# Essentials of Modern
# Telecommunications Systems

For a listing of recent titles in the *Artech House Telecommunications Library*, turn to the back of this book.

# Essentials of Modern Telecommunications Systems

Nihal Kularatna

Dileeka Dias

Artech House, Inc.
Boston • London
www.artechhouse.com

**Library of Congress Cataloging-in-Publication Data**
Kularatna, Nihal.
    Essentials of modern telecommunications systems / Nihal Kularatna, Dileeka Dias.
      p.  cm. — (Artech House telecommunications library)
    Includes bibliographical references and index.
    ISBN 1-58053-491-0 (alk. paper)
    1. Telecommunication systems. I. Dias, Dileeka.  II. Title.  III. Series.
    TK5101.K838  2004
    621.382—dc22
                                         2004047640

**British Library Cataloguing in Publication Data**
Kularatna, N.
  Essentials of modern telecommunications systems.
  —(Artech House telecommunications library)
  1. Telecommunications systems
  I. Title  II. Dias, Dileeka
  621.3'82

  ISBN  1-58053-491-0

**Cover design by Igor Valdman**

Chapters 7 and 8 are based on the contents of Chapter 12 of the IEE book *Digital and Analogue Instrumentation: Testing and Measurement*, Nihal Kularatna, 2003. (Reproduced with permission of IEE.)

© 2004 ARTECH HOUSE, INC.
**685 Canton Street**
**Norwood, MA 02062**

International Standard Book Number: 1-58053-491-0
Library of Congress Catalog Card Number: 2004047640

10 9 8 7 6 5 4 3 2 1

*This book is dedicated ...*

*...to my loving Amma and Thatha who, unfortunately, did not witness any of my humble writing activities—but for whose unselfish love and support none of this would have been possible, and to my cherished mother-in-law who left us recently...*

*...to the memory of John Robinson Pierce, the engineer who named the Transistor, who gave me the brief, but most valuable advice on how to commence on larger projects like books...*

*...and to my dear friend Martin Edwin Allard who shared the work related to my first published paper in telecommunications...and departed at a very unexpected moment...proving the uncertainty of life...*

*with loving appreciation to my wife Priyani, and daughters Dulsha and Malsha, who have always managed to tolerate my addiction to electronics and writing...and have always lent their unselfish assistance and encouragement. They are the remaining loves of my life and for them I am fortunate beyond measure.*

Nihal Kularatna

*...to the memory of my parents Kanakaratne and Dharma Subasinghe, who showed me the value of wholehearted dedication in all aspects of life.*

*...to the memory of my father-in-law Brigadier P.N.K. Dias, a truly practical engineer.*

*...to all my teachers who, in the past, have shown me the way to learn.*

*and*

*...to all my students, who give me now, the opportunity of passing it on to the future.*

Dileeka Dias

# Contents

## CHAPTER 4

### Cellular Systems                                                        143

## CHAPTER 7

### Transmission Techniques

# Foreword

In my own lifetime, telecommunications have transformed the planet beyond recognition.

Virtually everything we wish to do in the field of communications is now technologically possible. The only limitations are financial, legal, or political. In time, I am sure most of these will also disappear, leaving us with only limitations imposed by our own morality.

But at the same time, the communications revolution has bypassed tens of millions of people in many parts of the world. We are reaching the point in our technological evolution when we can—and must—commit more time and resources to solving the problems of poverty, deprivation, and inequality.

I discussed some of these concerns when addressing the UN General Assembly two decades ago, during the World Telecommunications Year 1983. I suggested that the 'A telephone in every village' would be one of the most effective social stimulants in history, because of its implications for health, weather forecasting, market information, social integration and human welfare. I added: "Each new telephone installation would probably pay for itself, in hard cash, within a few months. I would like to see a cost-effectiveness study of rural telephone systems for the developing countries of Asia, Africa, and Latin America. But the financial benefits, important though they are, might be insignificant compared with the social ones."

When I spoke my mind from that famous podium in New York, I did not imagine how quickly my words would be illustrated by actual developments. Just as satellite television swept across the globe during the 1980s, the Internet spread rapidly in the 1990s. (The father of the World Wide Web, Tim Berners-Lee, says he was inspired by my 1964 short story "Dial F for Frankenstein.")

In this book, Nihal Kularatna and Dileeka Dias help us to navigate through the complex world of telecommunications technology. They cover fundamentals, current applications, and future developments while avoiding complex mathematics and management jargon. I believe this book will provide valuable additions to our understanding and mastery of information and communications technologies that now shape our lives, livelihoods, and leisure.

*Sir Arthur C. Clarke*
*Fellow of King's College, London*
*Patron, Arthur C Clarke Institute for Modern Technologies*
*Colombo, Sri Lanka*
*April 2004*

# Preface

This book is principally a result of over 20 years of experience gained by us in conducting undergraduate and postgraduate courses, as well as continuing professional development (CPD) courses and seminars in electronics and telecommunications. One of us, early in our career, was able to work on real time and processor based message switching hardware, designed using early logic gates, and without monolithic processor chips, and was then responsible for maintaining early versions of digital switches with monolithic processor chips (AXE10 systems from L M Ericsson, Sweden) and the associated mobile switching subsystems of the early types. This was followed by the opportunity to work for the Arthur C. Clarke Institute for Modern Technologies (ACCIMT), an institute named in honor Sir Arthur C. Clarke, considered to be the "godfather of satellite communications." This created the opportunity to collect a fair amount of information on telecom developments, its enabling technologies, and conduct short courses in telephony and data communications.

During the last 10 years, the global telecommunications network has seen extraordinary growth in services and subscribers, spurred by many groundbreaking innovations. Rapid development imposes new demands on knowledge for all involved in related research, development, service creation, and provision. Correspondingly, in our day-to-day work, we experienced the necessity of continuously updating our knowledge, building up a massive collection, of reference and teaching material in the process. This book is an outgrowth of this material, compiled with the intention of passing on some of the essentials behind the fascinating world of telecommunications today. Fairly broad in scope as is inevitable, the book attempts to present the technology developments, particularly of the last decade, relevant to each chapter. The focus is on systems, and technology in context, rather than theory.

As a unique feature of this book, we track the developments in semiconductor technology in parallel with those in the communications technologies that they enable. Building up from the simple telephone instrument, the book provides a peek at today's complex integrated circuits including systems-on-chip (SoC) solutions for telecommunications systems. We attempt to highlight, through illustrative devices and applications, the role that human expertise and its integration into hardware and software plays in the global telecommunications network—mankind's largest and most complex engineering creation.

Many postgraduate and undergraduate students the at the University of Moratuwa, Sri Lanka, and participants of CPD courses at the ACCIMT gave us excellent feedback on the need for new knowledge and its half-life problem, which inspired us to seriously consider in this work. The work itself was a project founded on the global telecommunications network, which enabled the authors and the publisher to work together via the global telecom network, irrespective of geographic boundaries, and across many a *last mile*.

# Acknowledgments

Documenting the essential technology behind the marvels of modern telecommunications was indeed a pleasant and rewarding experience for us. Gratitude is due to many teachers, colleagues, friends, and family for helping us along the way to successfully reach our goal.

Our experience as teachers at university and industry levels lead us to seriously consider the publication of a book and collect relevant information and develop ideas over a considerable period of time. In this exercise, we are very thankful to the tireless attempts by Chandrika Weerasekera, Nimalasiri and Kumara of the Arthur C Clarke Institute for Modern Technologies (ACCIMT), and Thushara Dhammika and S. Wimalasiri of the University of Moratuwa, Sri Lanka, who helped over a period of many years in compiling collections of technical information. The authors also wish to thank all the students, undergraduate, postgraduate, and practicing engineers for guiding them along the way, technically.

Nihal Kularatna is also very grateful to the head of the department, Professor Allan Williamson, Professor John Boys, and the colleagues of the Deptartment of Electrical and Computer Engineering at the University of Auckland, for the encouragement given at departmental level and access to most needed communication facilities. Free availability and access to the departmental resources at the School of Engineering, University of Auckland was a great asset to the project. Also Nihal Kularatna is very grateful to the library team at the University of Auckland, Patsy Hulse and Susan Brookes. Susan, many thanks for your untiring efforts of locating many references, saving many of my valuable hours for reading and writing instead of searching for information! His gratitude is also extended to Thiranjith Weerasinghe for assistance with some urgent illustrations at the last moment.

As a postgraduate student at the University of California, Davis, Professor Kamilo Feher, Dileeka Diaz's research supervisor was about to publish his third book at the time she met him. During the course of the next 6 years, Professor Feher showed her how to avoid Low Information Content sentences (LICs) in writing, which has aided her greatly in research and publications. Dileeka Diaz thanks Professor Feher for the initial inspiration and guidance.

The University of Moratuwa, Sri Lanka, where Dileeka Diaz has worked since the completion of her Ph.D., has given her the environment and the freedom of mind to work on academic pursuits of her liking, even in spite of a heavy workload. She thanks the university, and particularly, the Department of Electronic & Telecommunication Engineering, for the environment and facilities they extended for this work, as well as for the opportunities given to handle many academic and administrative challenges, which were stimulants in undertaking a project of this nature.

Those who directly helped with the preparation of the text are Chandrika Weerasekera, Secretary to the CEO of the ACCIMT and Thanoja Rajapakse of the Lanka Education and Research Network (LEARN). Jayantha Perera, Senior Staff Technical Officer of the Department of Electronic and Telecommunication Engineering handled the most tedious task of figure preparation. Thank you, Chandrika, Thanoja, and Jayantha for your fast response to our requests, and the excellent quality of work you turned out.

Ethan Bordeaux, DSP, Software Engineer with Analog Devices Inc., added a unique chapter to this book. Ethan, thank you for your really hard work, in spite of your busy schedule at a difficult time in the industry, and the cooperation you willingly extended at various stages in editing the manuscript.

Sumith Kumaratunga, Director of David Pieris Information Technologies (DPIT), Sri Lanka gave us the idea of inserting a map of Sri Lanka into the book as part of the switching chapter. We thank Sumith and the engineers of DPIT for the idea of showing how a developing country can effectively adopt new technologies rapidly.

The aim of this book is to provide a broad approach to understanding the world of modern telecommunications systems, and to highlight developments over nearly half a century, particularly after the invention of the transistor in 1947 and the commercialization of the microprocessor in the early 1970s. In this book a large amount of published material, with special attention to it being state-of-the-art, from many industry and academic sources has been used. The individuals and organizations who have made this possible and deserve important acknowledgment are:

1. Analog Devices, Inc., for material in the final chapter, and permission for Ethan Bourdeaux to document some of his experiences in developing software for an adaptive multirate codec.

2. Dr. Robin Mellors, director of publishing of IEE, and commissioning editor Sarah Cramer for much of the material in Chapters 7 and 8 based on Chapter 12 of Volume 11 of the IEE Electrical Measurement Series, and for the copyright permission to many other illustrations from IEE publications.

3. William Hagen and Sandy Bjornsen of IEEE for providing information and permissions for material from many IEEE publications.

4. Pia Stenervall of Ericsson Editorial Services of Telfonaktiebolaget LM Ericsson of Sweden for some material in Chapter 3.

5. Joan M. Lynch of EDN Magazine, Mark David of Electronic Design Magazine, Rick Nelson and staff of T& M World Magazines for copyright permissions and also for documenting much of the state-of-the-art development information related to telecom integrated circuits, processors, and technology.

6. Dr. Bernard Arambepola, Mark Levi, and Des Byrne of Zarlink Semiconductor.

7. Douglas A. Balog of Intersil Corporation, USA.

8. Lillian Su, Emma Lee, Chris Geen, and Serene Chen of Texas Instruments for assistance, information, and copyrights.

9. Mary and the staff Pascual of IDT Corporation, U.S.A.

10. Tiffany Plowman of Silicon Laboratories for information and permissions for some items in Chapter 3.

11. Christoph Liedtke of Infineon Technologies for permissions to some items in Chapter 3.

12. The marketing and permissions departments of several publishers, including Sabrina Paris of Pearsons Education, Inc., and Tiffany Gasbarrini of Elsevier, U.S.A.

While compiling the manuscript and collaborating from Indian and Pacific Ocean islands, it would not have been possible to complete this work in a reasonable period of time without the wonderful support extended by the individuals and companies listed above, to whom the authors are greatly indebted.

Most sincere appreciation from the authors go to their colleague and close friend, Wayne E. Houser, Jr., retired USIA Foreign Service Officer and Voice of America Relay Station Manager, for untiring follow-up assisted by his lovely wife Gloria in the information collection and copyright clearance stages. And, *new* congratulations to Wayne who was recently named the first recipient of the Richard T. Liddicoat Scholarship for Gemology. Wayne will soon be setting engineering aside to be off to the Gemological Institute of America in Carlsbad, California to become a gemologist! We wish Wayne the very best of luck in this new endeavor!

The authors are also very thankful to the editorial and production staff at Artech House Publishers, in particular Mark Walsh, Barbara Lovenvirth, Jill Stoodley, Kevin Danahy, and the editorial and production staff. The speed of the publication process was highly impressive!

Nihal Kularatna is very grateful to many friends who encouraged him in this work, with special reference to Keerthi Kumarasena (who also taught the practical aspects of Buddhism for a peaceful and meaningful life), Padmasiri Soysa (ACCIMT librarian who maintains and manages a vast array of industrial documentation), and including friends such as Mohan Kumaraswamy (who helped write his first technical paper in 1977), Sunil, Kumar, Ranjith, Vasantha, Lakshman, Upali, Kithsiri, Lal/Chani, and Shantha/Jayantha. Also his heartfelt gratitude is to JJ and DJ Ambani, Erajh Gunaratne, Niranjan De Silva, and the corporate staff of the Metropolitan Group, in Sri Lanka, for providing services, facilities, and the opportunity to work with them. The opportunities provided by new wireless local loop providers such as Suntel and Lanka Bell in Sri Lanka deserves special credit, as they allowed him to seriously learn wireless local loop concepts. He gratefully remembers the support extended by Mahinda Ramasundara, Director-Technical of Suntel-Lanka in this regard.

Finally, this book would not have progressed beyond the title were it not for the authors' loving families.

Nihal Kularatna thanks Priyani for taking up the reins of all important family matters, and daughters Dulsha and Malsha who provided their love and support from the first page through to the last.

Dileeka Dias is indebted to her mother-in-law, Mrs. Maulee Dias, whose help allowed her the time to work on this book. She also thanks Gihan for his suggestions and criticisms at many stages of the manuscript preparation, and for tolerating endless hours spent on the book's work, and little Savidu and Kaneel, who unknow-ingly sacrificed a lot of time and opportunities for family activities.

   To Sir Arthur Clarke, a humanitarian above all …our most sincere gratitude for your inspiration, friendship and advice, and continued support of our activities, wherever they may be. And thank you once again for documenting your thoughts and experiences in the foreword of our book.

## CHAPTER 1

# The Communication Channel and the Communication Network

## 1.1 Development of Telecommunications

The first demonstration of Samuel Morse's electric telegraph in 1844 marks the birth of telecommunications as we know it today, the transfer of information across distances by electric means. The introduction and successful implementation of telegraph systems across the Atlantic formed the first telecommunication network. In 1876, Alexander Graham Bell obtained the first telephone patent (U.S. No. 174,465), from which point the global telecommunication network took off on an amazing adventure of technological inventions and innovations, which continues to this day. Today, the global telecommunication network integrates the world irrespective of geographical boundaries with an array of voice, data, and multimedia services.

Simple voice telephony within a bandwidth of 4 kHz was the reason for the initial remarkable success of the telecommunications network. Switching was next identified as a key element in telephone systems. Early systems were manual. As networks grew, the need was felt for automation of number reception and establishing channels. The automatic switching systems started with the mechanical Strowger systems installed in the early 1890s, followed by electromechanical crossbar and reed relay systems. The latter types of switching systems were in use well into the 1990s, due to their rugged, maintenance-free nature.

The invention of the transistor in 1947 changed the world of microelectronics and telecommunications. Proliferation of inexpensive solid-state components in the early 1960s helped the advancement of telecommunications systems. In mid 1960s, exchanges progressed towards stored program controlled (SPC) systems, where a mini computer was used for overall control and number processing, while the voice path was a physical channel.

In the late 1970s and early 1980s, low-cost digital components such as memories and microprocessors helped convert SPC systems to fully digital systems combining number processing by real-time processor systems and voice digitization by pulse code modulation (PCM).

In the early 1970s, satellite communications, a concept envisaged by Sir Arthur C. Clarke in 1945 [1], became a reality for international traffic, competing with submarine cables. Simultaneously, modular software helped the exchanges with billing, call handling, signaling, and special services. During the 1980s, fiber-optic transmission systems entered as a solution to the annoying delay (approximately 250 ms) in satellite communications and the high maintenance costs in transoceanic cables. Digital modulation and multiplexing techniques coupled with optical systems helped integrate the world's national telephone network as one global entity.

During 1960s to 1970s, with the development of computer systems, the need for data communications networks arose. Telephone systems were identified as inexpensive ways to couple computers using 300- to 1,200-bit/s modems, leading to the integration of data and voice traffic in the telephone network. These developments were supported by powerful semiconductors and software, and the cumulative data traffic over the global telecom systems developed well and exceeded voice traffic by 2000 [2], as illustrated in Figure 1.1(a).

Recognizing the inefficiency of the circuit-switched public switched telephone network (PSTN) for data communications, packet-switching techniques and systems evolved in the 1970s, giving rise to early X.25 packet switches. This laid the foundation for today's data networks. The early 1990s was a period of dramatic change. With computers and computer networks proliferating, data communications through public telecom systems grew, fueled by the development of Internet.

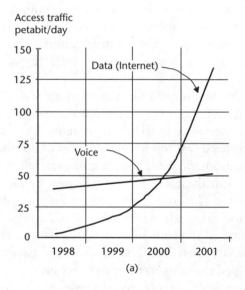

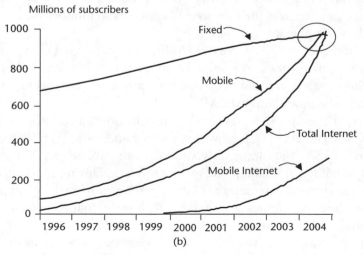

**Figure 1.1**　Growth of traffic and subscribers: (a) voice and data traffic, and (b) fixed and mobile subscribers. (*Source:* [2]. Courtesy of Telefoukatiebolaget L M Ericsson.)

In addition to the huge global copper cable base of over 900 million pairs by the 1980s, cellular phone systems were becoming popular, but restricted by call charges and hand phone costs. Cellular systems were also used in many parts of the world as a remedy against the long waiting list for land lines. Figure 1.1(b) shows the growth of mobile services compared to fixed services, predicting an interesting crossover by 2003.

To overcome the huge backlog of waiting subscribers, wireless local loop (WLL) technology was introduced in the early 1990s as a cheaper alternative to cellular. Derived from cellular technology, WLL systems were faster and less expensive to install. Throughout the 1990s, cellular systems flourished with decreasing call charges and phone prices, leading to an interesting mix of telecom services for voice and data.

With the continuing demand for bandwidth due to the advent of these new applications and services, the last mile over copper pairs or its radio equivalent, WLL for voice, posed insurmountable bottlenecks. Wideband access channels to subscribers became an essential requirement. A variety of wired and wireless wideband access techniques have been thus developed to cater for this need. Evolving from the integrated services digital network (ISDN), the bandwidth of the copper local loop was further exploited in digital subscriber loop (DSL) techniques, which started entering the market around 1993. In the wireless arena, cellular technology has also been extended for wideband access with code division multiple access (CDMA)–based third generation systems.

With the growth in Internet- or Internet protocol (IP)-based services in the latter part of the 1990s, creative design has resulted in transferring not only data, but also real-time traffic such as voice and video over such networks. Such techniques, called Internet telephony or voice over IP (VoIP), provide economic advantages to both service providers and end users by enabling an integrated network instead of separate networks for different types of traffic. However, meeting quality of service (QoS) demands of these services is an ongoing challenge for researchers. In summary, the trends in communications technology are wireless and wired wideband access, multimedia services, and integrated networks. These new possibilities have revolutionized the traditional thinking of regulators and telecom service providers. Telecommunications, which began with the plain old telephone system (POTS), has evolved to provide pretty amazing new services (PANS) today.

Powerful semiconductors evolving into systems on a chip (SoC) and the complex software for systems intelligence have been the key enabling technologies with diminishing costs. Software in communications devices has allowed such devices to be reconfigured or easily redesigned to suit different services and standards [4–6]. Parallel advances in radio frequency (RF) integrated circuits (ICs) have brought low-cost, high-speed, reliable, compact, power-efficient communications devices to the end users. Recent developments such as the gallium nitride transistor [7] show promise toward making future broadband wireless networks a practical reality.

The book provides the *essential engineering principles* applied in modern telecom systems, with Chapter 1 as an introduction to the telecommunications channel and the telecommunications network. The remaining chapters expand on the different telecommunications services and techniques, as applied to the evolving mix of advanced systems. Each chapter covers the basics, operation, and practical implementation using state-of-the-art components in the relevant area.

## 1.2  Introduction to Telecommunications

### 1.2.1  Information and Signals

Telecommunications systems transfer time-varying information over a distance using electrical, electromagnetic, or optical means. Information may be binary data as in a file transfer, text as in a fax message, voice, video, or multimedia (a combination).

The most common is speech. Figure 1.2 shows a typical speech waveform. Speech consists of a sequence of elementary sounds called phonemes, which originate in the voice box, acquire pitch by passing through the vocal cords, and excite resonances in the vocal tract. The result is the string of sounds that we call speech with unique identifying properties. A transducer such as a microphone converts voice into electrical form for transmission over the telecommunications system. Once converted to an electrical signal, the speech signal acquires technical descriptors: *frequency, phase,* and *amplitude,* corresponding approximately to pitch, intelligibility, and intensity, respectively [8]. The characteristics of the signal are source dependent. Speech signals have different characteristics from video signals and from computer data.

### 1.2.2  The Communications Channel

The communications channel acts as a conduit for carrying the information-bearing signal from the transmitter to the distant receiver. The goal is to deliver the message exactly as it was sent, and the channel must have sufficient capacity. However, the channel can also cause impairments to the signal. Different signals pose different demands on the channel. Basically, the channel should be capable of carrying the information ideally without loss. The impairments cause loss of information, which is seen as distortion. Communications theory quantifies information, the capacity of a channel, and conditions for distortion-free transfer.

Physically, a channel may be a pair of wires, a coaxial cable, an optical fiber, free space, or a combination of these. The channel may carry a single signal (direct or its amplified version) or a bundle of many signals, depending on its capacity and

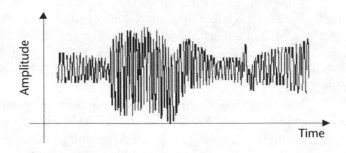

**Figure 1.2**  A speech waveform.

application. A pair of channels between two points for both-way communications is called a *circuit*.

### 1.2.3   Communications Systems, Networks, and Services

In a well-defined communications system, there are three essential components for effective information transfer: a transmitting device, a transport mechanism, and a receiving device. The transport mechanism may range from a simple single channel to a complex communications network consisting of many circuits, switching devices, and other components. A communications system may carry information only in one direction (simplex), in both directions sequentially (duplex or full duplex), or alternately in each direction (half duplex), as depicted in Figure 1.3(a). It may be point to point as in a telephone conversation, point to multipoint as in a radio or TV broadcast, or between many points simultaneously as in the case of a computer network as shown in Figure 1.3(b).

A communications network facilitates communication between users when the need arises. It may be a structured framework without geographical boundaries such as the PSTN, or it may be a network spanning over a small area serving a single organization, such as a private automatic branch exchange (PABX) or a local area network (LAN). A switched network such as the PSTN can set up a channel over the switched network on demand, as depicted simply in Figure 1.3(c). A switching device at each subscriber enables appropriate line selection on demand. Now the transport mechanism is a complex switch and circuit arrangement. This concept is extended to larger telecommunications networks with more subscribers.

In the PSTN, the number of switches (*exchanges*) and their locations are governed by the overall number and the geographical distribution of stations requiring service [9]. Interconnections among exchanges are multiple circuits, called *trunks*, with sufficient capacity to cater for the demand. These carry multiple conversations simultaneously using special *multiplexing* and *transmission* techniques. Multiplexing refers to the bundling of several conversations together, and transmission refers to the transport of these bundles.

*Signaling* is another important concept in the telecommunications network. It conceptually uses specialized messages to set up, maintain, terminate, and bill calls. Advances in switching, signaling, and transmission have enabled modern networks to handle a complex combination of subscribers, systems, and a variety of advanced services.

Applications that are implemented on telecommunications systems are referred to as communications services. Older examples are telegraphy and voice telephony. Modern examples are cellular telephony and multimedia. We are now in the *Information Age*, where information has become an important commodity for both the business community and the greater public. In today's service aspects, telecommunications systems can be primarily divided into PSTNs, public land mobile networks (PLMNs), and packet switched data networks (PSDNs).

## 1.3   Signals in Communications

### 1.3.1   Types of Signals

Signals can be classified on the basis of the way in which their values vary over time:

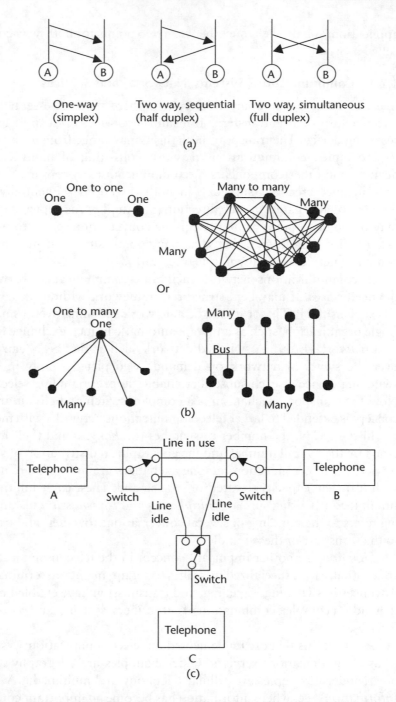

**Figure 1.3**  Communications channels and switching: (a) simplex channel, (b) channel classification as per direction, and (c) the concept of switching.

- *Analog:* This is a continuous function that assumes smooth changes in value with finite rates of change. The microphone output is an example of an analog signal. Figure 1.4(a) shows an analog signal.
- *Discrete:* This is a noncontinuous function whose values form a discrete set and occur at isolated points in time. A discrete signal is shown in Figure 1.4(b).

- *Digital:* A function that assumes a limited set of discrete values for constant durations of time is called a digital signal. Changes of values are instantaneous. A *binary* waveform is one that has only two allowable values and is commonly used for the communication of digital data. A binary digital waveform is shown in Figure 1.4(c).

### 1.3.2   Characteristics of Signals

A signal can be characterized in the time or frequency domain. In the time domain, the signal's characteristics are identified as a function of time, as in Figure 1.2 or as we would see on an oscilloscope. In the frequency domain, the signal's characteristics are represented as its constituent frequency components, as viewed on a spectrum analyzer. As most signals in communications are random, in addition to time and frequency domain characterization, statistical characterization is also possible.

### 1.3.2.1   Frequency Domain Characteristics of Signals

Any complex waveform can be decomposed into either a discrete set of sinusoids or a continuum of sinusoids, each having a different amplitude, frequency, and phase.

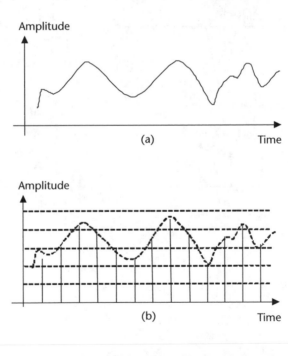

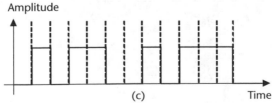

**Figure 1.4**   Different signals: (a) analog, (b) discrete, and (c) digital.

The tool for this characterization is *Fourier analysis*. Fourier analysis shows that any periodic signal is composed of a *dc component*, a *fundamental component,* and a series of *harmonics*. The fundamental is a sinusoid having the same frequency as the periodic signal, and harmonics are sinusoids at integer multiples of the fundamental frequency. Figure 1.5 shows two examples of periodic signals and their frequency content, known as the *Fourier series*.

Nonperiodic signals are composed of a continuous range of frequency components, called the *Fourier transform*. For example, Figure 1.6 shows the time and frequency domain representations of a rectangular pulse and a pulse train. The Fourier series (fundamental and harmonics) or the *Fourier transform* (continuous range of frequency components) is called the *spectrum* of the signal. Note that the spectral envelope in Figure 1.6(a) has the same shape as the envelope of spectrum lines, repeating at frequencies of *1/T* in part (b).

The range of frequencies that encompasses all of the energy present in a signal is known as the *bandwidth* of that signal. Hence, the bandwidth of the periodic signal

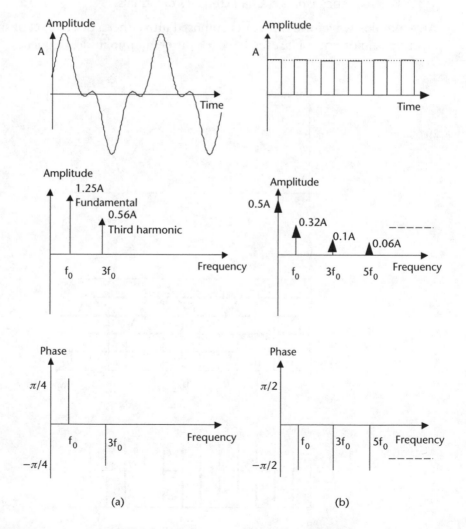

**Figure 1.5** Periodic signals and their frequency content: (a) a composite sinusoid with third harmonic, and (b) a square wave.

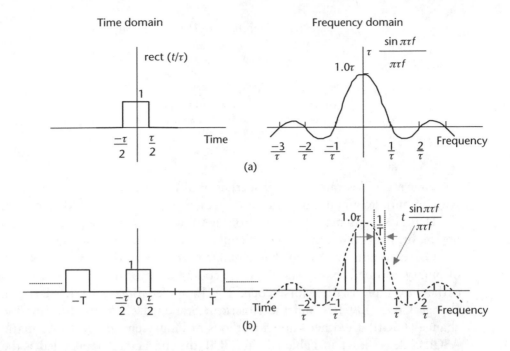

**Figure 1.6**    Time and frequency domain view of a single pulse and a pulse train: (a) single pulse, and (b) pulse train.

in Figure 1.5 is $3f_0$ and that of the square waveform is infinite. When we select a limited bandwidth for a signal or a channel, some distortion will occur. However, practical communications systems are designed to keep the distortion at an acceptable level by filtering the nonessential frequencies and allowing the signal to progress through limited bandwidth systems.

### 1.3.2.2    Statistical Characteristics of Signals

A deterministic signal can be expressed in the form of an equation, where all terms are completely known. However most signals in practical communications systems are not deterministic. To predict the behavior of random, or nondeterministic signals, statistical techniques such as probability density function, autocorrelation, and cross-correlation functions are used. Auto correlation represents a signal's similarity to a time-delayed replica of itself. Hence, it is a measure of how fast the signal changes and, correspondingly, its frequency characteristics. The cross-correlation function represents similar comparative properties for a pair of signals.

### 1.3.2.3    Practical Examples

Examples of analog signals are signals from a microphone (a low-audio-bandwidth case), musical performance (a high-bandwidth audio), signals from a video camera (video bandwidth), or signals from a sensor circuit to measure such variables as temperature, humidity, and acceleration. Analog signals in communications have wide-ranging characteristics. As an example, Table 1.1 compares voice and music.

**Table 1.1** Comparison of Electrical Parameters of Speech and Music

| Signal | Frequency Range | Frequency of Peak Energy | Dynamic Range* |
|--------|-----------------|--------------------------|----------------|
| Speech | 100–7,000 Hz | 250–500 Hz | 35–40 dB |
| Music | 20–20,000 Hz | 200–600 Hz | 75–80 dB |

*The dynamic range is the ratio of the power produced by the loudest and the softest portion of the signal.

Figure 1.7 illustrates some properties of different types of signals and compares with amplitude modulation (AM) and frequency modulation (FM) radio signals.

Digital signals may originate from a computer or other processor-controlled device, or through digitization or digital encoding of an analog signal.

One common way in which computers represent and output data is in the form of binary-encoded characters. Binary encoding of information is well known—the earliest example is the Morse code. The Extended Binary Coded Decimal Interchange Code (EBCID) and the American Standard Code for Information Interchange (ASCII) are other known methods of binary encoding of information. The ASCII code is shown in Table 1.2. In ASCII, the character A is encoded as the seven-bit binary sequence *1000001*.

Analog signals are digitized by *sampling*, approximating the samples to a discrete set of values (*quantization*) and then expressing the sample values as binary codes (*encoding* or *digitization*). This process and the advantages of carrying signals in digital form are discussed in detail in Chapter 2.

## 1.4  Characteristics of the Communications Channel

The communications channel carries signals from the transmitter to the receiver. In the ideal situation, signals must be carried without distortion. However, irrespective

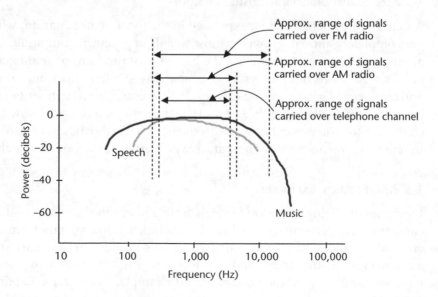

**Figure 1.7**   Comparison of speech and music signals.

**Table 1.2**   The ASCII Code

| | Bit Positions 5, 6, 7 | | | | | | | |
|---|---|---|---|---|---|---|---|---|
| | 000 | 100 | 010 | 110 | 001 | 101 | 011 | 111 |
| *Bit Positions* 0000 | NUL | DLE | SP | 0 | @ | P | ´ | p |
| *1, 2, 3, 4* 1000 | SOH | DC1 | ! | 1 | A | Q | a | q |
| 0100 | STX | DC2 | ,, | 2 | B | R | b | r |
| 1100 | ETX | DC3 | # | 3 | C | S | c | s |
| 0010 | EOT | DC4 | $ | 4 | D | T | d | t |
| 1010 | ENQ | NAK | % | 5 | E | U | e | u |
| 0110 | ACK | SYN | & | 6 | F | V | f | v |
| 1110 | BEL | ETB | ' | 7 | G | W | g | w |
| 0001 | BS | CAN | ( | 8 | H | X | h | x |
| 1001 | HT | EM | ) | 9 | I | Y | i | y |
| 0101 | LF | SUB | * | : | J | Z | j | z |
| 1101 | VT | ESC | + | ; | K | [ | k | { |
| 0011 | FF | FS | , | < | L | \ | l | | |
| 1011 | CR | GS | - | = | M | ] | m | } |
| 0111 | SO | RS | . | > | N | ^ | n | ~ |
| 1111 | SI | US | / | ? | O | - | o | DEL |

of the media used, the channel has a limited bandwidth and introduces impairments to the signal.

### 1.4.1   Bandwidth

The ability of the channel to carry signals (power or energy) at different frequencies is expressed as the channel's *frequency characteristics* or *frequency response*. The frequency characteristics of a local loop in a telephone network, by international agreement, are illustrated in Figure 1.8, indicating that the channel can carry a significant amount of power in the range 0.3 to 3.4 kHz, compared to the frequency ranges outside. The range that the channel can carry "reasonably well" is called the *passband* or the *bandwidth* of this channel. This is 3.1 kHz for the local loop.

### 1.4.2   Channel Impairments

An ideal channel will carry the signal undistorted, while the nonideal channel causes impairments to the signal causing distortion at the receiving end.

#### 1.4.2.1   Bandwidth Limitations

One of the main impairments caused by the channel is the elimination of some of the signal's frequency components due to the limited bandwidth. Figure 1.9 indicates the effect of different channel bandwidths on a square waveform of 500 Hz.

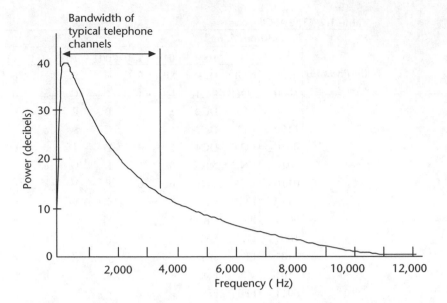

**Figure 1.8**   Characteristics of a local loop.

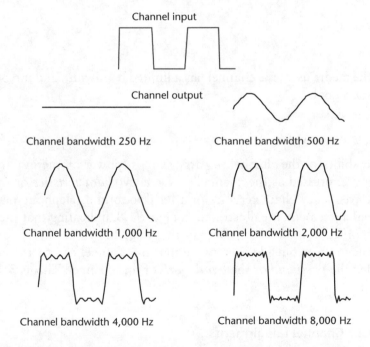

**Figure 1.9**   Distortion in a 500-Hz square waveform when transmitted over channels of different bandwidth.

### 1.4.2.2   Noise and Crosstalk

In any system, irrespective of the transmission media, there is unwanted electrical energy present, corrupting the information. This unwanted electrical energy is generally referred to as *noise*. Noise arises from different sources, and the most predominant type generated in electronic components is due to *thermal* noise. Noise may also be induced from natural sources such as thunderstorms and man-made sources such as car ignition or electrical/mechanical apparatus. When it is not possible to take account of noise sources individually, we assume that they produce a single random signal defined as additive white Gaussian noise (AWGN). Figure 1.10(a) shows AWGN and its additive effect on signals.

*Crosstalk* is a type of noise that arises due to parasitic coupling between different cables lying close to each other. This is caused by insufficient shielding, excessively large disparity between signal levels in adjacent circuits, unbalanced lines, or overloaded carrier systems [10]. The mechanism of how crosstalk occurs and the types *near-end* and *far-end* crosstalk are illustrated in Figure 1.10(b).

### 1.4.2.3   Attenuation

As signals travel along a channel, their amplitude or power is progressively reduced because of losses in the channel. These losses, called attenuation, may be due to (a) ohmic losses in conductors of the channel, (b) dielectric and other losses, or (c) distribution of signal power in space. Attenuation increases with the frequency. Hence, signals with large bandwidth will undergo unequal attenuation across their bandwidth, resulting in distortion of the signal in addition to attenuation.

### 1.4.2.4   The Cumulative Effect

The performance of communications channels is dependent on the combined effect of these impairments. Figure 1.11 shows the cumulative effect on signals.

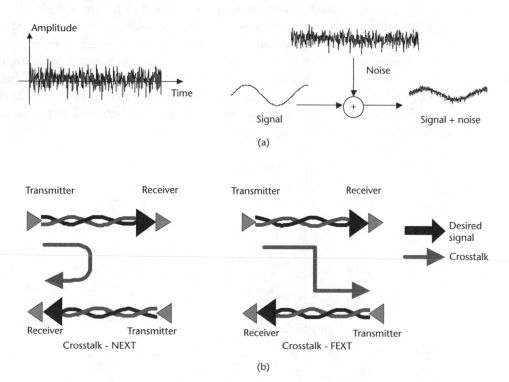

**Figure 1.10**   (a) AWGN, and (b) crosstalk.

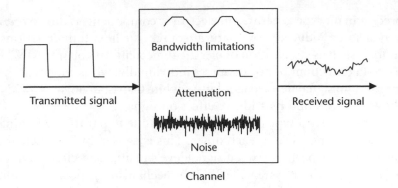

**Figure 1.11**  The combined effect of noise, attenuation, and bandwidth limitations.

A measure of quality of a communications system is given by the signal-to-noise ratio (SNR), defined as the power of the signal component (S) to the power of the noise component (N) in the composite signal. It is usual to express this ratio in decibels (dB) as:

$$SNR = 10\log\left(\frac{S}{N}\right) \tag{1.1}$$

Good quality communications systems achieve SNRs of 35 to 45 dB. For voice telephony, the SNR should be at least 30 dB [8, 10]. Many people have difficulty in recognizing words when the SNR in a telephone system drops below 15 dB. The SNR of a present-day telephone channel is estimated to be around 39 dB.

Whereas the SNR is a good measure of quality in analog communications, the bit error rate (BER) can be used in digital communications systems for a more quantitative measure. The BER gives the probability of a transmitted bit being in error at the receiver and is a measure of the overall quality of the system. Theoretical formulas are available to determine the BER performance of communications systems. These formulas are functions of the SNR, as illustrated in the example of Figure 1.12.

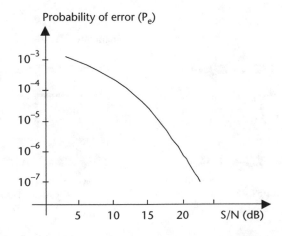

**Figure 1.12**  A typical BER performance curve of a digital communications system.

### 1.4.3    The Information Carrying Capacity of the Channel

In his landmark paper, *A Mathematical Theory of Communication* [11], Claude E. Shannon gave a precise mathematical meaning to the rate at which a communications channel can carry information. He proved a fundamental theorem that if the source information rate is less than or equal to the *channel capacity*, messages from the source can, in principle, be transmitted error free. The definition that Shannon gave to the capacity of the channel $C$ in bits per second is:

$$C = B \log_2 (1 + S / N) \qquad (1.2)$$

where $B$ is the channel bandwidth in Hz and the $S/N$ is the signal to noise power ratio at the receiving end of the channel.

This is known as Shannon's *channel coding theorem*. In this theorem, he demonstrated that the bandwidth *together* with the SNR determines the capacity (maximum achievable information rate) of a telecommunications channel. Shannon showed that error-correction methods exist, and that the possibility exists to attain virtually error-free transmission over noisy communications channels with error correction. Thus, Shannon predicted the future role of forward error correction (FEC) schemes.

In analog systems, the optimum system might be defined as the one that achieves the largest SNR at the receiver output subject to constraints such as channel bandwidth and transmit power. In digital systems, the theorem tells us whether information from a source can be transmitted over a given noisy channel. Figure 1.13 depicts the Shannon's limit by plotting $C/B$ versus carrier-to-noise ratio, a quantity closely related to the $S/N$. For comparison, the information carrying capacity of schemes such as phase shift keying (PSK) are also plotted.

For error-free transmission, encoding the message into a signal suited to the nature of the channel is necessary. Error-correcting codes add error-check digits to

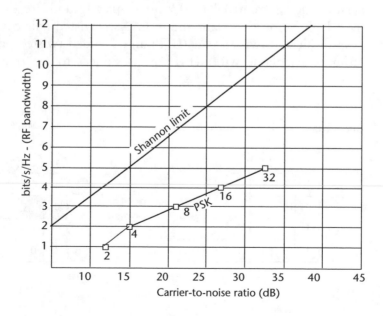

**Figure 1.13**   The Shannon's channel capacity limit.

the message to make up the signal that is to be transmitted. On reception, the correct message can be recovered even if a part of the message or the check digits are corrupted. A simple error-correcting code is given in Figure 1.14. The code uses 24 bits to transmit a message containing 16 bits. The 16 message bits are written into a 4 × 4 matrix. Check digits for each row and column are inserted such that there is an even number of 1s in each row and column. If one message digit is received in error, it can be located by the fact that the parity rule (i.e., even parity of 1s) is violated in one row and one column.

Complex FEC algorithms are common in communications systems today, aided by developments in semiconductors and software techniques. These are also common in other applications such as data storage and audio engineering [12].

### 1.4.4  Transmission Media

Transmission media can be classified as guided or unguided. Guided media such as twisted pair, coaxial cable, and optical fiber provide a physical path along which the signals travel. Unguided media employ an antenna for transmitting through air, vacuum, or water, and it is also referred to as wireless transmission.

### 1.4.4.1  Guided Transmission Media

For guided transmission media, the data rate or bandwidth depends critically on the distance. Resistance, capacitance, and the inductance of the medium cause the channel impairments in terms of bandwidth limitation and attenuation.

- *Twisted pair*. The most common guided transmission medium, twisted pair, consists of two insulated copper wires arranged in a regular spiral pattern. Typically, many pairs are bundled together into a cable with a tough protective sheath. Twisting decreases the crosstalk. The wires have thicknesses ranging from 0.4 to 0.9 mm typically and are exclusively used in the telephone outside plant and in LANs. Twisted pair comes in two varieties: unshielded twisted pair (UTP) and shielded twisted pair (STP). UTP is ordinary telephone wire. This type is subject to external electromagnetic

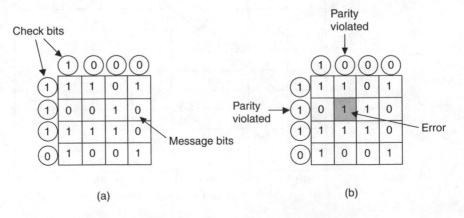

**Figure 1.14**  A simple error correcting code: (a) encoding, and (b) decoding.

interference and noise. STP provides better performance in terms of band-width, attenuation, and crosstalk, but it is more difficult to work with and more expensive.

The oldest type of voice-grade UTP used in telephony has a characteristic impedance of 100Ω. For high-speed LANs, two other types of UTP cables have been developed. The most commonly used of these are called Category 3 and 5 UTP, and have bandwidths of about 16 MHz and 100 MHz, respectively. While voice-grade UTP has about one twist per foot, Category 5 cables have about one to three twists per inch. Categories 5e (enhanced 5), 6, and 7 are also coming into use, having bandwidths up to 600 MHz.

- *Coaxial cable.* Coaxial cable consists of a hollow outer cylindrical conductor that surrounds a single inner conductor. A dielectric material fills the space, and the outer conductor is covered by a jacket or shield. See Figure 1.15(b). Coaxial cable is much less susceptible to interference and crosstalk than twisted pair and has a much larger usable bandwidth. In cable television systems and telephony trunks, multiplexing allows the simultaneous transmission of multiple signals over coaxial cable.

- *Optical fiber.* An optical fiber is a thin, reasonably flexible medium capable of guiding an optical ray. Various glasses and plastics are used. An optical fiber cable has a cylindrical shape and consists of three concentric sections: the core, the cladding, and the jacket, as depicted in Figure 1.15(c). The core is surrounded by the cladding, glass, or plastic coating, which has different optical properties than the core with a diameter in the range of 8 to 100 $\mu$m. The jacket protects the cable. Optical fiber transmits an encoded light beam by total internal reflection of the frequencies in the range of 100 to 1,000 THz ($10^{14}$ to $10^{15}$ Hz), covering portions of the visible and infrared spectra. Common wavelengths are 850 nm, 1,300 nm, and 1,550 nm.

Compared to hundreds of megabits per second over about 1 km for coaxial cable and up to 100 Mbit/s over a few tens of meters of UTP, optical fiber helps data rates of hundreds of gigabits per second over tens of kilometers. Smaller size and lighter weight, lower attenuation, and electromagnetic isolation are the attractive properties for long-haul, high capacity installations (over 1,500 km, 20,000 to 60,000 voice channels).

Extremely thin, large-bandwidth single-mode fibers are used in long-haul and high-speed trunks. Multimode fibers with lower cost and lower bandwidths are specified by the core/cladding diameters in $\mu$m (e.g., 50/125).

Fiber figure of merit (FOM) is given in megahertz-kilometers, the bit rate–distance product (i.e., the bandwidth scales with distance). Table 1.3 illustrates fiber characteristics. Figure 1.15(d) compares attenuation of guided transmission media.

### 1.4.4.2  Wireless Transmission

Wireless systems are based on unguided transmission, with radio waves of different frequencies propagating in different manners. The electromagnetic spectrum is a limited resource and must be shared by many. To provide some order and to minimize interference, the usage of the electromagnetic spectrum is regulated by

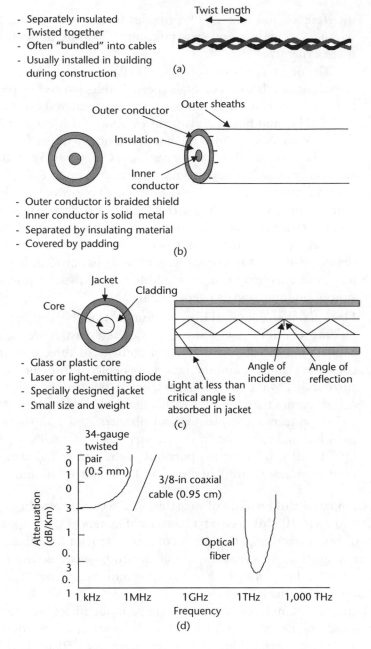

**Figure 1.15**   Guided transmission media: (a) twisted pair, (b) coaxial cable, (c) optical fiber, and (d) comparison of attenuation characteristics.

**Table 1.3**   Common Fiber Characteristics

| Fiber Type | Attenuation (dB/km) | | FOM(MHz-km) | |
| --- | --- | --- | --- | --- |
| | 850 nm | 1,300 nm | 850 nm | 1,300 nm |
| Single mode | – | 1.0 | – | Essentially infinite |
| 50/125 | 3.5 | 2.0 | 400 | 400 |
| 62.5/125 | 3.5 | 1.5 | 160 | 500 |
| 100/140 | 5.0 | 4.0 | 100 | 200 |

international and national governmental organizations. Table 1.4 summarizes the typical capacities of transmission links implemented in the world's telecommunications networks.

### 1.4.5   Modulation and Encoding

#### 1.4.5.1   Modulation

Modulation is a process of combining different properties of different signals for effective transmission over long distances by impressing the *information-bearing signal* on a *carrier*. Modulation allows multiple transmissions over single media, increasing the effectiveness of limited resources such as the spectrum and minimizing impairments. Simple examples are FM and AM used in radio broadcasting.

Shifting signals from their *baseband* (original) range of frequencies to another is accomplished through *modulation*. A baseband signal from 0 to 4 kHz, such as a signal from a microphone, might be shifted to the 60- to 64-kHz range. Analog baseband signals are rarely transmitted, with the most common exception being the local loop. Digital signals may be transmitted with or without modulation. In either case, binary data maybe *encoded* into multilevel digital signals to enable the transmission of more data in a given bandwidth, as shown in Section 1.4.5.2.

Communications systems most often do not carry signals in baseband form. There are several reasons for this. Baseband signals attenuate rapidly as they travel along channels, and the addition of noise renders them unusable. On the other hand, if a baseband signal is impressed upon a carrier, the higher frequency modulated signal can be made to travel further through directive antennas.

If we transmit unmodulated multiple baseband signals over a single channel, they will interfere with each other. Also, for economy, modulation allows long-distance transmission facilities of high capacity to be shared.

The wavelength of a radio signal in free space is the velocity of light divided by frequency. Baseband signals have low frequencies and therefore long wavelengths. For effective transmission and reception, antennas need to have lengths of a quarter of a wavelength or more. At 4 kHz, a wavelength of 75,000m makes the construction of antennas impossible. Further, radio transmission at long wavelengths is plagued by lightning and other disturbances.

**Table 1.4**   Transmission Media Used in Telephone Networks

| Transmission Medium | Capacity (Number of Voice Channels) |
|---|---|
| *Unguided:* | |
| Twisted pair | 12–96 |
| Coaxial cable | 600–10,800 |
| Fiber-optic cable | $672$–$6.25 \times 10^6$ |
| *Guided:* | |
| Terrestrial microwave | 600–6,000 |
| Communications satellites | 8,000–112,500 |

(*Source:* [14].)

### 1.4.5.2   Encoding

Most signals originate in analog form but are processed mostly in digital form. *Encoding* generally refers to the way that an analog signal is represented by a digital data stream (described in detail in Chapter 2) and the way that a binary data stream can be converted to a multilevel digital signal for better bandwidth efficiency.

A two- to-four-level encoding scheme is illustrated in Figure 1.16(a). The four-level signal encodes two bits in each of its levels. As a result, each level carries two bits of information at half the rate of the original binary signal. Hence, the bandwidth required for this four-level signal is half that required for the binary signal. This relationship holds whether the signal is transmitted with or without modulation. In higher level encoding techniques, $L$ bits are represented by one of $2^L$ possible levels. Figure 1.16(b) illustrates a four-level signal with amplitude shift keying (ASK) modulation. This is called four-level ASK.

### 1.4.5.3   Digital Modulation Techniques

In digital modulation techniques, the information-bearing signal is digital, either binary or multilevel. The techniques can be broadly classified as: ASK, frequency shift keying (FSK), phase shift keying (PSK), and a mixture of ASK and PSK

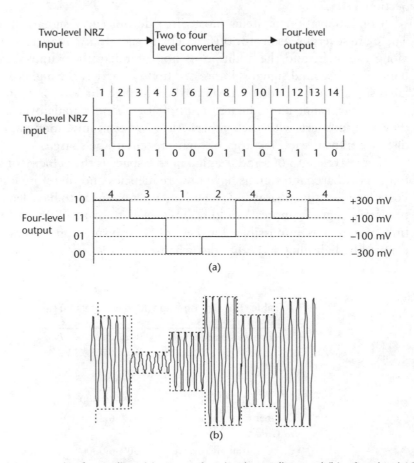

**Figure 1.16**   Example of encoding: (a) two- to-four-level encoding, and (b) a four-level ASK signal.

called quadrature amplitude modulation (QAM). The first three, illustrated in Figure 1.17, are analogous to AM, FM, and phase modulation (PM), with the difference of the modulation signal being digital. In binary ASK, the carrier amplitude has two discrete levels. In binary FSK, the carrier has two discrete frequencies. An early use of modulators/demodulators (modems) was for data transmission over the telephone network. The Bell System 103 modem used FSK to transmit data at 300 bps [13]. In practice, these modulation techniques use filtering to transmit digital signals in the limited bandwidth channel without significant degradation [14].

Complex multilevel digital modulation, coupled with powerful data-compression, error-correction, and channel-equalization techniques, is used in modern high-speed telecommunications systems. High-capacity digital backbone networks and DSLs are examples (see Chapter 6).

### 1.4.6    Multiplexing

In modern telecommunications systems, long-distance trunks are implemented through microwave or optical fiber links simultaneously used by tens of thousands of users. This deliberate sharing of a communications medium by a number of different signals is called *multiplexing*. The frequency-shifting characteristics of modulation and the digital encoding of signals using techniques such as PCM form the basis of multiplexing. The bundled signals are separated at the receiving end by a process called *demultiplexing*.

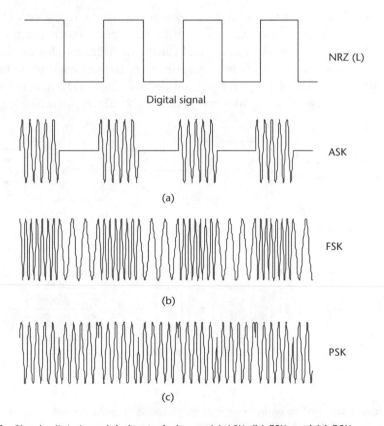

**Figure 1.17**    Simple digital modulation techniques: (a) ASK, (b) FSK, and (c) PSK.

There are two basic types of multiplexing, *frequency division multiplexing* (FDM) and *time division multiplexing* (TDM). In FDM, channel sharing is done in frequency, where each signal occupies a separate frequency range in the channel. FDM is the oldest approach to multiplexing in communications networks [15]. In TDM, each signal occupies a separate time interval in the channel. Figure 1.18 illustrates these basic concepts. A more recent development in multiplexing techniques is *code division multiplexing* (CDM) and will be discussed in Chapters 4 and 5.

## 1.5　The Communications Network

We have been discussing the communications channel, together with techniques for transmission, multiplexing, and handling impairments. In a communications network, channels are established between users when needed. The basic purpose is for one user (or subscriber) to exchange information in any form with any other user of the network.

The basic components for communication through a network can be categorized as:

- Access;
- Switching and signaling;
- Transmission (transport).

The basic structure of a network consists of switches, or exchanges, interconnected by the transport network. Subscribers are connected to the nearest exchange via the access network. The control functions necessary for establishing, maintaining, and disconnecting communications over the network are handled by signaling. Figure 1.19 shows the reference model for the modern network. Compared to the older networks, the modern systems use network intelligence (signaling) and

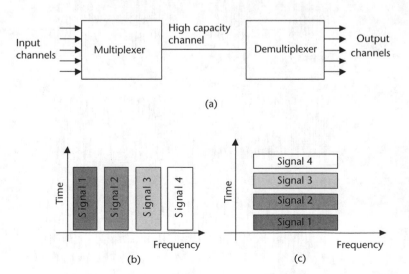

**Figure 1.18**　(a) The basic concept of multiplexing/demultiplexing, (b) channel sharing in frequency, and (c) channel sharing in time.

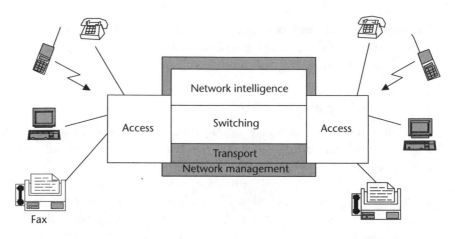

**Figure 1.19** Reference model for the modern communications network.

network management as more centralized functions, due to high levels of computing resources available.

### 1.5.1 Access

The access network provides the connection between the subscriber's premises and the local exchange. This connection is commonly referred to as the *last mile*. Technologies used in the access network are the basic local loop and the fixed wireless local loop, the digital subscriber loop (forming the PSTN), cellular (forming the PLMN), and fiber-based techniques such as fiber to home (FTTH) and other broadband access techniques.

The oldest type of access network is the *local loop* or *subscriber loop,* a pair of wires drawn from the exchange to each subscriber, carrying the subscriber's voice in analog form. The physical components that make up the local loop are called the *outside plant facilities* or the access network.

WLL, radio local loop (RLL), and fixed wireless access (FWA) are generic terms for an access system that uses a wireless link to connect subscribers to their local exchange in lieu of a copper pair. Depending on the existing telecommunications infrastructure, demand for services, and local market conditions, this technology can be both a substitute and a complement to the local loop. These techniques are described in Chapter 5.

With mobile access technologies, the WLL concept has been further extended to allow the subscriber to move freely while accessing the telecommunications network. The predominant technology for mobile access today is *cellular systems,* which are actually access networks for the public telecommunications system. These PLMN systems are described in Chapter 4.

Modern telecommunications service requirements are voice, together with such high-speed data transmissions as video. While the central facilities (switches, transport networks) have developed to cater to modern needs, the local loop creates a serious bottleneck. At the same time, it is not economic to install a dedicated fiber or a coaxial cable to every subscriber to exchange wideband signals. Short local loops have useful bandwidths as large as 30 MHz, and most can cater up to few megahertz easily. ISDN and DSL techniques described in Chapter 6 exploit this bandwidth to provide wideband access of several megabits per second.

### 1.5.2  Switching and Signaling

What makes services such as telephony unique is that they give us the ability to communicate with any other. Their selective reach makes it very different than broadcasting, in which everyone receives the same signal.

Each subscriber accesses the nearest exchange. All calls made by a subscriber are carried by the same access channel. However, the overall path traveled by a voice signal between one subscriber and another is different from call to call. The access channel is only a part of this *circuit*. To achieve this person-to-person service, transmission paths must be rapidly configured and reconfigured by switching and signaling.

Old manual switching systems have grown through several different technologies to become sophisticated software-controlled, distributed digital systems.

The intelligence to carry out call control functions is implemented through *signaling* mechanisms. This is the common thread that unites all components in a communications network. Signaling is the mechanism that allows network components (subscriber premises equipment, network switches) to establish, maintain, terminate, and bill sessions in a network. Signaling systems also access various databases to obtain information about connecting via more than one exchange. Chapter 3 discusses the evolutions in switching and signaling and their present status.

### 1.5.3  Transmission

The communications links connecting the switches in a communications system form the *transmission* or the *transport* network. They consist of high-capacity communications facilities capable of carrying thousands of multiplexed calls simultaneously, made up of a combination of copper cables, fiber-optic cables, and terrestrial and satellite microwave links. The transmission network simply provides a sufficient number of channels from one exchange to another. Exchanges use these channels for calls that they route from one exchange to another. Modern high-capacity transmission technologies are detailed in Chapter 7. Special measurement techniques on transmission systems are described in Chapter 8.

The major trend in telecommunications has been the digitization of facilities. While many of the chapters illustrate these developments within the network, Chapter 2 is specifically dedicated to the basic techniques in bringing this digitization out of the network to the subscriber.

## 1.6  The Future

In the 1980s, when the computer and networking industry reached maturity, the worldwide telephone network was used as the main information-carrying highway, making use of the well-developed exchange infrastructure and transmission systems. In the mid 1990s, the increasingly popular Internet resulted in an explosive growth in data transmission over the public telecommunications systems. The growth of traditional fixed-voice subscriptions is beginning to slow and may level off in coming years. By contrast, a continued strong growth in mobile communications is foreseen, where the number of worldwide subscribers will approach one billion by 2003–2004 [2].

With the advent of new, bandwidth-hungry multimedia services centered on the Internet, the capabilities of the PSTN for data services were soon exhausted. Though new packet-based switching and transmission techniques such as frame relay, asynchronous transfer mode (ATM), synchronous digital hierarchy (SDH), and synchronous optical network (SONET) have been developed, the basic user access, both wired or wireless, remains a bottleneck. Thus, broadband access via cable, fiber, or wireless is envisaged to be an area of major development in telecommunications in the future.

Packet-based networks, which were originally designed to carry nonreal-time data traffic, have today found applications in carrying real-time traffic such as voice and video. This has revolutionized the development of communications networks, which are evolving toward integrated networks based on packet switching and the transmission control protocol/Internet protocol (TCP/IP) suite of protocols. Figure 1.20 illustrates the multiplicity of access and transmission technologies in the complex telecommunications world.

Increased accessibility and a richer supply of services impose stringent demand on the network. This is recognized by the introduction of intelligent network (IN) capabilities by network providers throughout the world. The end result of IN concepts is that services and service provisioning can be separated from network infrastructure, signaling, and transport. These set the stage for the next evolutionary

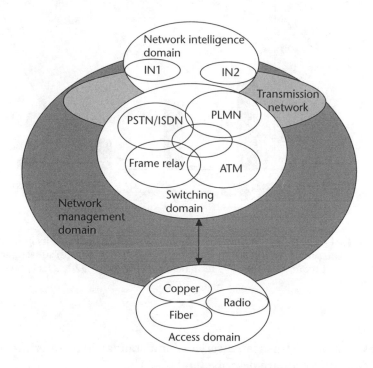

**Figure 1.20**   Reference model for the complex telecom network with a multiplicity of networks.

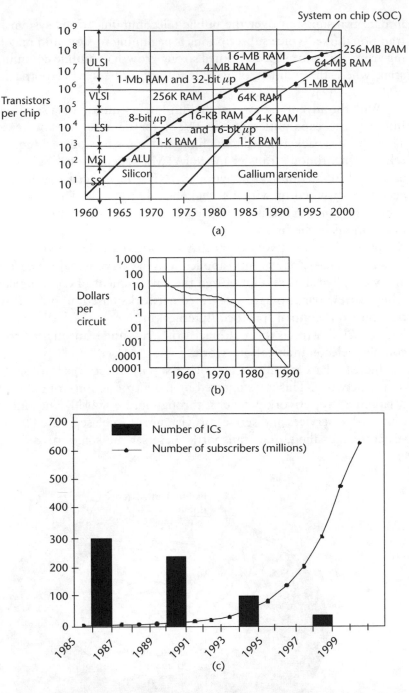

**Figure 1.21** Illustration of modern technology development and its impact on telecommunications: (a) the progress of integration, (b) cost path of ICs, and (c) subscriber growth and integrated circuit reduction in mobile terminals. (*Source:* [19]. ©2001 IEEE, reprinted with permission.)

steps in more sophisticated services and standardized interfaces. Evolutions along these technologies are described in [16–21].

## 1.7   Enabling Technologies

Advances in the following supporting technologies have spurred the growth of tele-communications infrastructure and services around the world:

- Silicon IC manufacturing processes;
- Radio frequency systems and products providing wireless connectivity using Gallium Arsenide (GaAs) and Silicon Germanium(SiGe) technologies;
- Digital signal processors (DSPs) and application-specific ICs (ASICs);
- Optical components and systems;
- Smart antenna systems;
- Software-defined radio and embedded intelligence in telecommunications systems;
- Intelligent signaling systems.

The result of these technologies have been:

- Reduction of the cost of services by improving system capacity;
- Reduction of the cost, size, and power consumption of subscriber equipment by improved semiconductor technologies;
- Development of a variety of communications devices capable of handling a variety of services;
- Development of communications services separately from the communications infrastructure.

These results are amply demonstrated in several aspects through Figure 1.21, which illustrates the progress made in the level of semiconductor integration of semiconductor devices, their cost path, and the impact on one particular industry, that of mobile communication, of these developments.

## References

[1]   Clarke, A. C., "Extra Terrestrial Relays—Can Rocket Stations Give World-Wide Radio Coverage," *Wireless World*, Vol. LI, No. 10, October 1945, pp. 305–308.

[2]   Nilsson, T., "Toward Third Generation Mobile Multimedia Communications," *Ericsson Review*, No. 3, 1999, pp. 122–131.

[3]   Guizani, M., A. Rayes, and M. Atiquazzaman, "Internet Telephony," *IEEE Communications*, April 2000, pp. 44–46.

[4]   Gatherer, A., et al., "DSP-Based Architectures for Mobile Communication: Past, Present and Future," *IEEE Communications*, January 2000, pp. 84–90.

[5]   Bassak, G., "DSPs for Next-Generation Cell Phones Balance Performance and Power," *EDN*, November 23, 2000, pp. 105–112.

[6]   Bing, B., and N. Jayant, "A Cell Phone for All Standards," *IEEE Spectrum*, May 2002, pp. 34–38.

[7]   Eastman, L. F., and U. K. Mishra, "The Toughest Transistor Yet," *IEEE Spectrum*, May 2002, pp. 28–33.

[8]   Carne, B. E., *A Telecommunications Primer*, Englewood Cliffs, NJ: Prentice Hall, 1995.

[9] Clark, M. P., *Networks and Telecommunications, Design and Operation*, New York: John Wiley and Sons, 1991.

[10] Reeve, W. D., *Subscriber Loop Signaling and Transmission Handbook—Analog*, New Jersey, IEEE Press, 1991.

[11] Shannon, C. E., "A Mathematical Theory of Communication," *Bell System Technical Journal*, Vol. 27, July 1948, pp. 379–423.

[12] Gappmair, W., and C. E. Shannon, "The 50th Anniversary of Information Theory," *IEEE Communications Magazine*, April 1999, pp. 102–105.

[13] Couch, L. W., *Digital and Analog Communication Systems*, Singapore: Addison Wesley Ongman, 2001.

[14] Feher, K., *Digital Communications: Satellite/Earth Station Engineering*, Englewood Cliffs, NJ: Prentice Hall, 1983.

[15] Martin, J., *Data Communication Technology*, Englewood Cliffs, NJ: Prentice Hall, 1988.

[16] Ericsson and Telia, *Understanding Telecommunications*, Vol. 1, Ericsson Telecom AB, Telia AB, Studentlitteratur AB, 1998.

[17] Johnson, J. T., "Rebuilding the World's Public Networks," *Data Communications*, December 1992, pp. 61–80.

[18] Hassan, M., A. Nayandoro, and M., Atiquzzaman, "Internet Telephony: Services, Technical Challenges and Products," *IEEE Communications*, April 2000, pp. 96–103.

[19] Bi, Q., G. I. Zysman, and H. Menkes, "Wireless Mobile Communication at the Start of the 21st Century," *IEEE Communications*, January 2001, pp. 110–116.

[20] Young, G., K. T. Foster, and J. W. Cook, "Broadband Multimedia Delivery Over Copper," *Electronics and Communication Engineering*, February 1996, pp. 25–36.

[21] Garrahan, J. J., et al., "Intelligent Network Overview," *IEEE Communications*, March 1993, pp. 30–36.

[22] Ericsson and Telia, *Understanding Telecommunications* Vol. 2, Ericsson Telecom AB, Telia AB, Studentlitteratur AB, 1998.

## CHAPTER 2

# Digital Telephony

## 2.1 From Analog to Digital

The term *digital telephony* refers to the use of digital technology in the message path of voice communications networks [1]. This implies digitization of voice and its subsequent switching and transmission. The first signals to be transmitted electrically were digital. A worldwide network of telegraph circuits existed at the time the telephone was invented, and a considerable body of knowledge existed on digital transmission. It was evident that with digital transmission, the received signal's quality need only be sufficient to detect the presence or absence of a pulse. Codes, such as the Morse code, were devised to ensure sufficient quality of the received information.

The introduction of voice telephony presented a completely new set of problems, as the absolute quality of the analog signal was of interest, rather than just a quality sufficient to detect the presence or absence of a signal. Thus, an entirely new body of knowledge grew related to speech transmission over long distances.

By the 1930s, long-distance carrier systems using FDM were widely in use, transporting groups of telephone circuits over long distances. The bandwidth of these trunks based on cables became a concern as telephony subscribers grew. Though the concept of TDM was introduced, the lack of suitable electronic components retarded its implementation. Microwave radio was introduced in the 1930s as a new medium for long-distance transmission. This medium offered a large bandwidth but with severe problems of noise and distortion. As a solution to this problem, Alec Reeves patented PCM technique in 1938, showing that by converting analog signals into digital code and transmitting them in place of analog signals, the problems of noise and distortion could be overcome. Shannon's information theory expositions in 1948 made the potential for PCM and digital transmission apparent. However, practical electronic devices did not exist at that time to implement and introduce PCM commercially.

In parallel, switching techniques too grew to support the growth of telephony by establishing continuous physical connections between incoming and outgoing circuits. Throughout these developments, the objective was to establish a complete path for the analog voice to travel from the caller to the called party.

The invention of the transistor in 1948 made practical the dormant ideas of PCM and TDM. This also provided a tool to combine transmission and switching.

A major milestone for telephony and its digitization was the installation of the first computer-controlled switching system in 1965. The stored program control (SPC) feature simplified many administrative and maintenance tasks for the operating companies. Thus, the world's telecom systems gradually readopted digital transmission.

Developments rapidly followed from these, in parallel with the development of IC design and fabrication technology, allowing voice digitization, fully digital switching, advanced control of exchange hardware, and transmission systems.

This chapter covers the essentials of transferring digitized voice over different types of communications networks: the PSTN, fixed and mobile wireless networks, as well as relatively new applications over packet-switched networks. The chapter also describes modern electronic components, devices, software, and related standards.

### 2.1.1  Digitization of the PSTN

Today, almost 100% of the world's telecom networks use digital transmission and switching. Figure 2.1 shows the evolution of the analog telephone network to digital in the following stages:

- Twenty-four-channel T1 systems were installed on relatively short-haul interoffice trunks within the exchange areas in the 1960s.

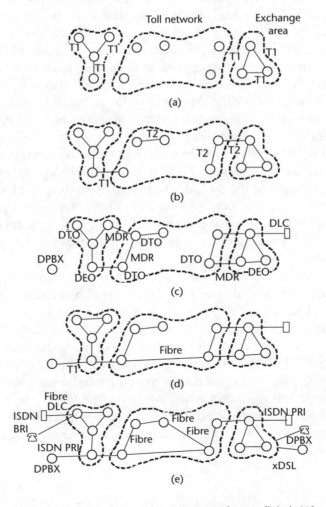

**Figure 2.1**  The evolution of the telephone network from analog to digital. (*After:* [1].)

- In the 1970s, higher capacity T2 systems were introduced for long-distance digital transmission.
- T1 coverage expanded greatly, and by the late 1970s digital loop carrier (DLC) systems and digital private branch exchanges (DPBXs) came into use. Integrated islands of digital switches (toll offices and end offices) interconnected by digital transmission links grew.
- Realization of fully integrated and interconnected digital network came about in the early 1980s with fiber-optic transmission for high-density, long-haul routes. Digital connectivity to customer premises equipment occurred as T1/E1 became the preferred interface for large PABXs.
- End-to-end digital connectivity for voice or data became a reality in the late 1980s with the introduction of ISDNs.

The trend continued with fiber-based technologies, SDH, and SONET. In the mid 1990s, DSL techniques brought bandwidths of several megahertz to individual subscribers, and digital techniques progressed from the network to the subscriber loop.

### 2.1.2 Digitization of Wireless Systems

Wireless systems, both fixed and mobile, have also evolved to encompass digital techniques in their subscriber interface (air interface), while adopting the same digital techniques as the PSTN for transmission and switching. The most important aspect of digital techniques in wireless systems is the development of sophisticated voice coding techniques, described in Section 2.5, which have been developed to satisfy the need for utilizing RF bandwidth efficiently and for operating in harsh environments.

### 2.1.3 Telephony over Packet-Switched Networks

As illustrated in Section 2.1.1, digitization of the PSTN grew from the inside out (i.e., from the network toward the subscriber). Currently, another similar evolution is taking place—that of packet-based transmission and switching techniques.

The most popular application of packet-based techniques for telephony services is IP telephony or Internet telephony, the transmission of voice over the Internet. The underlying principle is that the digitized voice is broken into packets and transmitted and switched individually, without an end-to-end circuit as in the traditional PSTN. These services not only bring telephony at a lower cost, but also enable a variety of services, the integration of many services, and the creation of new services [2]. As such, these techniques constitute an important component of digital telephony today and are discussed in Section 2.6. Section 2.7 discusses the devices and software that support the three types of digital telephony services introduced here.

## 2.2 Advantages of Digital Techniques

The introduction of digital technology was motivated by the desire to improve the quality, add new features, and reduce costs of conventional voice services.

Digitization did not arise from the needs of the data communications industry for better data transmission services. As more and more of the network became digitized, more and more support for direct use of the facilities became available for data applications [1]. The following section summarizes the advantages.

### 2.2.1   Integrated Networks

Techniques used for digital telephony can be used uniformly for any kind of digital traffic. A digitally encoded message presents a common format for transmission, irrespective of the source. Thus, it need not provide any special attention to individual services and can be indifferent to the nature and origin of the traffic. Signals from different sources can be multiplexed, switched, and transmitted in a uniform manner.

### 2.2.2   Modern Semiconductor Technology and Software Support

As the cost of microelectronics dropped, digital techniques gradually became more attractive. The circuitry used to multiplex and switch digital signals is much cheaper than the complex amplifiers and filters used to handle analog signals. Great advantages have been achieved by using very large-scale integrated circuit (VLSI) devices such as analog to digital (A/D) and digital to analog (D/A) converters, voice codecs, multiplexers, DSPs, and switching matrices.

As digital signal processing and switching is mostly software-controlled today, the result is ease of reconfiguration, upgrading, and integration. Many semiconductor manufacturers have introduced specialized telecom parts such as switch fabric ICs to help build fast switching systems, while optical switching systems are gradually coming out of research laboratories.

### 2.2.3   Regeneration

In digital transmission, the requirement is only to detect the presence or absence of a pulse to be detected in the presence of channel impairments. To transmit signals over long distances, intermediate devices that remove some of these impairments are therefore needed.

In analog transmission, such devices are amplifiers, which simply amplify the signals. In digital transmission, regenerators are used. These are devices that catch the bit stream before it is submerged in the noise and then remove noise and distortion from it by creating it afresh. Figure 2.2 compares an amplifier and a repeater. An amplifier cannot eliminate noise and distortion from a signal, while a repeater can.

The direct benefit of signal regeneration is that any distortion introduced in a segment of a transmission link between two repeaters is not carried over to the next segment unless the distortion is so great as to cause errors in the regeneration process. Thus, the quality of communication becomes independent of bit rate and distance if regenerators are placed appropriately.

### 2.2.4   Ease of Signaling

In digital transmission, control information is indistinguishable from message traffic. One means of incorporating control information (i.e., an on/off hook or address

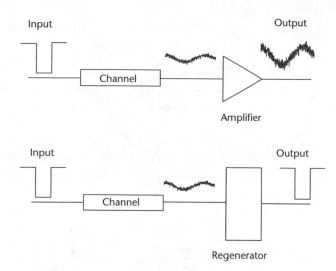

**Figure 2.2**   Amplification versus regeneration.

digits) involves time division multiplexing these as separate, but easily identifiable, control channels. This is called common channel signaling (CCS). Special control commands can also be inserted into the control channel, with digital logic in the receiver decoding them. The result is that the transmission and signaling equipment can be designed, modified, or upgraded independently of each other.

By contrast, analog transmission requires special attention for signaling. As a result, many varieties of control signal formats and procedures have evolved. The move to CCS removes most of the costs associated with interoffice trunks. DSL techniques extend this cost reduction to the subscriber lines.

### 2.2.5   Integration of Transmission and Switching

Figure 2.3(a) shows the multiplexing and switching functions at an exchange in an analog network. The transmission equipment is independent of switching equipment, with standardized interfaces. The incoming signals are demultiplexed, switched to the appropriate outgoing trunk, and then remultiplexed with other circuits routed to the same trunk. This particular switching technique is known as space division switching, as the operation involves the physical connection of a signal from one route to another.

Multiplexing can easily be integrated into the switching equipment in a digital system, as in Figure 2.3(b). Digital encoding/decoding is needed only at the end switches, while intermediate switches carry out the switching function simply by rearranging time slots taken from incoming trunks on the appropriate outgoing trunk. This not only reduces the equipment used, but also improves end-to-end voice quality by eliminating multiple A/D/A conversions and by using low error rate transmission links. This technique combines space division switching for physical connection between incoming and outgoing trunks and time division switching for placing and ordering time slots within each outgoing trunk. Packet switching, simply a form of statistical TDM, is increasingly used in transport networks.

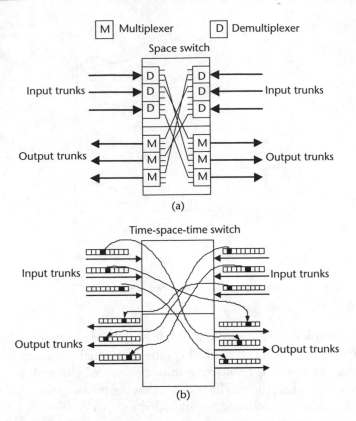

**Figure 2.3**   (a) Switching in FDM systems, and (b) switching in TDM systems.

### 2.2.6   Use of Optical Fiber Technology

The technological development that has had the single most significant impact on telecommunications networks is fiber-optic transmission. Digital transmission dominates fiber applications due to the ease of interfacing digital (on/off) signals to a fiber. Very large numbers of channels can be time division multiplexed to fill the capacity of optical fiber transmission facilities.

### 2.2.7   DSP

Once a signal is encoded, DSP enables many enhancements that result in the improvement of quality of services and efficient use of network and transmission facilities. They also enable high-quality communication in harsh environments.

   Some examples of DSP applications in transmission systems are error control, echo cancellation, and equalization. With respect to communication services or applications, data compression, encryption, noise, and interference cancellation are useful DSP functions.

## 2.3   Disadvantages of Digital Techniques

Digital techniques are, however, not without disadvantages.

   A digitized voice signal at a standard rate of 64 kbit/s requires at least eight times the bandwidth as the analog signal to be transmitted in binary form. Although more sophisticated voice encoding (compression) and modulation schemes exist, none of them can provide the same quality voice without some bandwidth or power penalty.

Especially in long-distance trunks, this bandwidth penalty is a significant disadvantage and is directly dependent on the encoding and modulation schemes used. With multilevel encoding and modulation, greater efficiency in terms of the bit rate in a given bandwidth is achievable. However, with limited transmit power, such a multilevel signal is no longer as immune to noise and distortion as before.

When digital information is transmitted, a timing reference is needed for the receiver to sample the incoming signal. The generation of such a timing signal is known as *synchronization*. When a number of digital transmission links and switches are interconnected, networkwide timing and synchronization must be established. In analog systems, synchronization requirements are much less stringent.

In addition to basic synchronization, when TDM is used in broadcasting, satellite, and mobile communications, an additional level of synchronization is needed. These systems use sophisticated synchronization techniques so that each Earth station or mobile unit times its transmission to arrive at the satellite or base station at precisely defined times, irrespective of their locations.

When digital techniques are used in telephone networks, it is necessary to provide standard analog interfaces to the rest of the network and the subscriber loop. These interfaces represent a significant cost of the digital systems. Digital loop interface devices are described in Section 2.7.

## 2.4   PCM Techniques and Standards

PCM is the first and the most basic of a wide range of voice digitization techniques, with different bit rates, complexity, and voice quality. Encoding a voice channel into a 64-kbit/s data stream provides excellent quality at a moderate cost. However, applications with strict bandwidth limitations such as cellular radio require much more sophisticated voice digitization techniques to operate in a harsh radio environment and to achieve significantly lower data rates, on the order of 8 to 16 kbit/s. Because of its usefulness in a variety of applications, the field of voice digitization continues to be an area of intense research. More details of advanced voice-coding techniques are found in Section 2.5.

### 2.4.1   PCM Basics

The PCM process takes time-separated samples of an analog signal (*sampling*) and represents each sample with a digital code related to its amplitude (*encoding*). These are the basic elements in A/D conversion. A sample from an analog signal may take on a continuous range of values. This requires the *quantization*, or the approximation of each sample into the closest among a number of discrete levels before digitally encoding. The necessity for this is seen by the fact that if a sample is to be encoded by an $n$-bit digital code, there can be only $2^n$ different codes. Hence, the signal must be quantized into this number of levels. The difference between two consecutive quantization levels $\Delta$ is known as the *quantization step size*. Figures 2.4(a) and (b) illustrate these basic concepts.

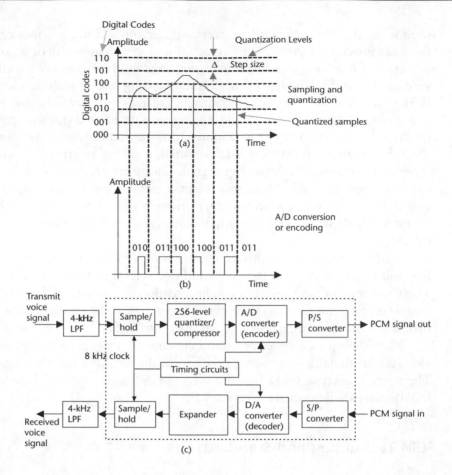

**Figure 2.4**  PCM encoding: (a) the basic steps, (b) the PCM data stream, and (c) a PCM codec.

At the receiver, the reverse process recovers the original analog signal by D/A conversion and low-pass filtering. The PCM encoding and decoding is carried out by a PCM codec (encoder/decoder), as shown in Figure 2.4(c). The terms compression and expansion in Figure 2.4(c) will be discussed later in this chapter.

The basic PCM concepts must be further refined for practical implementation. For instance, *how frequently must the signal be sampled?* And *how many bits are needed to encode each sample?* The more frequently the signal is sampled, and the higher the number of bits representing each sample, the better would be the approximation of the analog signal with the digitized version. However, this will also result in higher data rates and hence in higher bandwidth requirements for transmission and more space for storage. Therefore, the answers to the two questions are important in achieving a trade-off between quality, data rate (or transmission bandwidth), and storage space.

### 2.4.2 Sampling

In 1933, Harry Nyquist derived the minimum sampling frequency required to extract all information in a continuous, time-varying waveform. This result, the *Nyquist Criterion*, is given as

$$F_s \geq 2W \qquad\qquad (2.1)$$

where $f_s$ is the sampling rate or sampling frequency, and $W$ is the bandwidth of the analog signal to be sampled. Hence, a signal has to be sampled at a minimum frequency known as the *Nyquist rate*, which is equal to twice its bandwidth. The maximum possible signal bandwidth that can be sampled at the Nyquist rate is known as the *Nyquist bandwidth*.

It is a well-known fact that the spectrum of a sampled signal consists of images of the analog signal replicated at multiples of the sampling frequency [3]. If a signal is sampled at a rate below the Nyquist rate, a phenomenon known as *aliasing* occurs. This phenomenon is illustrated in the time and frequency domains in Figure 2.5 using a sinusoidal signal at frequency $f_a$. The figure shows the effect of aliasing by progressively reducing $f_s$ from $8f_a$ to $1.5f_a$.

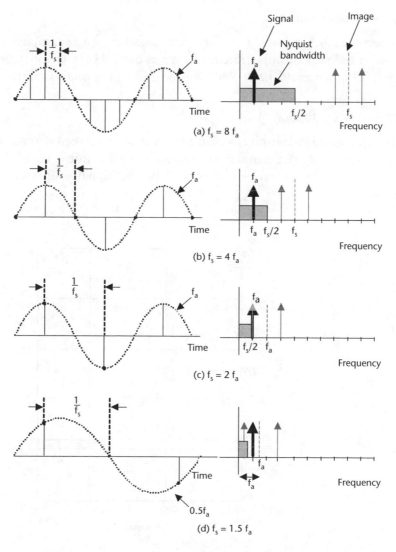

**Figure 2.5**   Aliasing in the time and frequency domains: (a) $f_a = 8f_s$, (b) $f_a = 4f_s$, (c) $f_a = 2f_s$, and (d) $f_a = 1.5f_s$.

In practice, a signal could contain a wide range of frequencies. For example, an audio signal contains frequencies up to 20 kHz. It is essential for an analog signal to be bandlimited before sampling to avoid aliasing. For basic voice, however, not only is the power content very low at high frequencies, it contributes very little toward the intelligibility of the speech. Therefore, it can be eliminated, reducing the required sampling rate and hence the transmission bandwidth. Hence in practical PCM systems, an *antialiasing filter* must precede sampling.

A filter template matching the ITU-T Recommendation G.712 for PCM channel attenuation characteristics in the range of 3,400–4,600 Hz is shown in Figure 2.6. The signal is sufficiently attenuated at 4 kHz by this filter to adequately suppress aliasing at a sampling rate of 8 kHz. Though the phase response of these band-limiting filters and the corresponding reconstruction filters are not critical for voice, they can be a serious impediment to data transmission over PCM systems.

### 2.4.3   Quantization

In Section 2.4.1 the necessity for approximation of the sampled values to a finite number of levels was illustrated. The errors caused by this approximation results in what is known as *quantization noise*, degrading the quality of the signal.

#### 2.4.3.1   Quantization Noise

The effect of quantization errors in a PCM encoder is treated as additive noise with a subjective effect that is similar to band-limited white noise. Hence, the quality of the signal can be quantitatively measured by the signal-to-quantization noise ratio (SQR).

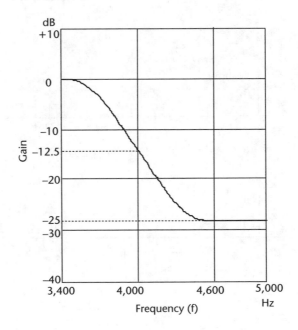

**Figure 2.6**   A filler template matching the channel attenuation characteristics for ITU-T/G.712 (*Source:* [58].)

Figure 2.7 illustrates the variation of quantization errors as a function of input signal amplitude. They can be assumed as being uniformly distributed between $\Delta/2$ and $-\Delta/2$ where $\Delta$ is the step size of the quantizer. It can be shown that if the number of quantization levels is $L = 2^n$, in what is called an $n$-bit quantizer, the SQR is given by [3]

$$SQR \leq 6.02n + 1.76 \, \text{dB} \tag{2.2}$$

Thus, an approximate 6-dB gain in SQR is achieved for each additional bit for quantization (i.e., for each doubling of the number of quantization levels). The above upper limit is valid for a full-amplitude sine wave input (i.e., one spanning the entire input range of the quantizer).

The higher the number of bits (i.e., the more levels that are defined), the less the quantization noise but the greater the bandwidth needed to transmit the extra bits. It can be shown that the SQR increases exponentially with the bandwidth as given here [4]:

$$\left( \frac{\frac{S_0}{N_0}}{\frac{S_i}{N_i}} \right)_{\text{max}} \alpha \, 2^{\frac{2B}{f_m}} \tag{2.3}$$

where $\dfrac{S_i}{N_i}$ and $\dfrac{S_0}{N_0}$ are the SQR at the PCM decoder input and output, respectively, $B$ is the transmission bandwidth, and $f_m$ is the information signal bandwidth. This extremely efficient trade-off results in a large improvement in SNR for a small increase of transmission bandwidth. CCITT recommendations for PCM voice specify eight-bit quantization. This results in a SQR of about 50 dB for a full-load sine wave input.

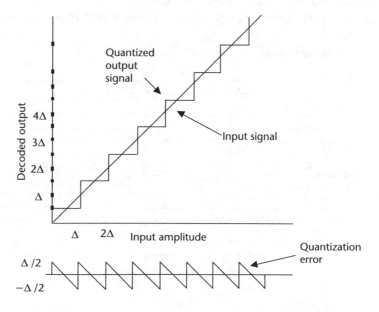

**Figure 2.7**  Quantization errors.

### 2.4.3.2  Nonuniform Quantization

The type of quantization so far discussed is called *uniform quantization* because the quantization levels are assumed to be equally spaced, as in Figure 2.4(a). This technique achieves the maximum SQR only for a full-amplitude signal. In practice, this cannot be the case, and the average SQR can be significantly lower than the maximum given in (2.2). Figure 2.8(a) depicts uniform quantization of a full-amplitude and a smaller sinusoidal signal. It can be seen qualitatively that the approximation of the smaller signal is inferior to that of the larger signal. How much lower an SQR we have in a practical system depends on the *dynamic range* (the ratio of maximum and minimum amplitudes) of the analog signal.

Equation (2.2) can be generalized to take into account smaller signal levels as [1]:

$$SQR = 1.76 + 6.02n + 20\log_{10}\left[\frac{A}{A_{max}}\right] dB \qquad (2.4)$$

where $A$ is the maximum amplitude of the signal being quantized, and $A_{max}$ is the full-load range of the quantizer. This implies lower SQR for small signals compared to large signals.

The probability of occurrence of small amplitudes in speech is much greater than large ones. Consequently, it seems appropriate to provide many quantization levels in the small amplitude range and only a few in the region of large amplitudes. As long as the total number of levels remains unchanged, no increase in transmission bandwidth will be required. However, the average SQR will improve. This technique is referred to as *nonuniform quantization*. Figure 2.8(b) illustrates the improvement in quantization of low-amplitude signals compared to uniform quantization.

The range of amplitudes that occur within a transmission system (i.e., the dynamic range) is considerable. The difference in level between two speech signals in telephony may easily exceed 30 dB. Thus, a nonuniformly quantized system is

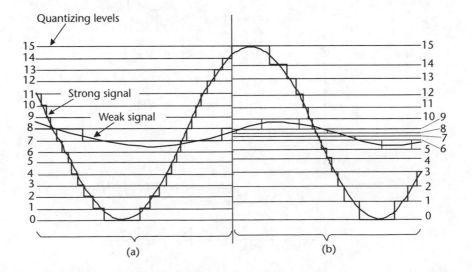

**Figure 2.8**  Quantization: (a) uniform, and (b) nonuniform.

inherently better able to accommodate a larger range of signal amplitudes than the uniform case. When the quantization intervals are not constant, the relationship between the codes and the samples they represent is nonlinear. In nonuniform quantization, the interval heights are increased with the sample value.

### 2.4.3.3   Practical Implementation of Nonuniform Quantization

The basic process of achieving nonuniform quantization is illustrated in Figure 2.9(a). The analog sample is first compressed and then uniformly quantized. Typical compression characteristics in Figure 2.9(b) show the effect of the compression. Successively larger input signal intervals are mapped into constant-length quantization intervals. An expander with inverse compression characteristics is used to recover the original sample value at the receiver. The processes of compression and expansion are called *companding*.

### 2.4.3.4   Companding Laws

The shape of the curve in Figure 2.9(b) is defined by a companding law chosen to match the statistics of the analog signal. Two companding laws are in existence today, the $\mu$-law used in North America and Japan (recommendation G.733 of ITU-T) and the $A$-law recommended by the CCITT used in most other parts of the world (recommendation G.732 of ITU-T). These laws are described by (2.5a) and (2.5b), respectively.

$$F_{\mu}(x) = \text{sgn}(x) \frac{\ln + \mu|x|}{\ln(1+\mu)} \tag{2.5a}$$

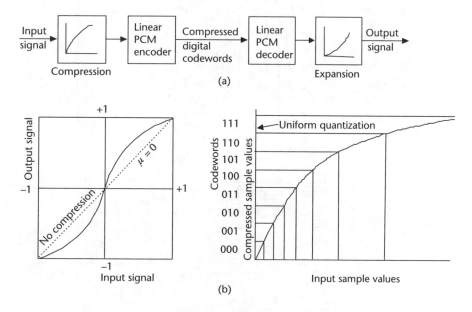

(a)

(b)

**Figure 2.9**   (a) Implementation of nonuniform quantization, and (b) typical compression characteristics.

$$F_A(x) = \begin{cases} \text{sgn}(x)\left[\dfrac{A|x|}{1+\ln(A)}\right], 0 \le |x| \le \dfrac{1}{A} \\ \text{sgn}(x)\left[1+\dfrac{\ln|Ax|}{\ln|A|}\right], \dfrac{1}{A} \le |x| \le 1 \end{cases} \qquad (2.5b)$$

where $x$ is the normalized input signal amplitude $(-1 \le \propto \le 1)$ and $\mu$, $A$ are the parameters, which define the amount of compression in each case.

The main difference between the two is that the $A$-law is perfectly linear for small signal values, while the $\mu$-law only approximates this linearity. Present systems use $\mu = 255$. The $A$-law characteristic is used in Europe and most other parts of the world with $A = 87.6$. Both the $A$-law and the $\mu$-law have the property of being closely approximated by a set of straight-line segments, also referred to as *chords*, with the slope of each successive segment being exactly one-half the slope of the previous one.

The $A$-law defines seven step sizes, equivalent to seven segments for both positive and negative excursions (i.e., a total of 14 segments). However, as the slope of the central region is the same for both positive and negative values, it is common to consider the two segments about the origin as one, resulting in 13 segments. The seven segments for the positive signals based on the $A$-law are shown in Figure 2.10(a). The $\mu = 255$ companding curve is segmented in a similar manner into

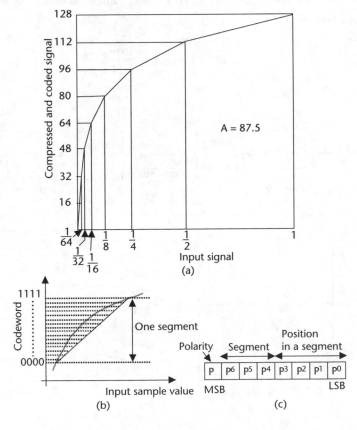

**Figure 2.10** (a) The segmented A-law characteristics, (b) quantization levels within a segment, and (c) the PCM word structure.

15 segments, with eight positive segments and eight negative, and the central region combined into one.

### 2.4.4   Encoding

For encoding, each segment of the compression characteristic is divided into equal quantization levels. In PCM, this number is 16. This is illustrated in the expanded compression curve in Figure 2.10(b), showing the uniformly spaced quantization levels within one segment (chord) of the compression characteristics. An eight-bit PCM codeword as shown in Figure 2.10(c) is composed of one polarity bit, three bits to identify the segment (S), and four bits to identify the quantization level (Q) within the segment. The encoding/decoding table for the segmented *A*-law is shown in Table 2.1. The values are scaled to a maximum of 4,096 for integral representations.

**Table 2.1**   Segmented *A*-Law Encoding/Decoding Table

| Input Amplitude Range | Step Size | Segment Code S | Quantization Code Q | Code Value | Decoder Amplitude |
|---|---|---|---|---|---|
| 0–2 | | | 0000 | 0 | 1 |
| 2–4 | | 000 | 0001 | 1 | 3 |
| ... | | | ... | ... | ... |
| 30–32 | 2 | | 1111 | 15 | 31 |
| 32–34 | | | 0000 | 16 | 33 |
| ... | | 001 | ... | ... | ... |
| 62–64 | | | 1111 | 31 | 63 |
| 64–68 | 4 | | 0000 | 32 | 66 |
| ... | | | ... | ... | ... |
| 124–128 | | | 1111 | 47 | 126 |
| 128–136 | | | 0000 | 48 | 132 |
| ... | 8 | 011 | ... | ... | ... |
| 248–256 | | | 1111 | 63 | 252 |
| 256–272 | | | 0000 | 64 | 264 |
| ... | 16 | 100 | ... | ... | ... |
| 496–512 | | | 1111 | 79 | 504 |
| 512–544 | | | 0000 | 80 | 528 |
| ... | 32 | 101 | ... | ... | ... |
| 992–1,024 | | | 1111 | 95 | 1008 |
| 1,024–1,088 | | | 0000 | 96 | 1056 |
| ... | 64 | 110 | ... | ... | ... |
| 1,984–2,048 | | | 1111 | 111 | 2016 |
| 2,048–2,176 | | | 0000 | 112 | 2112 |
| ... | 128 | 111 | ... | ... | ... |
| 3,968–4,096 | | | 1111 | 127 | 4032 |

(*Source:* [1].) Line encoding is done subsequent to PCM encoding.

## 2.5   Low Bit-Rate Voice Coding

Modern telephony applications make use of low-bit-rate voice coding techniques in order to conserve bandwidth. The primary motivations for low-bit-rate coding are the need to minimize transmission costs and storage and the need to transmit over channels of limited capacity, such as mobile radio channels. In addition, there are also needs to share capacity for different services, such as voice, audio, data, graphics, and images, in integrated services networks and to support variable-rate coding in packet-oriented networks [5].

### 2.5.1   Digital Encoding of Waveforms

Following the introduction of PCM, network operators soon realized that by using 32-kbit/s adaptive differential PCM (ADPCM), they could double the capacity of important narrow bandwidth links such as undersea cables. At that time, the goal of speech coding was to provide a compression technology that would enable copper cables to handle the continual growth in voice traffic.

Digital encoding of waveforms, however, entails the introduction of some kind of coding distortion, such as quantization noise. Speech coders compress speech by analyzing and then quantizing features of the speech waveforms in ways that attempt to minimize any audible impairment [6]. As such, the basic challenge in waveform encoding is to achieve the minimum possible distortion for a given encoding rate or, equivalently, to achieve a given acceptable level of distortion with the least possible encoding rate [7]. The process of speech compression is very computationally intensive, incurs delays, and requires powerful DSPs for implementation.

Some encoding techniques, such as PCM, are *lossless*, providing a reconstructed waveform that exactly matches the original signal sample for sample. Other methods achieve higher compression through *lossy* techniques, which do not allow exact reconstruction of the signal but instead seek to preserve its information-bearing characteristics [7].

While PCM encodes signal samples one by one independently of neighboring samples, other, more advanced, signal-encoding techniques make use of the inherent correlation that exists between adjacent samples to achieve high compression ratios. This requires aggregation of samples and introduces a delay into the encoding process. Trade-offs between the encoding delay and the compression ratio may be made, depending on the application. Computational complexity of the encoding/decoding algorithms is another factor that determines the choice of hardware needed in different applications. Speech quality is yet another factor used for selection of a speech codec for a particular application.

Basic requirements in the design of low-bit-rate coders can be summarized as [5]:

- Data rate;
- High quality of reconstructed signals;
- Low encoder/decoder delays;
- Low complexity and power consumption;
- Robustness to random and bursty channel bit errors and data losses;

- Robust concatenation of codecs;
- Graceful degradation of quality with increasing bit error rates.

Rapid progress has been made in the coding of audio bandwidth signals during the late 1990s, with advanced encoding algorithms that have better compression than was thought possible in the early 1990s. These techniques are discussed in the following sections. DSPs with low power consumption, high processing power, speed, and specialized architectures have helped the implementation of these codecs, as discussed in Section 2.7.

### 2.5.2   Overview of Speech Coding Techniques

Speech coding algorithms can be classified into the following four types: direct quantizers, waveform coders, vocoders, and hybrid coders.

#### 2.5.2.1   Direct Quantizers

The simplest and most widely used coders in standard telecommunications systems are called direct quantizers. PCM, defined by the International Telecommunications Union (ITU) Recommendation G.711, falls into this category. For a variety of input levels, these quantizers maintain an approximate 35-dB SQR. PCM is used in encoding telephone bandwidth voice (0.3–3.4 kHz) sampled at 8 kHz and encoded at 8 bits/sample.

#### 2.5.2.2   Waveform Coders

Waveform coders are those in which an analog input signal is approximated by mimicking the amplitude-versus-time characteristics of the signal. As these coders are nonspeech-specific, they can cater to many nonspeech signals, background noise, and multiple speakers without difficulty. The penalty is the relatively high bit rate.

In differential PCM (DPCM), the sample-to-sample similarity, or correlation, may be exploited to reduce the bit rate by predicting each sample based on previous samples, comparing the predicted sample value to the actual sample, and encoding the difference between the two. As this difference is much smaller than the sample value, fewer bits are needed to encode it. DPCM is illustrated in Figure 2.11. Relatively straightforward implementation of DPCM can provide savings of one to two bits per sample, compared to standard PCM.

The *linear predictor* uses the weighted sum of several previous samples to form a good prediction of the current speech sample.

The predicted sample is given by:

$$\tilde{x}(n) = \sum_{i=1}^{p} a_i \hat{x}(n-i) \qquad (2.6a)$$

where $\hat{x}(n-i)$ is the encoded and decoded $(n-i)$th sample as shown in Figure 2.11(a).

The difference $e(n)$ between the predicted sample and the actual sample is given by:

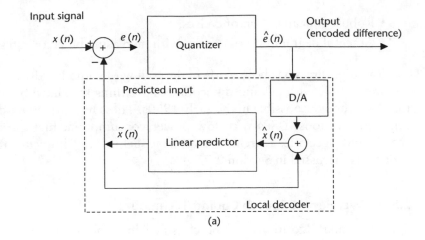

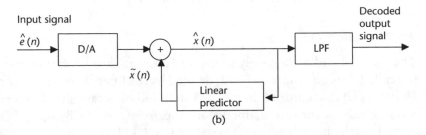

**Figure 2.11**   (a) DPCM encoder, and (b) decoder.

$$e(n) = x(n) - \tilde{x}(n) \tag{2.6b}$$

The quantizer produces $\hat{e}(n)$, the quantized prediction error, which is the encoder output.

The decoder in Figure 2.11(b) reconstructs the signal using the same linear prediction technique as in the encoder, as can be seen by comparing the signals at corresponding points in Figures 2.11(a) and (b). The encoder contains an embedded decoder, a common feature in many voice-encoding techniques.

Adaptive differential PCM (ADPCM), an extension of this technique, can provide further savings. The encoding is conceptually similar to DPCM. However, an *adaptive predictor* changes the quantization step size based on a shorter term average of signal statistics. The step size used to quantize $e(n)$ is a function of its amplitude.

Several ITU standards exist for ADPCM. The most important is Recommendation G.721, which forms the core of other ITU ADPCM standards. This defines a 32-Kbps bit rate speech-coding algorithm. ITU Recommendation G.726 extends ADPCM to include bit rates of 16, 24, and 40 kbit/s, giving good to excellent speech quality.

Recommendation G.721 is used extensively in the common air interface for the base stations in cordless telephony and WLL systems. It is also used commonly in digital circuit multiplexing equipment (DCME) to increase the capacity of long-distance trunks. The basic speech coding rate is 32 kbit/s, with 24 kbit/s and 16 kbit/s employed on channels at peak load periods. The 40-kbit/s rate is used for voice-band data transmission. The ITU-T Recommendation G.727 is a variant of G.726, suitable for use on packetized networks.

### 2.5.2.3   Vocoders

Vocoders try to preserve information and speech content without trying to match the waveform. They extract perceptually significant parameters from the input signal and use these to synthesize an output signal that is intelligible to the human ear, though not a waveform replica of the input [7]. A set of parameters is derived at the encoder. The parameter set includes the model of the speaker's vocal tract and a coarsely approximated excitation signal for this model. These parameters can be efficiently quantized for transmission. At the decoder, these are used to control a speech-production model. Hence, vocoders operate at very low bit rates. However, these require much more complex algorithms than waveform coders. Vocoders are also known as *parametric coders*.

The operation of vocoders is closely based on the speech-production model shown in Figure 2.12(a). This model uses a vocal tract representation that is excited by either a voiced or unvoiced signal. Linear predictive coding (LPC) is used to find a filter approximating the vocal tract. The vocal tract model is excited by a periodic pulse train and by a noise-like signal to produce voiced and unvoiced sounds, respectively. The amplitude and pitch of the former and the amplitude of the latter are determined in the analysis process. As the vocal tract model is used to generate sounds, it is also called the *synthesis filter*.

The encoding process is based on the principle shown in Figure 2.12(b). If the input speech signal is filtered with the inverse of the vocal tract filter, called the

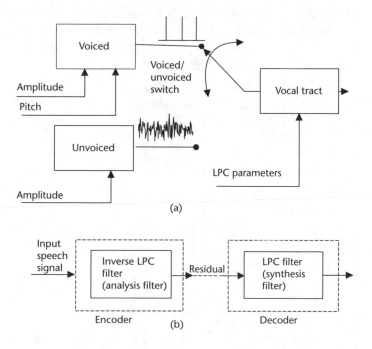

**Figure 2.12**   (a) Vocoder model of speech production, and (b) principle of operation of the vocoder.

*analysis filter*, and the output, called the *residual*, is used as the excitation for the synthesis filter (LPC model of vocal tract), the original input waveform is produced. In other words, the decoder synthesizes speech by passing the residual through an LPC synthesis filter. The analysis and synthesis filter are parts of the encoder and decoder, respectively.

Such analysis-synthesis provides the greatest economies in terms of bit rate. But they also provide fundamental limitations on subjective quality and great expenses in terms of coder complexity and cost. As vocoders are strongly based on the simple speech production model described in Figures 2.12, they perform poorly with high levels of background noise, multiple speakers, and nonspeech signals. Speaker identity information may also be lost in the process.

The actual implementation of vocoders is shown in Figure 2.13. The key factor here is the generation of the excitation signal parameters from the residual. If the residual signal is strongly periodic, the section of speech is declared as "voiced" with a measured pitch period, and the excitation is modeled as a series of unipolar pulses spaced at the pitch frequency. If not, the speech segment is declared as "unvoiced," and the excitation is approximated with random noise. The excitation signal is generated in the decoder from this relatively crude approximation and is fed to the synthesis filter. The transmitted information includes the excitation signal and the LPC coefficients representing the vocal tract, which are periodically updated. LPC analysis is an important technique in speech coding and is discussed in detail in [8].

LPC vocoders operate at around 2.4 kbit/s and give synthetic-quality speech, which is unacceptable for telephony applications due to its inability to provide speaker identification. It is, however, acceptable for military applications as a means of providing secure, low-bit-rate communications. FS-1015 LPC-10E is a U.S. federal standard algorithm widely used for military communications [9, 10]. Other vocoding techniques, such as sinusoidal transform coding, are discussed in [9–11].

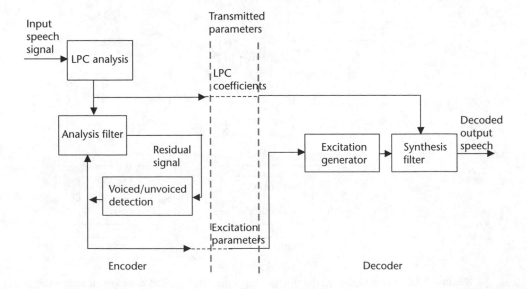

**Figure 2.13** Implementation of encoding and decoding in vocoders.

### 2.5.2.4  Hybrid Coders

Hybrid coders combine features from both waveform coders and vocoders. In common with vocoders, they make use of a speech-production model, but they use a more refined excitation signal than vocoders use. Hybrid coders operate at medium bit rates between those of vocoders and waveform coders and produce higher quality speech than vocoders. Many varieties of hybrid encoding schemes have been standardized and are widely used in telecommunication systems.

Instead of approximating the residual signal by a stream of periodic pulses or a noise signal, as in the case of vocoders, they generate a better approximation to the residual. Excitation parameters corresponding to this waveform are then transmitted with the LPC coefficients to the receiver. There are a number of hybrid encoding techniques, which differ in the way excitation parameters are generated. Some of these techniques quantize the residual directly, while others substitute approximated quantized waveforms selected from an available set.

*Hybrid Coders Based on Residual Quantization*
The simplest method of approximating the residual for digital transmission is direct quantization. However, this leads to relatively high bit rates, as in the case of PCM. The same adaptive techniques used in direct quantization can be used to reduce the bit rates, with varying degrees of success.

*Adaptive Predictive Coding*
In this technique, the adaptive techniques used in ADPCM are applied to encode the residual signal. However, the residual is noise like to a great extent, resulting in none of the differential encoding techniques being successful in reducing the bit rate.

*Residual Excited Linear Predictive Coding*
In this technique, the residual is lowpass filtered to around 1 kHz and resampled at a lower rate. As the high-frequency components of the residual are removed, the signal to be quantized has more correlated characteristics. Hence, it can be quantized at a lower bit rate. At bit rates less than 8 kbit/s, both adaptive predictive coding (APC) and residual excited linear predictive coding (RELP) become very complex to achieve good quality.

*LPC-Based Analysis-by-Synthesis Hybrid Coding Schemes*
In LPC vocoders, the LPC synthesizer model is used only at the decoder, as shown in Figure 2.13. However, LPC-based hybrid coding schemes use the synthesizer model at the encoder as well. At the encoder, the LPC coefficients of the synthesizer model are obtained directly from the input speech signal, and the excitation signal is derived using a closed loop analysis-by-synthesis technique, as illustrated in Figure 2.14. This approach to coding is also called linear-prediction-based analysis-by-synthesis (LPAS) coding [11].

The basic components of the LPAS codec in Figure 2.14 can be summarized as follows:

- *The basic decoder:* This receives data, which specify an excitation signal and a synthesis filter. The reproduced speech is generated as the response of the synthesis filter to the excitation signal.

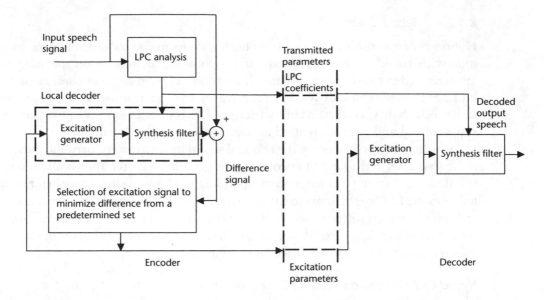

**Figure 2.14**  Principle of operation of LPAS codecs.

- *The synthesis filter*: The time-varying linear prediction (LP)-based synthesis filter in the encoder is periodically updated and is obtained by LP analysis of the signal, segment (frame) by segment.
- *Analysis-by-synthesis excitation coding*: The encoder determines the excitation signal one segment at a time, by feeding candidate excitation segments into a replica of the synthesis filter and selecting the one that minimizes the distortion between the original and reproduced speech.

Different algorithms used in the LAPS category in selecting the excitation signal are listed in Table 2.2.

### 2.5.2.5   Variable-Rate Speech Coding

For digital storage and for such telecommunications applications as packet voice, a variable-bit-rate (VBR) output is advantageous. VBR speech coders can exploit pauses and silent intervals in conversational speech and the fact that different speech segments may be encoded at different rates while maintaining a given reproduction quality. A common technique is to substitute silent periods with a noise-like signal known as *comfort noise*. Consequently, the average bit rate for a given reproduced speech quality can be substantially reduced if the rate is allowed to vary with time [11]. The rate may be controlled internally by the statistical character of the incoming speech signal or externally by the current traffic level in a multiuser communication network.

Applications that have motivated the development of variable-rate speech coding include speech storage, packetized voice, and digital speech interpolation (DSI) for DCME. Multiple access schemes for wireless communication, particularly CDMA, have recently become an important application for VBR coding. A variable-rate coding algorithm, QCELP, has been standardized by the Telecommunications Industries Association (TIA) for use in IS-95 CDMA cellular systems.

**Table 2.2**   LPAS Algorithms

| Name | Description |
| --- | --- |
| Multipulse excited LPC (MPE) | A series of nonuniformly spaced pulses is used as the excitation. No distinction is made between voiced and unvoiced speech. For high-quality speech, several pulses are required per pitch period. As all pulse amplitudes and positions must be transmitted, a quality versus bit rate trade-off has to be made. |
| Regular pulse excited LPC (RPE) | RPE uses regularly spaced pulse patterns for the excitation. The minimum-distortion pulse pattern is selected from a set of candidate patterns. The position of the first pulse and the pulse amplitudes are determined in the encoding process. The 13-kbit/s full-rate codec selected for the pan-European digital cellular system, Global System for Mobile (GSM), is essentially an RPE LPC algorithm selected by the European Telecommunications Standards Institute (ETSI). This is known as RPE with long-term prediction (RPE-LTP). |
| Code excited LPC (CELP) | The most important form of LPAS coding today is CELP. CELP improves MPE by using vector quantization (VQ). A predesigned set of excitation vectors is stored in a codebook, and for each speech segment the encoder searches for the vector that serves as the best excitation signal. The address of the selected vector is transmitted to the receiver. The selected excitation signal is generated by looking up the same codebook at the receiver. Performance is superior to MPE and earlier coding methods for bit rates ranging from 4.8 to 16 kbit/s. CELP refers to a family of algorithms. Almost all speech-coding standards recently developed belong to the LPAS/CELP category. These coders include U.S. Federal Standard 1016 (4.8 kbit/s), vector sum excited linear prediction (VSELP) (8 kbit/s for North American digital cellular systems), Japanese digital cellular standard (6.7 kbit/s), and ITU Recommendations G.723.1, G.728, and G.729. These are described in detail in [12]. G.723.1 and G.729 are preferred for packetized voice applications. |
| Algebraic codebook excited LP (ACELP) | If the codebook contains a number of possible sequences, the excitation for each frame can be described completely by the index to the appropriate vector in the codebook. In ACELP, a simplified codebook search technique is used [12]. ITU Recommendations G.729 and G.729A use the ACELP technique. |

An important component in variable-rate speech coding is *voice activity detection* (VAD), which is needed to distinguish active speech segments from pauses when only background noise is present. This is particularly challenging for mobile phones due to vehicle or other environmental noise. The most common technique used in VBR coding is to use a fixed rate for active speech and a lower rate for background noise. Because human conversations are essentially half duplex in the long term, using VAD can realize an approximately 50% reduction in bandwidth requirements. Figure 2.15 illustrates VAD.

Front-end clipping and holdover distortion shown in Figure 2.15 are two possible causes for speech quality degradation with VAD. Excessive front-end clipping can make it difficult to understand the speech, while excessive holdover time can reduce network efficiency. Too small holdover time can cause speech utterances to sound abrupt and disconnected. The decoder generates local noise (comfort noise) that it presents to the listener during silent periods. Decoder comfort noise generation (CNG) complements encoder voice activity detection.

### 2.5.3   Speech Coder Attributes

Speech quality as produced by a speech coder is a function of bit rate, complexity, and delay. There is a strong interaction between all of these attributes, and they can be traded off against each other.

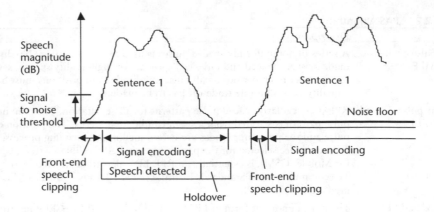

**Figure 2.15** VAD.

### 2.5.3.1 Bit Rate

The primary factor affecting the bit rate is the speech coding algorithm itself. Simple waveform coders that quantize and encode time-domain parameters generally result in higher bit rates than vocoders and hybrid coders, which encode parameters of the voice-production model. The different hybrid coding techniques discussed in Section 2.5.2.3 illustrate the compromise between coder complexity and bit rate.

### 2.5.3.2 Delay

Speech coding techniques process one frame of speech at a time, the length of which depends on the algorithm. The speech parameters are updated and transmitted once every frame. This *frame time* is one component of delay in the speech coding process. Additionally, in most speech coding techniques, except the simplest ones, it is also necessary to look beyond the present frame to take advantages of correlation properties of speech to reduce the bit rate. The adaptive predictor in (2.6a) and (2.6b) is an illustration of this. This is referred to as *look-ahead delay*. The *algorithmic delay* is the sum of the frame time and the look-ahead delay and cannot be changed by changing the implementation. The decoding delay is typically half the algorithmic delay.

The second major component of delay comes from the time it takes the encoder to analyze the speech frame and the decoder to reconstruct it. This is referred to as *processing delay* and depends on the speed of the hardware used to implement the coder. The sum of the algorithmic and processing delay is the *one-way codec delay*. Table 2.3 illustrates the algorithmic delay components of standard speech codecs.

The third component is the communication delay or *propagation delay*, which is the time it takes for an entire frame to be transmitted from the encoder to the decoder. The total of these three components is the *one-way system delay*.

The delay of the system affects the user's perception of QoS. For transmission applications, delay is important for two reasons—first, because excessive delay (more than about 400 ms, one way) causes conversational difficulties, and, second, because if there are any sources of echo in the transmission path, delay can cause the echo to become objectionable. For ease of communication, it is preferable if the one-way delay is below 200 ms. If there are echoes, the maximum tolerable delay is

**Table 2.3**   Algorithmic Delay Components of Standard Speech Codecs

| Standard | Bit Rate (Kbit/s) | Frame Size/ Lookahead (ms) |
|---|---|---|
| *ITU Recommendations:* | | |
| G.711 (PCM) | 64 | 0.125/0 |
| G.726 (G.721, G.723)[1] G.727 (ADPCM) | 16, 24, 32, 40 | 0.125/0 |
| G.722 Wideband coder | 48, 56, 64 | 0.125/1.5 |
| G.728 (LD-CELP) | 16 | 0.625/0 |
| G.729 (CS-ACELP) | 8 | 10/5 |
| G.723.1 (MPC-MLQ) | 5.3 and 6.4 | 30/7.5 |
| G.729A (CS-ACLELP Annex A) | 8 | 10/5 |
| *Cellular Standards:* | | |
| RPE-LTP (GSM) | 13 | 20/0 |
| IS-54 VSELP (U.S. digital cellular) | 7.95 | 20/5 |
| PDC VSELP (Japanese digital cellular) | 6.7 | 20/5 |
| IS-96 QCELP (U.S. CDMA cellular) | 8.5, 4, 2, 0.8 | 20/5 |
| PDC PSI-CELP | 3.45 | 40/10 |
| *U.S. Department of Defense Standards:* | | |
| FS-1015 LPC-10E | 2.4 | 22.5/90 |
| FS-1016 CELP | 4.8 | 30/7.5 |
| MELP | 2.4 | 22.5/23 |

(*Source:* [10].)
[1]Note: G.721 and G.723 were merged into G.726 in 1990.

25 ms. Hence, the use of an echo canceler is often necessary. The effect of delay on speech quality is further discussed in Section 2.6.5.2 with reference to packetized voice systems.

### 2.5.3.3   Speech Quality

Voice quality is a way to describe and evaluate speech fidelity, intelligibility of the analog voice signal itself, and is defined as the qualitative and quantitative measure of the sound and conversation quality of a telephone call [13]. One of the most important contributions to speech quality comes from the speech codec.

The quality objective for speech codecs used in public telecommunications systems is usually to achieve the same speech quality as in a long-distance telephone call on the analog PTSN [9]. This is often referred to as *toll quality*. Because such calls are variable in quality, toll quality is not a very specific definition. A more precise target would be to meet the transmission performance standards agreed by the ITU for PSTNs. However, there are also services where it would be uneconomical to meet the ITU quality criteria with currently available technology and where lower speech quality is deemed acceptable. Analog cellular services are such an example. The full-rate speech codecs in second generation digital cellular systems were chosen to give quality as good as the analog cellular systems.

For speech codecs forming part of a telecommunications system, the important factors that affect the speech quality are transmission errors, wide dynamic range of input signals, background noise, multiple/tandem codecs, and delay.

Traditionally, the mean opinion scores (MOS) based on subjective testing under carefully controlled conditions have been used for speech-quality testing. This test rates speech quality on a scale of 1 to 5. The assessment of speech quality from low-bit-rate voice codecs poses problems quite different from the assessment of wave-form codecs such as *A*-law PCM used at higher bit rates. This is because the distortions produced by low-bit-rate codecs are diverse in character, and subjective evaluations of these degradations vary significantly from person to person. Quality of speech classified according to the MOS for a number of standard codecs is given in Figure 2.16. This picture represents the quality for a single encoding without background noise.

### 2.5.3.4 Complexity

Speech coders often share or are implemented on special-purpose hardware such as DSPs and microprocessors. Their complexity can be assessed in terms of computing speed in millions of instructions per second (MIPS), RAM, and ROM requirement. Having to put more RAM and ROM on a chip results in a more expensive chip. Implementation structures of speech codecs is discussed in [8]. Algorithmic complexity can be qualitatively appreciated by examining Figures 2.11, 2.13, and 2.14. Higher complexity also results in greater power usage. For portable devices, this means reduced time between battery recharges (more expense) or using larger batteries (more weight).

### 2.5.3.5 Tandem Connections and Bridging

When a speech codec is connected in tandem with the same or a different codec, distortion accumulates at each step. Such interconnections can occur, for example, when low-rate coding is used (as shown in Figure 2.17) in the national/international links or mobile-to-mobile connections. In applications such as teleconferencing, it is necessary to bridge several callers so that each person can hear all others. This means

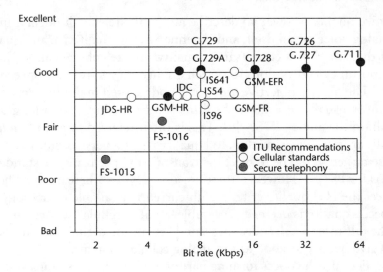

**Figure 2.16** Quality comparison of standard speech codecs. (*Source:* [12]. ©1996 IEEE, reprinted with permission.)

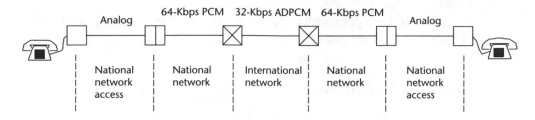

**Figure 2.17**   Tandem connection of codecs in a mixed analog/digital network.

decoding each bit stream, summing the decoded signals, and then reencoding. This process not only doubles the delay, but also reduces speech quality. For certain types of speech codecs, tandem connections with other speech codecs (identical or different type) can result in a small amount of distortion introduced by the first codec, leading to a large increase in the distortion introduced by the second codec. Reference [14] describes how the distortion budget is managed to ensure end-to-end transmission quality.

A comparison of codecs using a broader classification of quality and complexity are presented in Table 2.4. As there is no universally accepted measurement of complexity for speech coding algorithms, complexity is presented relative to PCM, which has been assigned an arbitrary value of 1. Table 2.4 illustrates the general trade-offs between speech quality, bit rate, and complexity.

The three quality categories used in Table 2.4 are:

1. Toll quality—quality of long distance PSTN connections;
2. Communications quality—lower than toll quality but speaker identity and intelligibility maintained;
3. Synthetic quality—speech sounds synthetic in nature and speaker identity information is largely lost.

While the complexity of new speech coding algorithms continue to increase, the computational power available with digital signal processors is correspondingly

**Table 2.4**   Complexity and Quality Comparison for Several Popular Speech Codecs

| Speech Coding Technique | Bit Rate (kbit/s) | Speech Quality | Complexity |
|---|---|---|---|
| PCM (G 711) | 64 | Toll | 1 |
| ADPCM (G721) | 32 | Toll | 10 |
| CELP (G728) | 16 | Toll | 450 |
| RPE-LPC (GSM) | 13 | Communications | 100 |
| VSELP (U.S. Cellular Standard) | 8 | Communications | 250 |
| FS -1016 (U.S. Department of Defense) | 4.8 | Synthetic/ Communications | 400 |
| FS-1015 LPC-10 (U.S. Department of Defense) | 2.4 | Synthetic | 100 |

(*Source:* [9].)

increasing to keep pace with the demand. Modern DSPs used in the implementation of speech codecs are discussed in Section 2.7.

## 2.6   Voice over the Internet

VoIP comprises several interconnected processes that convert a voice signal into a stream of packets, transmit them on a packet-switched network, and convert them back into voice at the destination. It allows the traditional services of the PSTN, voice, and fax information to travel over a packet data network concurrently with data packets.

VoIP is being adapted for transmission over technologies such as frame relay, ATM, and media such as cable and twisted pair using DSL techniques. These variations are referred to as VoFR, VoATM, VoCable, and VoDSL and are generally referred to by the umbrella names voice over packet (VoP) or VoX. VoP provides rich benefits for all levels of users, from networking equipment manufacturers and service providers to businesses and home users.

This section will describe the basic principles, practical implementation, architectures, standards, and technical difficulties in VoP technologies, with emphasis on VoIP. The terms VoIP and Internet telephony will be used interchangeably in the discussion.

### 2.6.1   Introduction to Voice over Packet-Switched Networks

Traditionally, the networking world has been divided along two lines: circuit switching and packet switching. The PSTN is based on circuit switching, which offers a guaranteed QoS to subscribers by establishing a dedicated end-to-end circuit for each call. By contrast, data networks comprise LANs and wide area networks (WANs) that are intertwined in the form of the Internet and many partitioned intranets, carrying data on logical circuits for which resources are not reserved, but allocated when available. Hence, these networks do not provide the same QoS to subscribers as circuit-switched networks. Chapter 3 provides a summary of packet switching and circuit switching.

However, in recent times, we have witnessed the merging of these two worlds, both in terms of services as well as underlying transport and switching techniques. The circuit-switched networks designed for point-to-point communication of real-time voice have been adapted for the growing needs of data communications via modem technologies, ISDN, digital carrier, and, most recently, integrated services frame relay and ATM backbones [15]. The proliferation of the TCP/IP-based Internet in the 1990s has given rise to a strong interest in carrying real-time traffic such as voice and video over the Internet. With the revolution of multimedia in the computer industry, voice, video, and data are now being carried on both networks. Supplementary services such as call forwarding and transfer, which originated in private telephone networks and later migrated to the PSTN, are now being developed for packet networks [16].

ITU-T Recommendation H.323 [17, 18] applied to multimedia communications over packet-switched networks has greatly aided the development of IP telephony and other associated services. H.323 has application in a variety of network media, including LANs, enterprise networks, wide area networks, dial-up connections to LANs, and the Internet.

Because the Internet was not initially designed for carrying real-time traffic, voice over the Internet presents a number of challenges and technical issues. In the IP community, QoS has not been considered with the same importance as it was in the telecommunications community. The current Internet service model is flat, offering a classless, "best-effort" delivery service [15]. However, the next generation of IP, version 6, and other new technologies developing in its support, provide different levels of QoS, where customers requiring the assurance of better QoS and willing to pay more should be able to get better service than customers paying the basic rate. These developments predict a favorable future for real-time services over the Internet.

In the near term, Internet telephony is motivated by lower costs. In the longer term, telephony over the Internet brings in a number of services and their integration, which are not possible using traditional circuit-switched telephone networks [19]. It affords an opportunity to rethink many of the fundamental design choices that have governed the architecture of the world's most widely used telecommunications service, telephony [2].

### 2.6.2   Basics of VoIP

The basic steps involved in originating an Internet telephone call are conversion of the analog voice signal to digital format and translation of the digital data into packets for transmission over the Internet. The process is reversed at the receiving end. Voice coding usually includes some form of compression to save bandwidth. The digital speech data is processed in units known as frames, with each frame containing a portion of a speech signal of a specific duration. These frames are inserted into IP packets, which contain additional information (overhead) such as packet sequence numbers, IP addresses, and timestamps, all of which are necessary for the packet to traverse the network successfully. To reduce the inefficiencies caused by this overhead, it is common to pack several voice frames into one IP packet. Packetization of voice is illustrated in Figure 2.18(a). The IP packets are received in a play out buffer at the receiver, decoded in sequential order and played back. These basic processes are shown in Figure 2.18(b).

Transferring voice over a packet network designed to transport time-insensitive data requires VoIP to be much more complex than this simplified overview. The major constraints that the Internet poses on voice transmission are all due to the absence of an end-to-end circuit through the network. This results in packets being delayed at nodes, the delay being variable, and loss of packets in the network. These issues, their effect on voice quality, and some techniques used to overcome these limitations are discussed in Section 2.6.5.

### 2.6.3   VoIP Configurations

VoIP systems can allow communication between two multimedia PCs over a data network, between two telephones over a data network, and between PCs and phones.

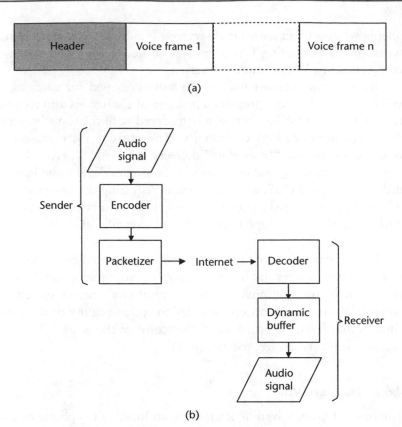

**Figure 2.18**  (a) Packetization of voice, and (b) basic processes in VoIP.

### 2.6.3.1  Basic Configurations

The two basic models for VoIP are the PC-to-PC architecture and the phone-to-phone architecture shown in Figure 2.19(a) and (b).

The PC-to-PC configuration is based on the assumption that users have access to multimedia computers that are connected to the Internet. These computers can be on a LAN, as in a corporate environment, or connected via telephone lines to Internet service providers (ISPs), as in home use. All encoding and packetization of the voice signal occurs in codec hardware and software on the sender's PC, while play out of the received signal occurs in the sound card on the receiver's PC. Alternately, the codec could be implemented in hardware as part of a modem, network interface card, or sound card. A user places a call by specifying the IP address of the recipient or by looking up the recipient's name in a public directory such as the dynamic host configuration protocol (DHCP) service provided by ISPs.

In the phone-to-phone architecture, the user calls the Internet telephony gateway that is located near a central office (CO) switch or a PABX. The gateway prompts the user to enter the phone number of the recipient and initiates a VoIP session with the gateway that is closest to the recipient. This gateway then places a call to the recipient's phone. End-to-end communication can proceed subsequently, with voice sent in IP packets between the two gateways. Encoding and packetization occurs at the sender's gateway while decoding, reassembly, and replay occur at the

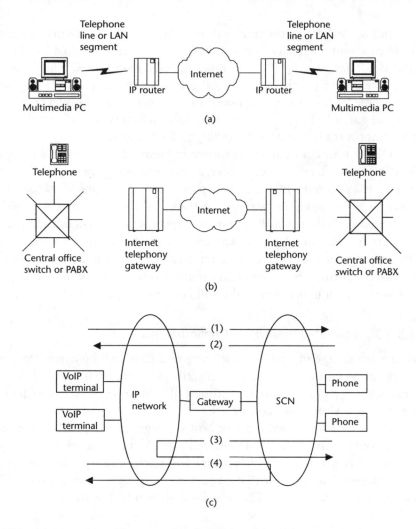

**Figure 2.19** (a) PC-to-PC architecture of VoIP, (b) phone-to-phone architecture of VoIP, and (c) connection types in hybrid IP/switched communications network (SCN) networks.

recipient's gateway. The CO or PABX may digitize the message or pass it on to the gateway for digitization. The signal on both users' local loops is analog. The gateways are implemented in hardware using powerful DSPs and ASICs capable of processing many VoIP calls at the same time.

### 2.6.3.2 Hybrid VoIP Configurations

Hybrid VoIP configurations, where a PSTN subscriber will call a PC user and vice versa, have naturally risen. Such services have attracted a great deal of interest in recent years. Hybrid configurations also include PC-PC paths with intermediate transport over the switched-circuit network (SCN) and phone-to-phone paths with intermediate transport over IP or other packet networks. These configurations are illustrated in Figure 2.19(c). SCN refer to all circuit-switched networks including ISDN, fixed wireless, and cellular networks.

In Figure 2.19(c), the terminals attached to the IP network are assumed to have VoIP capability and may be connected to the IP network via a LAN or a dial-up connection. They may be PCs or stand-alone IP telephones. The IP network and the SCN are connected through a VoIP gateway. This gateway transports signaling as well as media information between the two networks. However, it should be noted that the gateway has many other functions, which will be discussed in Section 2.6.4. One important function is the billing of local users.

Of the four configurations shown in Figure 2.19(c), type 1 is a call initiated by VoIP terminal to an SCN customer, as exemplified by various *PC-to-phone* services. Type 2 is a *phone-to-PC* service. Type 3 is a call from one SCN customer to another, where, for example, the two local networks are circuit switched, and long-distance transmission is done via IP networks. Type 4 services will occur as service providers migrate traditional telephony services to an IP infrastructure. In this type, the long-distance carrier uses an existing circuit-switched route, while the local networks are IP based. During the evolutionary phase from circuit-switched to packet-switched networks, such hybrid networks will be commonplace.

### 2.6.3.3 Integrated Voice/Data Configurations

A key advantage of VoIP is that voice and data can be carried over the same network, eliminating the need for subscribers to maintain two or more services and bear several types of communication bills. An integrated voice/data network configuration is shown in Figure 2.20.

Figure 2.20 is an example of how an organization with branches in two locations could implement an integrated voice/data network. Gateways installed at either end convert voice to IP packets and back. Voice packets are transported over the IP network together with data packets between the two locations. The PABX can route calls to the local PSTN as well (as shown in Location B).

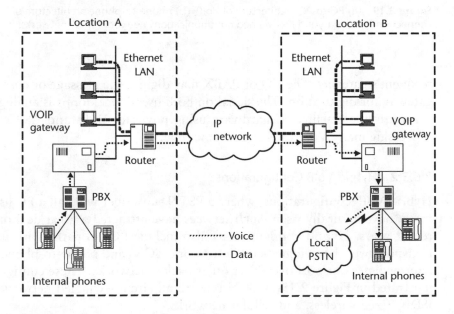

**Figure 2.20**   An example of an integrated voice/data network.

### 2.6.3.4   Value-Added Services

From the users' perspective, the key advantage of Internet telephony at present over conventional telephony is the call cost. However, current price models of both types are evolving, and there may not be any significant difference in the future. In order to survive and thrive, Internet telephony will need to offer not only the same high quality for voice calls, but a set of advanced services. Some examples of advanced services are free phone, split charging, credit card calling, and toll-free calling. How Internet telephony can support these supplementary services are discussed in [16, 20].

## 2.6.4   VoIP Standards

Signaling is one of the most important functions in the telecommunications infrastructure because it enables various network components to communicate and interwork with each other. Advanced signaling systems such as SS7 and Q.931 have been developed for the PSTN (see Chapter 3). Similar efforts have been undertaken in the latter part of the 1990s to define signaling protocols for VoIP.

### 2.6.4.1   Standardization Efforts for VoIP

The ITU started work on standardizing signaling protocols for VoIP in 1995. In 1996, ITU-T Study Group 16 decided on H.323 version 1, referred to as a standard for "Real-Time Video Conferencing Over Nonguaranteed QoS LANs" [17]. By the end of 1996, most PC client software vendors were developing H.323-compliant products.

H.323 was heavily weighted towards multimedia communications in a LAN environment. The rapidly growing popularity of VoIP paved the way for a revision of the H.323 specification. From the beginning, interworking of IP networks with the PSTN was one of the main focuses of this revision. Version 2 of H.323, defined to accommodate these additional requirements, was accepted in January 1998 [18]. This was called "Packet-Based Multimedia Communication Systems." New features being added to the next revision underway will include fax-over-packet networks and fast connection mechanisms.

In parallel with H.323, the Internet Engineering Task Force (IETF) standardized the session initiation protocol (SIP), which provides advanced signaling and control functionality for a wide range of multimedia services [21]. SIP, standardized in early 1999, is part of the overall IETF multimedia architecture that has emerged over the past few years. At the present time, H.323 is in more widespread use than SIP and, hence, is described in detail in the following sections. Details of SIP can be found in [21, 22].

### 2.6.4.2   The H.323 Standard

H.323 is part of a family of ITU-T recommendations called H.32x, which provides multimedia communication services over different networks. This family of standards is summarized in Table 2.5. Multimedia communication services include real-time audio, video, and data communications.

H.323 can be applied in a variety of mechanisms—audio only (IP telephony); audio and video (video telephony); audio and data; and audio, video, and data. It can also be used in multipoint-multimedia applications.

**Table 2.5**   The H.32x Series of Standards

| Standard | Network |
| --- | --- |
| H.324 | SCN |
| H.320 | ISDN |
| H.321, H.310 | B-ISDN |
| H.322 | LANs that provide guaranteed QoS (ISO-Ethernet) |
| H.323 | Nonguaranteed QoS LANs |

One of the primary goals in the development of the H.323 standard was interoperability with other multimedia networks. This interoperability is achieved through the use of a gateway, which performs any signaling and media translation required for interoperation.

### 2.6.4.3   The H.323 Architecture

H.323 specifies four kinds of components, which, when networked together, provide the point-to-point and point-to-multipoint communication services. The four components are *terminals*, *gateways* (GWs), *gatekeepers* (GKs), and *multipoint control units* (MCUs). The functional architecture of H.323 is shown in Figure 2.21. Gateways, gatekeepers, and MCUs are logically separate components but can be implemented as a single physical device.

A typical H.323 network is composed of a number of *zones* interconnected via a WAN. Each zone consists of a single H.323 GK, a number of H.323 terminal endpoints, a number of H.323 GWs, and a number of MCUs interconnected via a LAN. A zone can span a number of LANs in different locations or just a single LAN. The only requirement is that each zone contains exactly one GK, which acts as the administrator of the zone.

*Terminals*

An H.323 terminal, or *client*, can either be a personal computer or a stand-alone device such as an IP phone, video phone, or a terminal adapter that connects an analog phone or fax machine to the H.323 network. It is an endpoint in the network

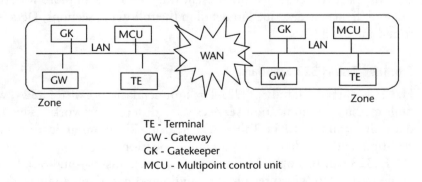

TE - Terminal
GW - Gateway
GK - Gatekeeper
MCU - Multipoint control unit

**Figure 2.21**   The architecture of an H.323 network.

that provides for real-time two-way communications with another H.323 terminal, GW, or MCU. This communication consists of control, indications, audio, video, or data between the two terminals. A terminal may set up a call to another terminal directly or with the help of a GK. H.323 terminals may be used in multipoint conferences. The H.323 recommendation defines call signaling, control message multiplexing, audio codecs, and data protocols. Figure 2.22(a) shows an example of an H.323 terminal.

H.323 does not specify audio or video equipment, data applications, or the network interface. It does specify certain capabilities for a minimum level of interoperability with other terminals. G.711 audio (PCM) is mandatory, and if video is

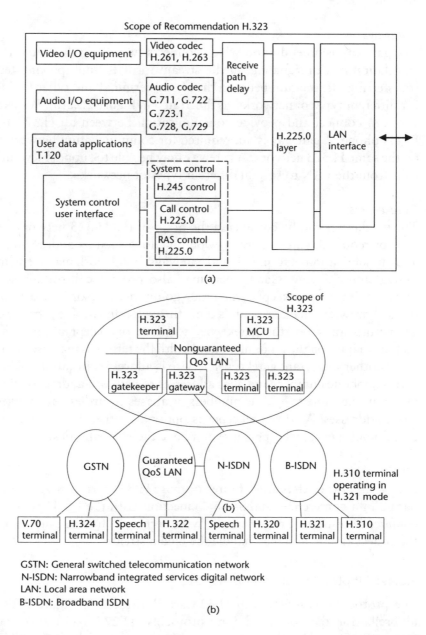

GSTN: General switched telecommunication network
N-ISDN: Narrowband integrated services digital network
LAN: Local area network
B-ISDN: Broadband ISDN

(b)

**Figure 2.22**   (a) An H.323 terminal, and (b) interoperability of terminals through a gateway.

supported, H.261 is mandatory. H.323 provides a variety of options for audio and video codecs. These provide a choice of lower bit rates, lower delay, or improved quality. The choice of codecs also provides compatibility with other terminal types. H.245 specifies messages used for capability exchange, opening and closing logical channels for media streams, and other commands and requests. H.225 specifies the messages for call signaling, registration and admission, and packetization and synchronization of the media streams. Receive path delay is optional for both audio and video streams and may be used to provide lip synchronization or jitter control. Jitter and its control are discussed in Section 2.6.5.4. For data applications, H.323 recommends the T.120 series of protocols. The protocols will be described more in Section 2.6.4.4.

### Gateways

The gateway is an endpoint, which connects two dissimilar networks. It provides translation of call signaling, media stream formats, multiplexing techniques, and transferring information between the H.323 terminal and other ITU terminal types for different types of networks, as illustrated in Figure 2.22(b). For example, a gateway can connect and provide communication between an H.323 terminal and a PSTN phone. A gateway is not required for communication between two terminals on the same H.323 network. A gateway may be able to support several simultaneous calls from the LAN to several different types of networks.

### Gatekeepers

The gatekeeper can be considered the brain of the H.323 network. Although they are not required, gatekeepers provide functions such as addressing, authorization and authentication of terminals and gateways, bandwidth management, accounting, billing, and charging. Gatekeepers may also provide call-routing services. Clients register with the gatekeeper when they go online. One of its most important functions, bandwidth management, is done through admission control. Terminals must get permission from the gatekeeper to place or accept a call. This permission includes a limit on the amount of bandwidth the terminal may use on the network.

Another important gatekeeper service is address translation. This function converts telephone numbers and alias addresses to network addresses, allowing users to maintain the same telephone numbers or aliases regardless of changes to their network addresses. As the gatekeeper is optional in the H.323 architecture, systems that do not have gatekeepers may not have these capabilities.

### MCUs

MCUs support conferences of three or more H.323 terminals. All terminals participating in a conference establish a connection with the MCU. The MCU manages conference resources, negotiates between terminals for the purpose of determining the codecs to use, and may handle the media stream.

### 2.6.4.4   Protocol Relationships in H.323

The protocol relationships in H.323 are illustrated in Figure 2.23. H.323 is an umbrella for the following four protocols: H.225 registration admission and status (RAS) and call signaling, H.245 control signaling, Real-Time Transport

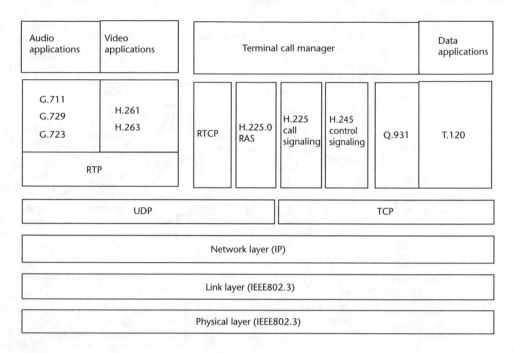

**Figure 2.23**   Protocol relationships in H.323.

Protocol (RTP)/Real-Time Control Protocol (RTCP), and Q.931. Table 2.6 summarizes the seven phases of an H.323 call and the protocols involved in each step. More details on these protocols can be found in [23].

### 2.6.4.5   An Example of an H.323 System

An example of an H.323 network is shown in Figure 2.24 [16]. Domain X is an H.323 network, and domain Y provides connectivity to the SCN. The H.323,

**Table 2.6**   The Seven Phases of an H.323 Call

| Phase | Protocol | Functions |
| --- | --- | --- |
| 1. Call admission | RAS | Request permission from GK to make/receive a call. At the end of this phase, the calling endpoint receives the Q.931 transport address of the called endpoint. |
| 2. Call setup | Q.931 | Set up a call between the two endpoints. At the end of this phase, the calling endpoint receives the H.245 transport address of the called endpoint. |
| 3. Endpoint capability negotiation and logical channel setup | H.245 | Negotiate capabilities between two endpoints. Determine master-slave relationship. Open logical channels between two endpoints. At the end of this phase, both endpoints know the RTP/RTCP address of each other. |
| 4. Stable call | RTP/RTCP | Two parties in conversation. RTP is the transport protocol for packetized voice. RTCP is the counterpart that provides control services. |
| 5. Channel closing | H.245 | Close down the logical channels. |
| 6. Call teardown | Q.931 | Tear down the call. |
| 7. Call disengage | RAS | Release the resources used for this call. |

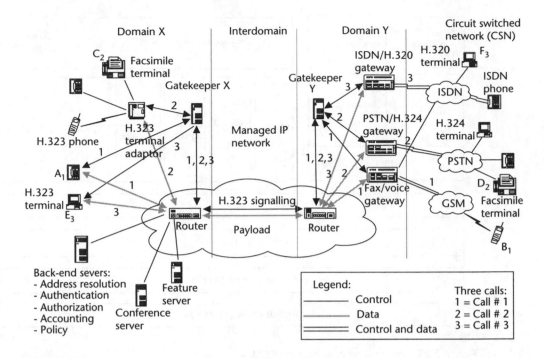

**Figure 2.24**   An example of an H.323 network. (*After:* [16].)

PSTN, ISDN, and GSM networks are interconnected through gateways to allow seamless interoperability among devices on these different networks. Traditional telephony devices can be used on an H.323 network by connecting them to an H.323 terminal adapter. Such a terminal adapter is a fax/voice gateway and may provide the functionality of a CO to these devices. H.323 signaling is routed via a gatekeeper, but the payload (media) is transferred directly between endpoints.

The two domains, X and Y, have three terminals each, A, C, E and B, D, F, respectively. The subscript of the terminal name indicates the call it is involved in. For example, A1 indicates terminal A is involved in call number 1.

The calls are as follows:

- Call 1—voice only call between an H.323 endpoint (A) and a GSM phone (B);
- Call 2—fax call between fax machines C and D on H.323 network and PSTN, respectively;
- Call 3—multimedia call between H.323 and H.320 (ISDN) clients E and F, respectively.

It is assumed that A and the gateway representing B have discovered and registered with their respective gatekeepers X and Y. It is also assumed that through interdomain signaling, gatekeeper X knows that gatekeeper Y will complete calls destined for client B. So, when client A sends a call setup message to X with the address (telephone number) of client B, gatekeeper X uses this to find the IP address of gatekeeper Y. Gatekeeper X then forwards this message to gatekeeper Y, which then sends it to B. If B answers the call, it returns a connect message via the two gatekeepers to client A. The connect message triggers the gatekeepers to start the accounting for the call. Then the two clients start sending multimedia packets to each other.

### 2.6.5  Technical Challenges of VoIP

The technical challenges that face VoIP services are mainly related to the lack of guarantee in terms of bandwidth, packet loss, delay, and jitter, which affect the quality of voice over packet switched networks in general and the Internet in particular. This section looks at these technical issues and solutions that are being developed for IP telephony.

#### 2.6.5.1  Packet Loss

Unlike the PSTN, no end-to-end physical circuits are established in packet-switched networks. Rather, only virtual circuits may be available, which incorporate a specified path, but no resource reservation for a call on that path. Hence, no guaranteed bandwidth is available for virtual circuits. This results in *packet loss* being a common phenomenon in all such networks.

Packets from many sources are queued for transmission over an outgoing link in a packet switch (router) and transmitted one by one from the head of the queue when capacity on the link becomes available. A packet is lost in the network if there is no space in the queue when it arrives at a router. As the network traffic increases, the possibility of routers becoming congested increases, leading to higher packet loss.

Packet loss can cause severe damage to voice quality in IP telephony. As each IP packet contains 40 to 80 ms of speech, approximately in units called phonemes (identifiable units of speech), when a packet is lost, a phoneme is lost. While the human brain is capable of reconstructing a few lost phonemes in speech, too much packet loss makes voice loose its continuity and hence intelligibility. Figure 2.25(a) shows how voice quality degrades with packet loss. The MOS, a measure for voice quality, was described in Section 2.5.3.3. A higher score implies better quality.

Because packet loss is a direct consequence of insufficient link bandwidth or packet processing capability of network routers, an obvious solution is upgrading the network infrastructure. In the last few years, several promising techniques such as ATM, SDH, SONET, and WDM have been developed for gigabits per second–terabits per second transmission speeds. To complement these, high-speed, switching-based router technologies have also evolved.

However, network upgrade is an expensive and long-term solution. It may be more feasible in a privately managed network such as an enterprise network than in the public Internet. While network upgrade attempts to reduce packet loss, several other techniques focus on repairing the damage caused by it. Some techniques used to address this problem are summarized in Table 2.7. Further details on these techniques are found in [13, 22].

#### 2.6.5.2  Packet Delay

Timing is an important characteristic of voice, resulting in the need for real-time communication. The syllables of speech are as important as the intervals between

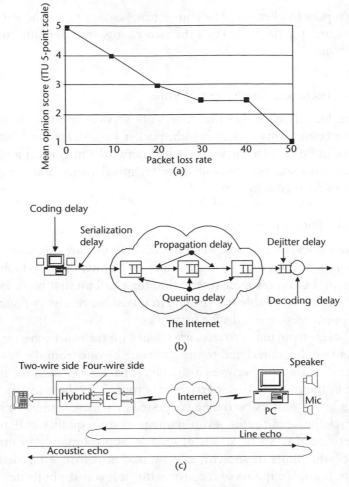

**Figure 2.25** (a) Voice quality as a function of packet loss rate (*Source*: [19]), (b) sources of delay in IP telephony, and (c) the two types of echo.

**Table 2.7** Techniques for Repairing Damage Caused by Packet Loss

| Technique | Description | Remarks |
|---|---|---|
| Silence substitution | Substitutes silence in place of missing packets | Causes clipping; successful for small packets (<16 mS) and low loss rates (<1%) |
| Noise substitution | Substitutes white background noise for lost packets | Performs better than silence substitution; also known as CNG |
| Packet repetition | Replays last received packets in place of lost ones | Repeated packets are faded for better quality |
| Packet interpolation | Uses characteristics of packets in the neighborhood of the lost one to generate a replacement | Sound quality is better than silence substitution or packet repetition |
| Frame interleaving | Interleaves voice frames across different packets | Loss of a packet results in multiple short gaps in the voice stream; increases system delay |
| FEC | Transmits information redundantly in several packets | Introduces overheads to the system; good performance even at bit error rates as high as 3% |

them. Delay can have several negative effects on speech performance. If additional delay is inserted between syllables, the rhythm of the voice is lost. Delays above 150 ms in each direction can interfere with the dynamics of speech conversation [23]. Long delays can cause two speakers to enter a half-duplex communications mode. They also exacerbate echo because any reflected signal coming back to the sender is now noticeable.

The basic guideline for voice transmission is that delays below 150 ms are acceptable for most applications. However, delays up to 400 ms are still acceptable for long-distance communications, as users are mentally prepared for this. Voice quality deteriorates significantly above 400 ms [19]. Measurement results for delay and packet loss on several routes on the Internet are presented in [15].

The primary source of delay in circuit-switched networks such as the PSTN is the signal propagation delay, which depends on distance. It therefore is constant. Because the signal travels at the speed of light, it can be kept well below 400 ms except in the case of long-distance satellite links.

Delay is one of the biggest technical challenges for IP telephony. In packet-switched networks, there are many factors contributing to the delay that are not present in circuit-switched networks. While some of these delays are constant and known in advance, some are unpredictable. Figure 2.25(b) illustrates the types of delay that are encountered in IP telephony. The total delay can easily exceed 400 ms. A typical domestic circuit-switched call suffers just 50 to 70 ms of delay in comparison.

### Codec Delay

The codec converts the voice from analog to digital, and vice versa, and also performs voice compression to reduce the bandwidth requirement for digital voice transmission. Voice coding techniques and their delay characteristics are described in Section 2.5.2 and Section 2.5.3, respectively.

Speech codecs in the ITU-T G.700 series described in Section 2.5 have been tested extensively and shown to have good quality under a wide range of conditions. In particular, the ITU Codecs G.723.1, G.728 and G.729, and G.729A have shown to work well in the presence of constraints found on the Internet. Though these codecs were designed with different applications (mostly wireless) in mind, they all are candidates for enabling VoIP. Table 2.8 lists the characteristics of these codecs and compares them with the characteristics of PCM encoding defined by ITU-T Recommendation G.711.

While G.723.1 provides the lowest bit rate, it also suffers from the largest delays. G.729 trades off a slightly higher bit rate and more complexity for significant decrease in delay. G.729A provides the same performance as G.729 with lower complexity.

### Serialization Delay

This is the time taken to place a packet on the transmission line and depends on the packet size and the transmission speed. As each packet contains some amount of overhead in the form of a header, if packets are small, inefficiency will result. Therefore, it is common to pack more than one voice frame into one IP packet. Packetization delay is caused while compiling frames into a packet.

**Table 2.8**   Codec Characteristics

| Codec | G.711 | G.723.1 | G.729 | G.729A |
|---|---|---|---|---|
| Bit rate (kbit/s) | 64 | 5.3/6.4 | 8 | 8 |
| Frame size (ms) | 125 | 30 | 10 | 10 |
| Processing delay (ms) | 0 | 30 | 10 | 10 |
| Lookahead delay | 0 | 7.5 | 5 | 5 |
| Total encoding delay | 0 | 37.5 | 15 | 15 |
| Typical decoding delay | 0 | 18.75 | 7.5 | 7.5 |
| Frame length (bytes) | 1,000 | 20/24 | 10 | 10 |
| DSP MIPS | Not available | 16 | 20 | 10.5 |
| RAM (words) | Not available | 2,200 | 3,000 | 2,000 |

### Network Delay

IP-network delay is primarily the result of buffering, queuing and switching, or routing delay of IP routers. Specifically, IP-network delay comprises packet capture delay, switching/routing delay, and queuing time [13].

Packet-capture delay is the time necessary to receive the entire packet before processing and forwarding it through the router. Packet length and transmission speed determine this delay. Using short packets over high-speed trunks can shorten this delay but potentially decrease network efficiency. Switching/routing delay is the time the router takes to switch the packet. This time is necessary to analyze the packet header, check the routing table, and route the packet to the output port. This delay depends upon the architecture of the routing engine and the size of the routing table. New IP switches can speed the routing process by making routing decisions and forwarding traffic via hardware rather than software [13].

Queuing delay occurs at the various switching points of the network, such as routers and gateways. At these points, voice packets wait behind other packets waiting to be transmitted on the same outgoing link. Because the size of the queue depends on the amount and the statistical nature of network traffic, the queuing delay varies significantly from packet to packet. Queuing delays can be reduced by faster links in cases where uses have control of the infrastructure, as in the case of a corporate IP network. The IETF is working on mechanism such as integrated services (IntServ), differentiated services (DiffServ), and resource reservation protocol (RSVP) to prioritize voice packets over data packets to minimize queuing delay for voice and other delay-sensitive applications. More details on these techniques are found in [24].

### Propagation Delay

This is the time required by signals to travel from one point to another and is determined by the velocity of signal propagation in the transmission medium and the distance. Long, fixed delay is apparent in telephone calls that are routed over geostationary satellite, also causing a problem for conventional telephony.

### Other Sources of Delay

In certain implementations of VoIP, other delay sources are present. For example, in dial-up applications, modems cause delays. Such delays can be avoided by digital lines. Packet voice systems using multimedia PCs also incur delays due to operating system inefficiencies and sound card delays. Such problems can be reduced by using gateway interfaces that use powerful, specialized DSPs.

### 2.6.5.3    Echo

Generally echo exists in many PSTNs. However, due to the location in which echo originates and the relatively short end-to-end delays in these networks, a user often does not notice echo. However, with VoIP levels of delay, echo often becomes noticeable [13].

From a telephony perspective, echo is the sound of the talker's voice returning to the talker's ear via the telephone's speaker. If the time between the original spoken phrase and the returning echo is 25 to 30 ms, or if the echo's level is approximately –25dB or less, no annoyance or disruption to voice conversations will probably occur. However, when the echo is loud enough and has enough delay—usually around 30 ms and more—so that the speaker perceives the echo, the quality of a voice call becomes problematic.

In most cases, echo results from an electrical mismatch between the subscriber loop and the hybrid circuit at the subscriber line interface circuit (SLIC). End users will hear their own voice signal bouncing off the remote CO's SLIC. This is called *line echo*. The second form of echo occurs when a free-air microphone and speaker are used, as in the case of most PC end points. The remote user's voice signal is picked up by the microphone and echoed back. This is called *acoustic echo*. The lower part of Figure 2.25(c) illustrates the two types of echo.

To deal with unwanted echo, network designers deploy functional components known as *echo cancelers* in the local exchange, VoIP gateway, or the VoIP PC terminal, usually as close as possible to the hybrid that causes the echo, as shown in the upper part of Figure 2.25(c). In this figure, the echo heard by user A is most likely caused by user B's hybrid, and, hence, the echo canceler attempts to remove this. The echo canceler looks at the signal coming from user B, and if it identifies a part of user A's voice in it, it attempts to cancel this echo. Modern echo cancelers intelligently adapt to changing signal and hybrid circuit conditions with the aid of DSPs.

### 2.6.5.4    Network Jitter

Variation in the interframe arrival times at the receiver is called jitter. Jitter is potentially more disruptive for IP telephony than delay. Jitter occurs due to the variability of network delays, especially queuing delays in the network. IP packets belonging to the same stream may even undergo different delays. Network jitter can be significant even for average network delays. If an IP packet is inordinately delayed, it will be considered lost by the receiver. If this happens often, the quality of the voice will be affected.

To allow for variable packet arrival times and still achieve a steady stream of packets, the receiver holds the first packet in a *jitter buffer* for a while before reproducing the voice. This allows time for subsequent packets to arrive and be stored in the jitter buffer. The size of the jitter buffer is selected based on the expected jitter in the network.

The jitter buffer adds to the overall delay. Therefore, for high jitter, the overall perceived delay is high even if the average network delay is low. For example, for a moderate delay of 50 ms and a 5-ms jitter buffer, the total delay is 55 ms. By contrast, if a network has low average delay of 15 ms, but occasionally packets are delayed by 100 ms, the jitter buffer would have to be 100 ms, and the overall delay would be 115 ms.

Selection of the jitter buffer size is crucial to IP telephony systems to balance jitter removal and overall delay. If the buffer is too small, some packets may be lost. If set too high, the delay will be high. Ideally, the jitter buffer should be modified dynamically to suit network conditions. Some manufacturers of IP telephony equipment offer such intelligent buffers.

Consider as an example, two telephones connected via gateways to the Internet. The round-trip delay is approximately 200 ms (around 30 ms at each gateway multiplied by two for travel time, plus serialization and packetization delay). Encoding/decoding delays add another 200 ms. Finally, counting in a jitter buffer of at least 50 ms, the total round-trip delay adds up to 450 ms. Internet delays are known to vary up to about 2 seconds. No matter how well designed the VoIP devices and networks are, a fundamental delay exists that simply cannot be eliminated. The codec, packetization, and serialization delay components form this fundamental limit.

### 2.6.6   Voice Quality Measurement and Standards in VoIP

Voice quality in the PSTN has evolved over the years to be consistently high and predictable. It is now an important differentiating factor for new VoP networks and equipment. Although the PSTN does not exhibit perfect quality, users have become accustomed to PSTN levels of voice quality and compare other services with it. That is, PSTN voice quality is relatively standard and predictable. If PSTNs and IP networks are to converge, IP networks must include enhancement mechanisms that ensure the QoS necessary to carry voice [13]. Providing comparable service quality in IP networks will drive the initial acceptance and success of VoP services. Consequently, measuring voice quality in a relatively inexpensive, reliable, and objective way becomes very important. Section 2.5.3.3 introduced briefly the concept of speech quality.

#### 2.6.6.1   Factors Affecting Voice Quality

Three elements emerge as the primary factors affecting voice quality, particularly for VoP technologies: bandwidth, end-to-end delay, and computational complexity, as shown in Figure 2.26.

This figure shows the general relationships between the three factors. However, no known mathematical relationship exists that can be used to determine a single number or vector quantifying voice quality. The relationship between these three components can be quite complex. Breaking voice quality into three distinct areas makes its evaluation manageable.

Any given solution for packetized voice can be mapped into the space shown in Figure 2.26. Each point on this surface results in decoded speech of the same quality. The high point on the bandwidth axis represents traditional digital telephony based on 64 kbit/s PCM, which requires a large bandwidth but low computational complexity and delay. The point near the delay axis represents VoIP, which may suffer seconds of delay. Gateways that process many calls at the same time are based on

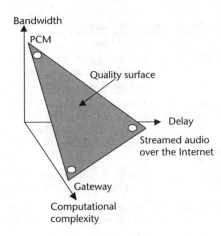

**Figure 2.26**  The relationship between the components of voice quality. (*After:* [15].)

powerful DSP hardware. These devices lie towards the high end of the computational complexity axis and have intermediate bandwidth requirements and delay.

### 2.6.6.2  Voice Quality Testing Standards

Traditional voice quality testing techniques involve comparing waveforms on a screen and measuring SNR and total harmonic distortion (THD), among others. These methods assume that changes to the voice waveform represent unwanted signal distortion, and that telephony circuits are essentially linear [19].

With low-bit-rate voice codecs such as G.723 and G.729, neither waveform preservation nor circuit linearity can be assumed, rendering these traditional testing methods ineffective. The bursty and time-insensitive nature of packet networks exposes the need for other testing methods as well. In VoP systems, the performance testing of echo cancelers, VADs, and other processors are also important.

Standardization efforts for speech transmission performance in VoP and hybrid networks by organizations such as ITU, TIA, and ETSI, as well as some testing techniques and practical testing techniques are discussed in detail in [13, 23, 25, 26]. Table 2.9 summarizes these.

## 2.7  Devices for Digital Telephony

Having discussed the key concepts, techniques, and applications of digital telephony, it is interesting to see how IC technology has helped the implementation of these systems in an economical manner, passing on the benefits to end users. The following sections describe the implementation of the concepts discussed in previous sections.

The following discussion presents devices relevant to digital telephony in the following categories, with examples from representative IC families.

- Subscriber line interface circuits (SLICs) interfacing the analog subscriber loop to the digital switches;

**Table 2.9**  Summary of Voice Quality Testing Standards

| Standard | Description |
| --- | --- |
| MOS; ITU-T Recommendation P-800 | This test quantifies quality by having large numbers of human listeners rate voice quality in a controlled and well-defined testing process. The rating is on a scale of 1 to 5. |
| Perceptual speech quality measurement (PSQM); ITU-T Recommendation P.861 | This test was originally used for testing voice codecs. It is an automated human listener. It uses an algorithm to compare the original speech and the received signal and provides a relative score that indicates the difference from a human listener's perspective. |
| PSQM+ | This is an enhancement of PSQM to take into account network impairments such as packet loss. |
| Perceptual analysis-measurement system (PAMS) | This test is similar in concept to PSQM but uses different types of models and scores. It provides *listening quality* and *listening effort* scores. |
| E-model; ITU-T Recommendation G.107 | This is a tool for assessing the impact of transmission planning decisions on speech performance. It consists of additive terms that capture noise, echo, propagation delay, and attributes of speech codecs. It introduces an equipment impairment factor (EIF). |

- Low-bit-rate voice codecs;
- Processors for VoIP systems.

### 2.7.1  Subscriber Line Interfacing

The two-wire local loop connects the customer premises equipment with the COs and through it to the worldwide telephone network. Almost all switches and trunks are digital, while the telephone handsets used are predominantly analog, sending and receiving analog signals via the local loop. The device or the module that interfaces the handset and the local loop to the digital switching system at the CO is the SLIC.

#### 2.7.1.1  The Telephone Instrument

Subscriber line interfacing is governed by the requirements of the analog telephone handset because of its widespread use. This conventional handset is described in detail in [27]. A block diagram showing the major functions of the analog telephone instrument and some of its components having a significant impact on subscriber line interfacing is shown in Figure 2.27.

Early phone instruments used electro-mechanical ringers and used ac voltage sources in excess of 70V in the frequency range of 20 Hz. The ringer circuit is always connected across the line so it can detect and signal an incoming call. Though the modern ringers have replaced the electro-mechanical ones, the high ac voltage principle was maintained throughout the modernization of telephone systems. The remainder of the telephone set is isolated from the line by the open contacts of the switch hook when the handset is on hook. No appreciable dc current flows in the local loop in the on-hook state. When the handset goes off hook, the loop closes, and the loop current flows from the CO dc power source through a relay coil at the CO to the telephone set. The relay is energized by the loop current and its closed contactssignal to other CO equipment that the subscriber is off hook. These circuits

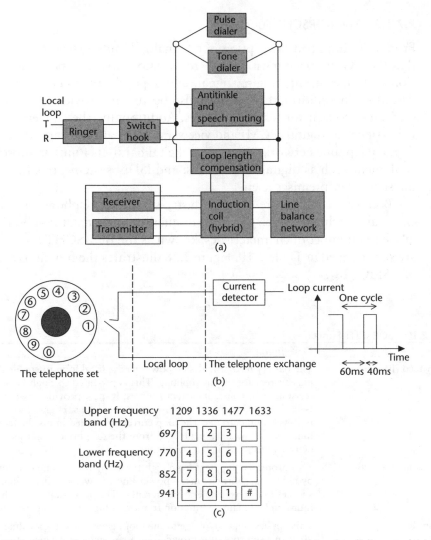

**Figure 2.27**   The analog telephone instrument: (a) block diagram, (b) pulse dialing, and (c) tone dialing.

then prepare to receive the dialed digits coming from the subscriber's dialer and to connect the call to the switch. In pulse dialing [Figure 2.27(b)], the dialed digits are encoded as a series of pulses by interrupting the loop current a number of times equal to the digit being dialed. The rotary dial in telephone instruments helped generate this pulse pattern, and current detectors at the CO detected these pulses. Tone dialing, which is most commonly used today, was developed to make the dial time shorter and thus utilize the CO equipment more efficiently. Each digit is encoded as a pair of tones, as shown in Figure 2.27(c). Digit receivers at the exchange consist of narrow bandpass filters.

Hybrid circuits are used to interface a two-wire circuit to a four-wire circuit to permit full-duplex operation. The local loop is a two-wire circuit. Inside the telephone handset, the hybrid provides two separate paths for the transmitted and received signals. A similar hybrid at the CO separates the two paths in the local loop to the four-wire transmission circuits to the switch and beyond.

### 2.7.1.2 The BORSCHT Functions

From this simplified description of the analog handset, several important facilities that the CO interface must provide to the subscriber can be identified—the provision of loop current, ringing signal, and supervisory functions such as off-hook detection. In addition, the CO must also be able to provide protection to the local loop and access it for testing purposes. Additionally, the conversion of the analog voice signal to digital (PCM) and vice versa must also be carried out by the CO in digital telephone networks. In modern digital handsets found in subscriber premises equipment such as digital PBXs, DSLs, and ISDN systems, this function is done at the subscriber premises itself.

Because of the nature of the industry-standard telephones and the electromechanical switches to which they were originally connected, the subscriber loop interface has a number of characteristics, known as the BORSCHT functions [1, 28] and are summarized in Table 2.10. Figure 2.28 illustrates the functions and components of a SLIC [28].

**Table 2.10** BORSCHT Functions

| Function | Description |
| --- | --- |
| Battery feed (B) | The battery feed provides dc power to the local loop to enable dc signaling and bias current for the microphone. This is provided through balanced feed resistances or constant current sources. It must provide power to the local loop (−48V) and 45 to 75 mA on low-resistance loops. Higher loop resistances result in lower loop currents. The loop current provided limits the loop length. It must also allow signaling to and from the telephone (dial tone, dial digits), low dc resistance, and high ac resistance. |
| Over voltage Protection (O) | This protects equipment from high transient voltage and current surges caused by lightning, short circuits to power lines, power line induction, and other sources of electromagnetic interference. This is in addition to the protection found in CO main distribution frames and in customer premises. |
| Ringing (R) | This applies ringing voltage to the loop, generally a high voltage, typically a 90-V rms sinusoid. It is provided by a ringing voltage generator. The ringing cadence (the pattern of ringing and silent intervals) may be controlled by the SLIC. The answer detection function is also included to initiate ring trip (cessation of ringing). |
| Supervision (S) | This is the detection of off-hook condition, dialed digits, and supervision of calls in progress. The basic principle used is the detection of the presence or absence of current flow in the loop. This function is required to be independent of loop length and should operate satisfactorily in the presence of noise and leakage current. |
| Coding (C) | This is PCM encoding of the transmit signal together with the simultaneous decoding of the received signal. After coding, the digital bit steam is sent to a PCM "highway," placed into a time slot, and transmitted via a trunk. The codec is interfaced to transmit/receive filters for the purpose of band limiting and reconstructing, respectively. |
| Hybrid (H) | The local loop is two-wire, carrying signals in both directions simultaneously. However, PCM and other transmission trunks are four-wire systems, requiring unidirectional transmit and receive circuits. The hybrid function converts the two-wire local loop to a four-wire path. The hybrid provides impedance matching to the two-wire loop, which acts to control echo. Old hybrid circuits used transformer-based techniques, while modern ones carry out the conversion electronically, sometimes coupled with digital signal processing techniques. |
| Test (T) | This provides access to the loop and the SLIC from an external test bus. |

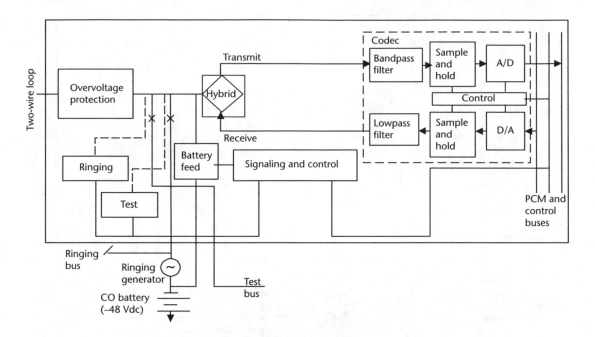

**Figure 2.28** Functions and components of a SLIC. (*Source:* [28]. ©1995 IEEE, reproduced with permission.)

### 2.7.1.3 IC Implementation of the Subscriber Line Interface

Because the subscriber loop interface has evolved through the days of magneto ringers, rotary dials, and step-by step switches, this has been more difficult to replace with modern electronics than any other component of the telephone network. While adopting modern technologies, it is necessary for SLICs to still meet the standards and characteristics that have evolved over the years. Early implementations of BORSCHT functions involved transformers, relays, and triacs [27]. To appreciate the role of modern ICs in SLICs, it is useful to study an early electronic SLIC and use it for comparison with state-of-the art SLIC ICs described later in the section. Figure 2.29 shows such a circuit in AT&T's No. 10-A remote switching system (RSS).

This is a circuit that is intermediate between a fully integrated circuit and no electronics (as in still earlier SLICs). It combines electronics with a transformer to provide, among other facilities, isolation from the local loop, application of ringing voltage, and access to the circuit for testing.

Due to the need to support high ringing currents, dc currents depending on loop length, proper impedance balance for echo cancellation, and 2- to 4-W conversion, different types of devices for SLICs have evolved, which support the BORSCHT functions to different levels. Some devices, called ringing SLICs, incorporate on-chip ringing current generators. SLICs specifically for short loops have the capability to feed low-range dc currents, while long-loop devices can feed higher currents. Some devices incorporate on-chip PCM codecs, while others need separate codecs to be interfaced. Some devices provide support for multiple local loops.

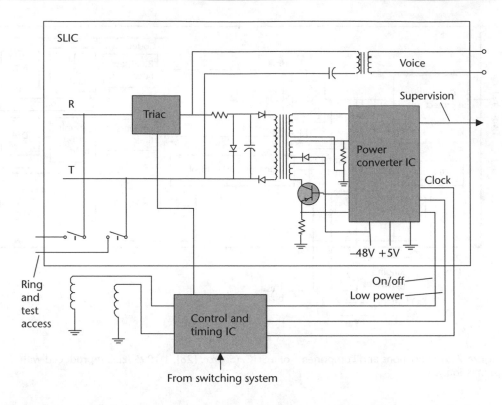

**Figure 2.29** Subscriber line interface circuit in the AT&T No. 10 A RSS switch. (*Source:* [27]. Reproduced with permission of Butterworth-Heinemann.)

The SLIC has to have a defined input impedance towards the subscriber line for maximum signal power transfer and return loss. The requirements for the input impedance vary from country to country and demand impedance matching to the different environments. Country-specific adaptations are also required for transhybrid loss, which is the loss between the transmit and receive ports of the 2- to 4-W hybrid.

To support these requirements, some modern devices have programmable parameters, such as dc current, ringing current impedance, and codec characteristics, offering flexibility for different country specifications. Such devices are often based on microcontrollers and DSPs having programming interfaces. Software controlled integrated test and diagnostic functions (ITDF) is another feature found in modern SLICs.

Advanced SLIC devices centered on DSPs are especially advantageous in performing the functions of 2- to 4-W conversion, encoding/decoding, and the associated filtering functions efficiently as well as supporting miniaturization and multiple line handling.

In the evolutionary process of SLIC ICs, the following additional features have been added to some devices due to practical and economical advantages:

- *Generation of metering signals.* In many countries, metering signals are sent to the subscriber for billing purposes. This is a sinusoidal burst.

- *Dual tone multifrequency (DTMF) generation.* Integrated DTMF decoders in the SLIC help integrate the calling number identification function into the SLIC.
- *On-hook transmission.* On-hook transmission is used in caller ID transmission. The caller ID information is sent using either FSK or DTMF. Some SLICs include this function. On-hook transmission is also used in equipment supporting direct inward dialing (DID).
- *Line echo cancellation.* Some SLIC devices contain an on-chip adaptive line echo cancellation unit for the cancellation of near-end echoes.
- *Universal tone detection.* Integrated universal tone detectors are useful in detecting special tones in the receive or transmit path, such as fax or modem tones. This is used in activation of proper filter coefficient sets for different applications such as voice, data modem, or fax.

Commercially available SLICs are categorized based on application environments such as CO, access network, and customer premises.

### 2.7.1.4   Illustrative ICs for SLICs

A few illustrative ICs for SLICs are discussed in this section, representing basic SLICs, ringing and nonringing SLICs, SLICs with on-chip codecs, and chip sets for SLICs. Also described are samples of codecs available for interfacing with simpler SLICs that do not have this function on chip. The reader is encouraged to identify the basic SLIC functions in each of the devices as well as the additional features available and their evolutionary path.

*Basic SLICs*

An example of a simple commercially available SLIC IC is the MH88612BV-K from Zarlink Semiconductor [29]. A functional block diagram of this is shown in Figure 2.30. This SLIC performs transformerless 2- to 4-wire conversion, employs feedback circuits to supply a constant current feed, or, if necessary, a constant voltage feed to the line. With a nominal −48-V battery, the device will supply a constant current of typically 25 mA to a line up to 1,100 ohms. For longer lines, the device automatically switches to constant voltage mode where the supplied current will be reduced according to loop resistance. The maximum line resistance supported is 2,000 ohms. The device has facilities to connect to an external ringing generator. The power denial (PD) function allows powering down of subscriber loop through the PD control input. It is useful for removing the battery voltage from the loop driver circuitry in the case of a faulty loop.

A typical application circuit for this device is shown in Figure 2.31. A comparison with Figure 2.29 may be useful at this point to understand the role of SLIC ICs in reducing the complexity and size of the subscriber loop interface.

The basic operation of this device is as follows: The switch hook (SHK) pin will go high to indicate that a telephone connected to the line has gone off hook. The SHK output will toggle to indicate dial pulses on the line. A ringing signal [e.g., 90-V root mean square (RMS) and −48-V dc] is applied to the line by disconnecting pin 15 (RV) from pin 11 (RF) and connecting the ringing voltage at pin 11 by use of the

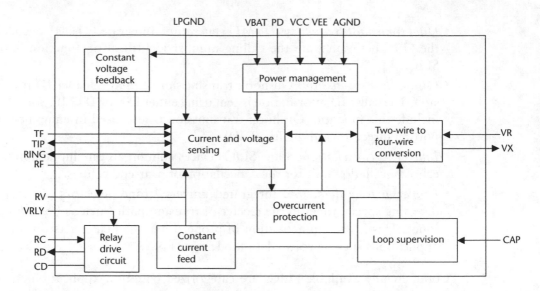

**Figure 2.30** The functional block diagram of the MH88612BV-K SLIC from Zarlink Semiconductor. (*Source:* [29]. Courtesy of Zarlink Semiconductor.)

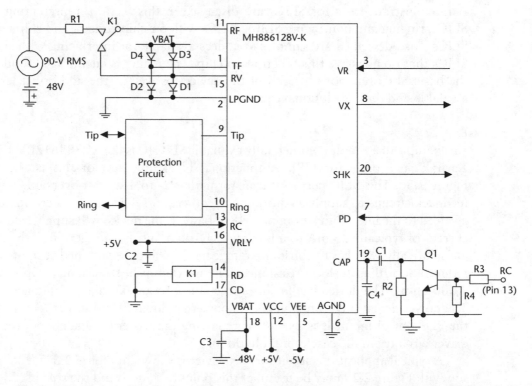

**Figure 2.31** A typical application circuit for the MH88612 SLIC. (*Source:* [29]. Courtesy of Zarlink Semiconductor.)

relay K1. Once an off-hook condition has been detected in response to the ringing signal, a logic low should be applied to pin 13 (RC), which will deactivate the relay K1 to disconnect the ringing voltage and reconnect pin 11 to pin 15.

*Ringing and Nonringing SLICs*

The UniSLIC14 family from Intersil Americas, Inc., is an example of low-power, nonringing SLICs. These devices have automatic single and dual battery selection, based on line length, to ensure the maximum loop coverage on the lowest battery voltage [30]. This device also provides a constant current for short loops and constant voltage with variable current, depending on loop resistance for long loops. A block diagram of this family of ICs is shown in Figure 2.32(a), and its loop current versus loop resistance curve is shown in Figure 2.32(b).

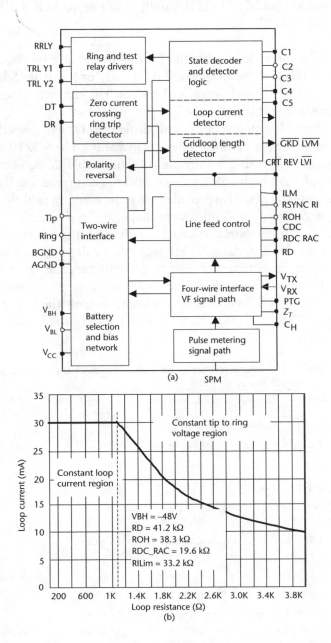

**Figure 2.32** (a) Block diagram, and (b) loop current versus loop resistance characteristics of the UniSLIC14 family of SLICs. (*Source:* [30]. ©2004. Intersil Americas Inc., all rights reserved. Reproduced with permission.)

Networks of many countries require the analog SLIC to terminate the subscriber line with an impedance that is complex rather than resistive in the voiceband frequencies. The UniSLIC14 accomplishes this impedance matching with a single network connected to the *ZT* pin [31].

The RSLIC family of ringing SLICs from Intersil offers flexibility for designs with high ringing voltage and low power consumption requirements [32]. This family operates up to 100V, which translates directly to the amount of ringing voltage supplied to the end subscriber. Loop lengths of over 5,000 ft can be supported. Intersil's nonringing SLICs (UniSLIC family) and ringing SLICs (RSLIC family) are summarized in Table 2.11.

*Single/Multichannel SLICs with On-Chip Codecs and DSP Functionality*
The ProSLIC® and the Dual ProSLIC® family of ICs from Silicon Laboratories are examples of SLICs with integrated codecs. The characteristics of these families are summarized in Table 2.12.

The ProSLIC integrates a ringing generator, DTMF decoder, and dual-tone generator, reducing the number of devices in a SLIC. The Si3210 generates battery voltages dynamically using a software programmable dc-dc converter from a 3.3- to 35-V supply. All high-voltage functions are performed locally with a few low-cost discrete components. Also provided are programming facilities for SLIC impedance, ringing frequency, amplitude, wave shape, and cadence.

The dual-tone generator enables support for a variety of signaling functions, such as caller ID FSK data, DTMF, and other call progress tones. The dc-to-dc converter controller is typically used to generate large negative ringing and off-hook battery voltages. The ProSLIC can generate battery voltages up to –94.5V from a single 3.3- to 35-V input and dynamically controls the battery voltage to optimize line power. Figure 2.33 shows a block diagram of the ProSLIC devices.

**Table 2.11**  Characteristics of Selected Devices of the UniSLIC and RSLIC Families from Intersil Americas, Inc.

| Device | DC Supplies | Loop Current Limit (mA) | Comments |
|---|---|---|---|
| *UniSLIC family (nonringing SLICs)* | | | |
| HC55120 | –48, –24, 5 | 20 to 30 | Central office (CO)/ digital loop carrier (DLC) applications |
| HC55130/40/42/50 | –48, –24, 5 | 20 to 45 | |
| *RSLIC family (ringing SLICs)* | | | |
| ISL5585A | –100, –24, 3.3 | 15 to 45 | 3.3-V device |
| ISL5585B | –85, –24, 3.3 | 15 to 45 | |
| ISL5585E | –75, –24, 3.3 | 15 to 45 | |
| ISL5586B | –85, –24, 5 | 15 to 45 | Balanced codec interface |
| ISL5586C | –100, –24, –15 | 15 to 45 | |
| ISL5586C | –75, –24, 5 | 15 to 45 | |
| HC55185A | –100, –24, 5 | 15 to 45 | Low transient currents |
| HC55185B | –85, –24, 5 | 15 to 45 | |
| HC55185E | –75, –24, 5 | 15 to 45 | |

From: [33, 34].

**Table 2.12** Characteristics of ProSLIC and Dual ProSLIC Family Devices from Silicon Laboratories

| | ProSLIC family | | Dual ProSLIC family | | |
| --- | --- | --- | --- | --- | --- |
| Feature | Si3210 | Si3211 | Si3220 | Si3225 | Si3232 |
| Number of channels | 1 | 1 | 2 | 2 | 2 |
| Integrated SLIC and codec | Yes | Yes | Yes | Yes | No |
| Internal Ringing to 65-V RMS | Yes | Yes | Yes | No | Yes |
| External ringing support | No | No | No | Yes | No |
| On-chip dc-dc converter | Yes | No | No | No | No |
| On-chip DTMF decoder | Yes | Yes | Yes | Yes | No |
| Subscriber line diagnostics | Yes | No | Yes | Yes | Yes |
| Audio and line card diagnostics | No | No | Yes | Yes | No |

*From:* [35].

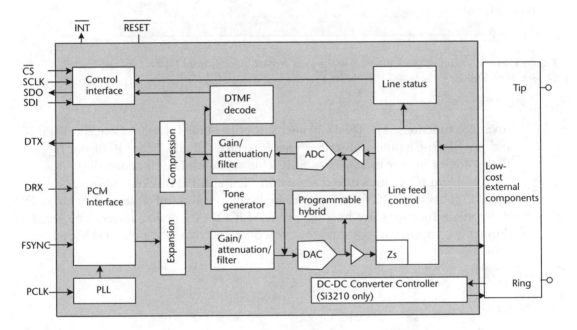

**Figure 2.33** A functional block diagram of the ProSLIC devices from Silicon Laboratories. (*Source:* [36]. Courtesy of Silicon Laboratories.)

The Dual ProSLIC family provides a complete dual-channel telephony interface. The Si3220 includes internal ringing to eliminate centralized ringing and ringing relays. The Si3225 supports centralized ringing for long-loop applications [37]. The Si3220 is ideal for short loops up to 12,000 ft, while the Si3225 is suited for longer loops up to 18,000 ft. A functional block diagram of the Dual ProSLIC is shown in Figure 2.34. This figure shows many SLIC functions that can be performed by an integrated DSP.

*Chip Sets for SLICs and DSP Functionality*
During the last decade, with the DSP techniques becoming implementable in chip sets within reasonable costs, the SLIC families have gained some flexibility and

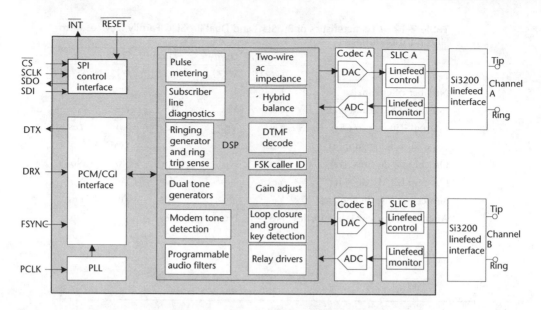

**Figure 2.34** A functional block diagram of the Dual ProSLIC devices from Silicon Laboratories. (*Source:* [37]. Courtesy of Silicon Laboratories.)

advanced functions. The Dual Channel Subscriber Line Interface Concept (DuSLIC) and the Multi-Channel Subscriber Line Interface Circuit (MuSLIC) from Infineon Technologies AG are examples of SLIC chip sets with DSP functionality.

The DuSLIC architecture is shown in Figure 2.35. Unlike traditional designs, DuSLIC splits the SLIC functions into high-voltage and low-voltage functions. The low-voltage functions are handled in the SLICOFI-2x codec device. The DuSLIC chip set is comprised of one dual-channel codec and two single-channel high-voltage

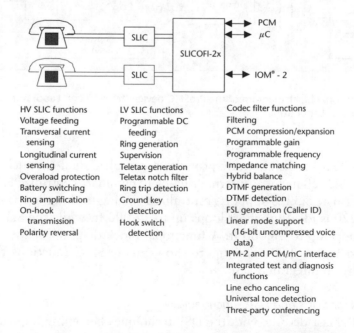

| HV SLIC functions | LV SLIC functions | Codec filter functions |
|---|---|---|
| Voltage feeding | Programmable DC | Filtering |
| Transversal current | feeding | PCM compression/expansion |
| sensing | Ring generation | Programmable gain |
| Longitudinal current | Supervision | Programmable frequency |
| sensing | Teletax generation | Impedance matching |
| Overload protection | Teletax notch filter | Hybrid balance |
| Battery switching | Ring trip detection | DTMF generation |
| Ring amplification | Ground key | DTMF detection |
| On-hook | detection | FSL generation (Caller ID) |
| transmission | Hook switch | Linear mode support |
| Polarity reversal | detection | (16-bit uncompressed voice data) |
| | | IPM-2 and PCM/mC interface |
| | | Integrated test and diagnosis functions |
| | | Line echo canceling |
| | | Universal tone detection |
| | | Three-party conferencing |

**Figure 2.35** The DuSLIC architecture. (*Source:* [38]. Courtesy of Infineon Technologies AG.)

SLIC chips. This chipset integrates the entire BORSCHT functionality based on a fully programmable DSP concept, offering flexibility for different country and line requirements with one single hardware.

The MuSLIC chip set consists of a 16-channel signal processor (PEB31665/6) together with four four-channel ADCs and DACs (PEB3465) and 16 single-channel ringing SLICS (PEB4165/66). The MuSLIC architecture is shown in Figure 2.36. This chip set is an example of enabling support for high-density SLICs, reducing space requirements and enhancing reliability in modern digital COs.

### 2.7.1.5   Codecs for SLICs

In some of the SLIC ICs described in Section 2.7.1.4, there are no integrated codecs. To fill this requirement, codec ICs specially designed for SLIC applications have evolved. This section illustrates some of these devices.

The types of codec ICs available include fixed $A$-law or $\mu$-law codecs, programmable $A/\mu$ or linear codecs, codecs with built-in transmit and receive filters, and

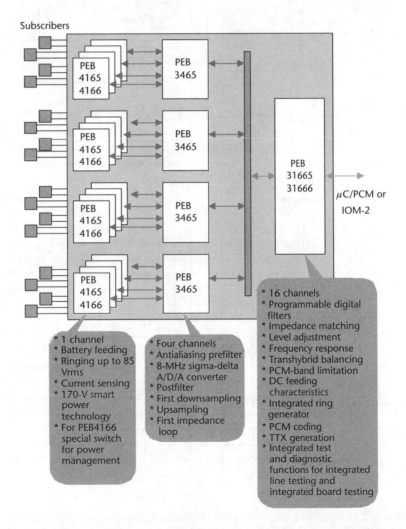

**Figure 2.36**   The MuSLIC architecture (*Source:* [39]. Courtesy of Infineon Technologies AG.)

multichannel codecs. Some codec ICs include on-chip, one or more of the following: echo cancellation, impedance matching, and transhybrid balancing through DSP-based loop filters, which use both transmit and receive signals for processing. These techniques produce significantly better performance compared to older analog techniques for achieving the same objectives.

*Fixed Encoding-Law Codecs*

An illustrative example is the MT89xx family of fixed encoding codecs from Zarlink Semiconductor. The MT8960/62/64/66/67 devices in this family provide $\mu$-law companding, and MT 8961/65 devices provide *A*-law companding. Figure 2.37 shows a block diagram of the MT 8960 and its encoding characteristics.

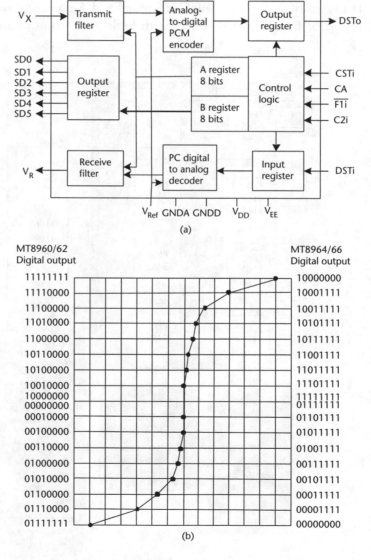

**Figure 2.37**  (a) Block diagram of the MT8960 $\mu$-law codec, and (b) encoding characteristics. (*Source:* [40]. Courtesy of Zarlink Semiconductor.)

Analog (voiceband) signals in the transmit path enter the chip at $V_x$ and are sampled at 8 kHz. The samples are assigned eight-bit digital values defined by the logarithmic (nonlinear) PCM encoding laws. Analog signals in the receive path leave the chip at $V_R$ after reconstruction from digital eight-bit words. Eight-bit PCM-encoded digital data enters and leaves the chip serially on DSTi and DSTo pins, respectively. The eight-bit digital word is output at DSTo at a nominal rate of 2.048 Mbit/s via the output buffer as the first eight bits of the 125-$\mu$S sampling frame. Separate switched capacitor filter sections are used in the transmit path prior to digital encoding and after digital decoding in the receive path. All filter clocks are derived from the 2.048-MHz master clock input C2i. Chip size is minimized by the use of common circuitry performing A/D and D/A conversion. A successive approximation technique is used with capacitor arrays to define the 16 steps and eight chords (linear segments in the nonlinear characteristics) in the signal conversion process.

The first bit of the serial data stream, bit 7 (MSB) represents the sign of the analog signal. Bits 6–4 represent the chord that contains the analog sample value. Bits 3–0 represent the step value of the analog sample within the selected cord, as discussed in Section 2.4.4 and illustrated in Figure 2.10. (Refer to Sections 2.4.3. and 2.4.4 to review the principles implemented in this device and compare Figures 2.9 to 2.10 with Figure 2.37(b).

Devices in the MT89xx series produce several options for output code formats. The device in Figure 2.37(b) shows two of these formats.

- MT8960-63: Sign plus magnitude PCM output code shown on the left side of Figure 2.37(b);
- MT8964/66: AT&T D3 PCM output code (true sign bit and inverted magnitude bits) shown on the right side of Figure 2.37(b);
- MT8965/67: ITU-T PCM output code (even bits inverted).

### Programmable Codecs

The MT91xx series is a family of programmable codecs from Zarlink Semiconductor. The family includes 9160/61/62.../67 devices [41]. Figure 2.38 shows a block diagram of the MT9160 programmable codec.

This device incorporates a built-in filter/codec, gain control, and programmable side tone path as well as on-chip antialias filters, reference voltage, and bias source. The device supports both ITU-T and sign-magnitude A-law and $\mu$-law requirements. Complete telephony interfaces are provided for connection to handset transducers. Each of the programmable parameters within the functional blocks is accessed through a serial microcontroller port compatible with Intel MCS-51®, Motorola SPI®, and National Semiconductor Microwire® specifications. The discussion of these interfaces are beyond the scope of this chapter. Readers are referred to the Web sites of these organizations for more details. Optionally, this device may be used in a controllerless mode utilizing power-on default settings.

### Multichannel Codecs

The 8210xx series of codecs from Integrated Device Technology, Inc. (IDT) and the PEB/PEF 2466 from Infineon Technologies AG are examples of multichannel codecs.

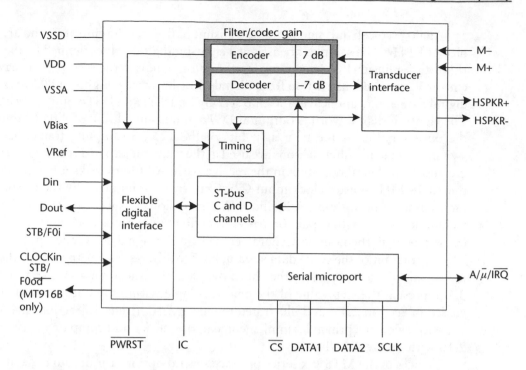

**Figure 2.38** Block diagram of the MT9160 programmable A/$\mu$-law codec. (*Source:* [41]. Courtesy of Zarlink Semiconductor.)

Some members of the IDT 8210xx family are the 821024/34 quad PCM codecs, the 821054/64 programmable quad PCM codecs, and the 821068 octal programmable codec. Figure 2.39 shows a block diagram of the IDT 821054 device.

This is a four-channel PCM codec with on-chip filters. A-law/$\mu$-law, as well as linear encoding/decoding, are supported. The SLIC signaling block includes an FSK generator and two programmable tone generators per channel, which can also generate ring signals, together with two programmable clocks for SLICs. The DSP core provides ac impedance matching, transhybrid balance, frequency response, and gain setting of the digital transmit and receive filters. The device also offers test capability with several analog/digital loop-back and level-metering functions. It supports two PCM buses in which the clock edge as well as time slot assignment are programmable.

Reference [31] describes a reference design for the UniSLIC14 discussed in Section 2.7.1.4 and IDT821054/64 programmable quad codec discussed in this section, forming a four-channel SLIC.

The PEB/PEF 2466 is a four-channel filter/codec combination from Infineon Technologies with PCM and microcontroller interfaces [43]. Figure 2.40 shows a simplified diagram of one channel of the PEB2466. This device, too, is based on a DSP architecture, where impedance matching and transhybrid balancing functions are performed by digital programmable loop filters between the transmit and receive path. The filter characteristics are adjusted according to local requirements. The DSP also performs frequency response corrections and level adjustments to enable the design of universally applicable SLICs.

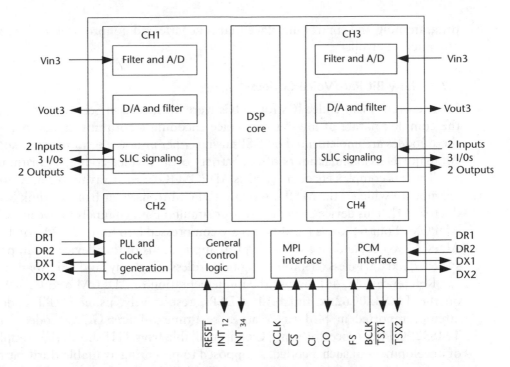

**Figure 2.39**  Block diagram of the IDT821054 programmable quad PCM codec. (*Source:* [42]. Reproduced with permission of Integrated Device Technology, Inc.)

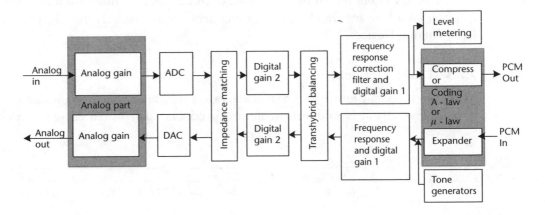

**Figure 2.40**  Simplified diagram of the PEB 2466 codec showing one channel. (*Source:* [43]. Courtesy of Infineon Technologies AG.)

Key features of this device are the impedance matching and transhybrid balancing functions implemented through DSP techniques. These functions complement similar functions implemented in the SLIC to which the codec is attached. In both of these functions, digital loop filters feed portions of the transmit signal back to the receive path to allow accurate and stable impedance matching and hybrid balancing.

The serial microcontroller interface is used to program the codec function as well as to monitor and control other SLIC functions. This interface provides a programming and control interface that is generic and nonproprietary for use with any microcontroller.

### 2.7.2 Low-Bit-Rate Voice Codecs

Some current PCM codec ICs for SLICs were illustrated in Section 2.7.1.5. Due to the complex nature of low-bit-rate voice encoding algorithms, a vast majority of these codecs are implemented on DSPs using either in-house or third-party software. All major DSP manufacturers offer a variety of such voice-coding algorithms.

The less complex encoders such as ADPCM (G.726) are available in IC form, an example of which is the MT9126 quad ADPCM transcoder from Zarlink Semiconductor [44]. This device can convert four simultaneous channels between PCM and ADPCM. Four 64-kbit/s PCM octets are compressed into four 32-, 24-, or 16-kbit/s ADPCM words, and vice versa. Typical applications of this device are in pair gain systems, fixed wireless telephony, digital cordless telephony, and voicemail systems.

References [45, 46] discuss the implementation of ADPCM and G.729 codecs on the TMS320C62xx and C54xx DSPs, respectively, using ITU-T codes. The attempt reported in [46] to create a real-time 8-kbit/s G.729 codec with the TMS320C54xx processor using C code available from ITU shows the complexities of development of such a codec, as opposed to procuring available third-party software, even with the added cost involved. Table 2.13 shows several DSPs commonly used for voice codecs and the standards supported by each. Chapter 9 details the development of a voice codec on DSP hardware.

Table 2.13 does not show variable-rate codecs such as those used in GSM. An example of a DSP for which these voice codecs are available is the Tiger SHARC® from Analog Devices, Inc.

### 2.7.3 Devices for VoIP

VoIP products fall into two broad categories: terminals that handle one or perhaps a few channels and infrastructure products such as voice gateways that

**Table 2.13** Voice Codecs

| Company | Device | Voice Codecs Supported | | | | | | | Software | |
|---|---|---|---|---|---|---|---|---|---|---|
| | | G.711 | G.722 | G.723.1 | G.726 | G.727 | G.728 | G.729 A/B | In house | Third party |
| Analog devices | ADSP-218x, ADSP-219x, ADSP-2106x, ADSP-2116x | x | x | x | x | x | x | x | | x |
| Audio codes | AC481xsA-C | x | | x | x | x | x | x | x | |
| DSP group | CT8020 DA | | | x | | | | | | |
| | CT8021 AB | x | x | x | | | x | x | | |
| 8 × 8 | VCPex | x | x | x | | | x | x | | |
| Motorola | 56311 | x | x | x | x | | x | x | x | |
| Texas Instruments | TMS320C5xxx | x | | x | x | | | | | x |
| | TMS320C5/6xxx | x | x | x | x | x | | x | x | |
| ZSP | 16402 | x | | x | x | | x | x | x | |

simultaneously handle many channels. High density is the goal of infrastructure products, and this density drives the main hardware features for VoIP devices: high speed, compactness, and low power.

Almost all VoIP products rely on DSPs for the heavy processing that is required for the voice codec as well as the IP call stack. Figure 2.41 shows the algorithms that a VoIP product must in general support. DSP vendors provide these algorithms either directly or through third parties. Many protocols exist for voice, fax, and data modems to make these devices applicable in products of all common telephony applications. DSP products supporting VoIP provide several ports with the flexibility of providing one of several protocols on each port. This way, a device can support any combination of fax, voice, and modem on one piece of equipment.

In most products, the voice processing and the IP call stack are handled by different devices. This is particularly true of gateways that carry hundreds or thousands of channels. The intense computational requirements of the voice codec as well as other voice-related processing mean that even a powerful DSP can handle just a few channels. Even in terminals, processing the call stack differs so much from voice processing that separate chips (i.e., voice processors and host processors) are often required.

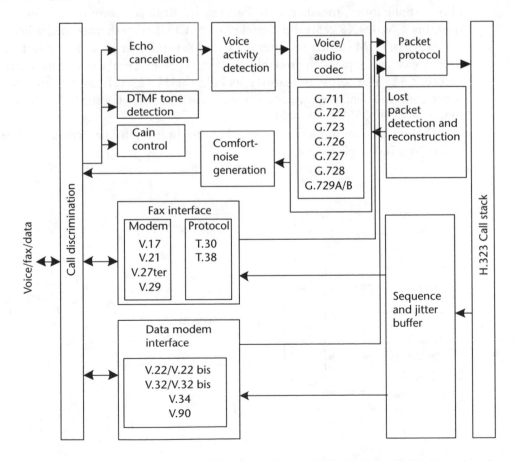

**Figure 2.41**   Algorithms needed in a VoIP product (*Source:* [47]. Courtesy of EDN. Reproduced with permission of Reed Business Information.)

The centralized nature of the IP call stack as shown in Figure 2.41 also means that one processor can manage many channels, enabling one device to manage the call stack and control many voice-processing devices.

### 2.7.3.1   Illustrative Devices for VoIP Terminal Equipment

Texas Instruments TMS320C54CST (client side telephony) DSP [48] is a device that supports a large number of industry's most demanded software algorithms for customer premises telephony devices. This device is based on the TMS320C5000 DSP platform and significantly minimizes the design and investment risk by integrating hardware and software needed to create a wide range of telephony applications, including VoIP. The on-chip telephony library includes data algorithms (such as modem, compression, and error correction), telephony algorithms (such as DTMF and CLI), and voice algorithms (such as VAD, CNG, voice coding, and echo cancellation). The chip combines software, on-chip data access arrangement (DAA) for modem applications, 16K words SRAM, 16K words ROM, a UART, two multichannel buffered serial ports, two timers, and a host processor interface. As a result of the C5000 DSP architecture, the 54CST DSP operates at less than 25-mW power consumption and up to 120 MIPS.

Netergy Microelectronics produces the Audacity-T2 processor for IP telephone applications, together with the Veracity VoIP software suite. This processor contains a MIPS-X5 central processing unit (CPU), an Ethernet media access controller, two TDM bit stream interfaces, and several host processor interfaces [49]. This device can implement an IP phone using a few external devices, as illustrated in Figure 2.42. The software features include the H.323 call control protocols and other common function implementations, such as voice coding, echo cancellation, tone generation, VAD, and CNG.

This device can implement a small media hub for home or small office/home office (SOHO) applications.

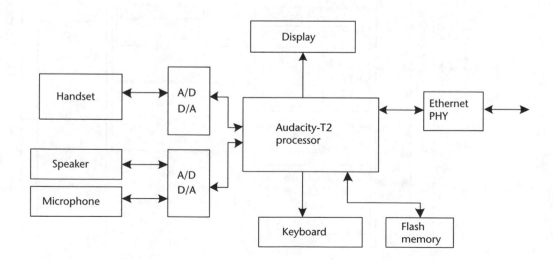

**Figure 2.42**   IP telephone built around the Audacity-T2 processor.

### 2.7.3.2 Illustrative Gateway Devices

In a VoIP system, gateways perform functions such as translation of protocols and media formats between networks and voice digitization when this facility is not available in the terminal equipment. Gateway functions were described in detail in Section 2.6.4.3.

Density is important in gateway devices because an application could have hundreds of telephone lines entering a gateway unit that could be physically a small box. High-speed processing supports many connections in one chip, helping to reduce device numbers in a card. Low-power requirements reduce heat buildup and also enable existing equipment to take in additional VoIP support cards using its power supply. All three work together to maximize the number of simultaneous calls a system can handle in a small space [47].

The same requirements of speed, power, and size drive wireless communications—hence the early devices used for VoIP products especially used devices developed for wireless communications. However, a major difference between today's VoIP products and wireless products is that the former supports several voice-coding algorithms as necessitated by the applications and standards, as described in Section 2.6. Products for wireless communications support only one voice codec.

Texas Instruments provides a range of devices for handling from two to 2,016 (OC-3) channels on a single chip. These devices are categorized as residential/SOHO, enterprise, and infrastructure gateways, depending on the number of channels supported. Table 2.14 summarizes these devices. These devices integrate Telogy Software™ products. More details of these devices are found in [49–54].

The TNETV1010 hardware and software architectures are shown in Figure 2.43(a) and (b), respectively.

Figure 2.44 illustrates the TNETV2010 high-density gateway processor. This device finds application in high-density carrier class VoIP gateway products. These offer up to 216 PCM channels on a single processor. The TNETV3030 has six fixed-point DSP cores based on Texas Instruments' TMS320C55 DSP. Each of the six cores operates at 300 MHz and shares multiple on-chip resources. The device supports multiple low-bit-rate codecs with other functions, such as echo cancellation, and value-added features, such as CLI.

The ADSP-21mod family of Internet gateway processors from Analog Devices, Inc. [56], is capable of implementing up to 40 communications channels in a single package. This family is summarized in Table 2.15.

This family employs all modem and VoP protocol functions and can be implemented without using external memory. The ADSP-21mod family offers a single-point interface for multiple lines to transmit voice, data, or fax. It addresses many of

**Table 2.14** Illustrative VoIP Gateway Products from Texas Instruments

| Device | Application/Type | Channel Density |
|---|---|---|
| TNETV1010 VoIP gateway solution | Residential/SOHO | 1–2 |
| C54X customer premises gateway solution | Residential/SOHO/enterprise | 1–12 |
| TNETV2020 enterprise VoIP gateway solution | Enterprise | 8–13 |
| TNETV2840 VoIP gateway | Enterprise/infrastructure | Up to 48 |
| TNETV30xx high-density VoIP gateway solution | Infrastructure | 192–2,016 |

(*Source:* [50].)

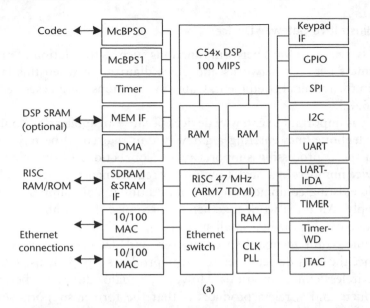

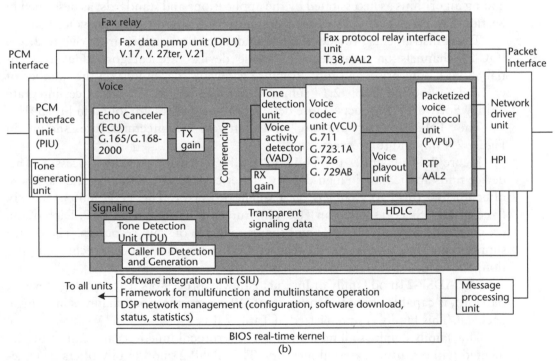

**Figure 2.43** TNETV1010 residential gateway architecture: (a) hardware, and (b) software. (*Source:* [50]. Courtesy of Texas Instruments.)

the Internet access challenges, optimizing voice quality through high-quality speech codecs, advanced jitter buffer management, and robust echo cancellation. The ADSP-21mod 90 device implements 16 modem channels or 40 voice channels in one package, using a pool of eight ADSP 218x DSP cores and 2 MB of on-chip memory, together with a number of host processor interfaces and serial interfaces [57].

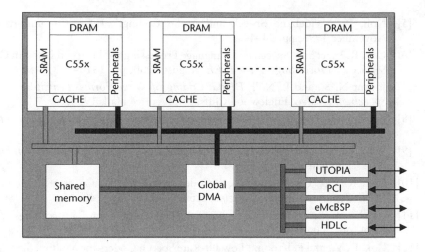

**Figure 2.44**   TNETV3010 high-density gateway processor architecture. (*Source:* [55]. Courtesy of Texas Instruments.)

**Table 2.15**   ADSP-21mod Family from Analog Devices, Inc.

| Processor | Number of Ports | Software Features |
|---|---|---|
| ADSP-21mod980-240 voice solution (600 MIPS) | 40 | G.711 |
| ADSP-21mod980-2416 voice solution (600 MIPS) | 16 | G.729 A/B, G.711 |
| ADSP-21mod980-110 integrated modem solution (600 MIPS) | 8 | Modem |
| ADSP-21mod970-110 integrated modem solution (312 MIPS) | 6 | Modem |
| ADSP-21mod870-110 integrated modem solution (52 MIPS) | 1 | Modem |

(*From:* [56].)

Reference [47] reports the Motorola 56307 DSP that can process both voice and the call stack. Nortel produces 80-channel daughter cards with nine Motorola 56307 DSPs on each side, one for the call stack and eight for the telephony algorithms. Each DSP handles 10 channels simultaneously, giving a total of 160 channels in one card. Ten such cards fit into a motherboard in a 1,600-channel gateway.

## References

[1]   Bellamy, J., *Digital Telephony,* New York: John Wiley and Sons, 2000.

[2]   Polyzois, C. A., et al., "From POTS to PANS: A Commentary on the Evolution of Internet Telephony," *IEEE Network Magazine,* May/June 1999, pp. 58–64.

[3]   Gibson, J. D., *Principles of Digital and Analog Communications,* New York: Macmillan Publishing Co., 1989.

[4]   Owen, F. F. E., *PCM and Digital Transmission Systems,* New York: McGraw-Hill Book Co., 1982.

[5] Noll, P., "Wideband Speech and Audio Coding," *IEEE Communications Magazine*, November 1993, pp. 34–44.

[6] Cox, R. V., et al., "Speech and Language Processing for Next-Millennium Communications Services," *Proceedings of the IEEE*, August 2000, pp. 1314–1337.

[7] Jayant N. S., and P. Noll, *Digital Coding of Wave Forms: Principles of Applications to Speech and Video*, Englewood Cliffs, NJ: Prentice Hall, 1984.

[8] Ackenhusen, J. G., *Real-Time Signal Processing: Design and Implementation of Signal Processing Systems*, Englewood Cliffs, NJ: Prentice Hall, 1999.

[9] Westall F. A., and S. F. A. Ip, (Eds.), *Digital Signal Processing in Telecommunications*, London: Chapman and Hall, 1993.

[10] Cox, R. V., "Three New Speech Coders from the ITU Cover a Range of Applications," *IEEE Communications Magazine*, September 1997, pp. 40–47.

[11] Gersho, A., "Advances in Speech and Audio Compression," *Proceedings of the IEEE*, June 1994, pp. 900–918.

[12] Cox, R. V., and P. Kroon, "Low Bit-Rate Speech Coders for Multimedia Communication," *IEEE Communications Magazine*, December 1996, pp. 34–41.

[13] Pracht, S., and D. Hardman, "Voice Quality in Converging Telephony and IP Networks," *EDN*, September 1, 2000, pp. 89–104.

[14] Dimolitsas, S., "Standardizing Speech-Coding Technology for Network Applications," *IEEE Communications Magazine*, November 1993, pp. 26–33.

[15] Kostas, T. J., et al., "Real-Time Voice Over Packet-Switched Networks," *IEEE Network*, January/February 1998, pp. 18–27.

[16] Korpi, M., and V. Kumar, "Supplementary Services in the H.323 IP Telephony Network," *IEEE Communications Magazine*, July 1999, pp. 118–125.

[17] ITU-T Rec. H.323, "Visual Telephone Systems and Terminal Equipment for Local Area Networks Which Provide a Non-Guaranteed Quality of Service," 1996.

[18] ITU-T Rec. H.323, "Packet-Based Multimedia Communications Systems," Geneva Switzerland, January 1998.

[19] Mahbub, H., and A. Nayandoro, "Internet Telephony: Services, Technical Challenges, and Products," *IEEE Communications Magazine*, April 2000, pp. 96–103.

[20] Glitho, R. H., "Advanced Services Architectures for Internet Telephony: A Critical Overview," *IEEE Network Magazine*, July/August 2000, pp. 38–44.

[21] Handley, M., et al., "SIP: Session Initiation Protocol," RFC 2543, March 1999.

[22] Henning, S., and J. Rosenberg, "The Session Initiation Protocol: Internet-Centric Signaling," *IEEE Communications Magazine*, October 2000.

[23] Perkins, M. E., et al., "Speech Transmission Performance Planning in Hybrid IP/SCN Networks," *IEEE Communications Magazine*, July 1999, pp. 126–131.

[24] Mathy, L., C. Edwards, and D. Hutchison, "The Internet: A Global Telecommunications Solution," *IEEE Network*, July/August 2000, pp. 46–57.

[25] GL Communications, Inc., "Voice Quality Testing," http://www.gl.com/voicequalitytesting.html.

[26] Agilent Technologies, Inc., "The Agilent Voice Quality Tester," http//onenetworks.comms.agilent.com/VQT/J1981/A_product.asp.

[27] Bigelow, S. J., *Understanding Telephone Electronics*, New York: Butterworth-Hienemann, 1997.

[28] Reeve, Witham D., *Subscriber Loop Signalling and Transmission Handbook: Digital*, New Jersey: IEEE Press, 1995.

[29] Zarlink Semiconductor, "MH88612BV-K Subscriber Line Interface Circuit (SLIC)" Data Sheet, January 1999.

[30] Intersil Americas, Inc., "HC55120, HC55121, HC55130, HC55131, HC55140, HC55141, HC55142, HC55143, HC55150, HC55151," Data Sheet FN4659.10, June 2002.

[31]   Intersil Americas, Inc., "UniSLIC14 and the IDT821054/64 Programmable Quad CODEC," Application Note AN9999.0, May 2002.

[32]   Intersil Americas, Inc., "HC55185," Data Sheet FN4831.9, May 2002.

[33]   Intersil Americas, Inc., "Non-Ringing SLICs," http://www.intersil.com/design/VoIP/unis-lic.asp.

[34]   Intersil Americas Inc., "Ringing SLICs," http://www.intersil.com/design/VoIP/rslic.asp.

[35]   Silicon Laboratories, "ProSLIC Products," http://www.silabs.com/products/proslic.asp.

[36]   Silicon Laboratories, "Si3210/Si3211/Si3212 ProSLIC Programmable CMOS SLIC/CODEC with Ringing/Battery Voltage Generation," Data Sheet Si3210-DS141, Rev. 1.41 6/03, 2003.

[37]   Silicon Laboratories, "Si3220/Si3225 Dual ProSLIC Programmable CMOS SLIC/CODEC," Data Sheet Si3220/Si3225-DS 096, Preliminary Rev. 0.96 7/03, 2003.

[38]   Infineon Technologies AG, "DuSLIC Dual Channel Subscriber Line Interface Concept," Product Overview, DS2, November 2000.

[39]   Infineon Technologies AG, "MuSLIC Product Brief PS09985," September 1998.

[40]   Zarlink Semiconductor, "MT8960 /61/62/63/64/65/66/67 Integrated PCM Filter Codec," Issue 10, May 1995.

[41]   Zarlink Semiconductor, "MT9160B/61B 5V Multi-Featured Codec (MFC), Advance Information," DS5145, Issue 3, March 1999.

[42]   Integrated Device Technology, Inc., "IDT821054 Quad Programmable PCM Codec with MPI Interface," February 2002.

[43]   Infineon Technologies AG, "SICOFI$^R$4-$\mu$C Four Channel Codec and Filter with PCM and Microcontroller Interface," Product Overview, DS1, February 2002.

[44]   Zarlink Semiconductor, "MT9126 Quad ADPCM Trascoder," Data Sheet, April 2002.

[45]   Texas Instruments, "Implementation of G.726 ADPCM on Texas Instruments TMS320C62xx Processors," Application Report BPRA066A, May 1999.

[46]   Texas Instruments, "Implementation of G.729 on the TMS320C54x," Application Report SPRA656, March 2000.

[47]   Ellis, G., "Internet Protocol: The Future Route for Telephony?" *EDN*, November 19, 1998, pp. 62–64, 66, 68, 70, 72, 74, 76–78.

[48]   Texas Instruments, "Client Side Telephony Solution," Product Bulletin SPRT228, 2001.

[49]   Netergy Microelectronics, "IP Phone Evaluation and Development Platform for the Audacity-T2 Processor," 000-0059-003 REV C 04/02, 2002.

[50]   Texas Instruments, "TNETV1010 VoIP Gateway Solution," Product Bulletin SPET010 V1.9R 4/04, 2002.

[51]   Texas Instruments, "VoIP Gateway Solution Summary," Product Bulletin SPET005 V1.2R 4/20, 2002.

[52]   Texas Instruments, "TNETV2020 Enterprise VoIP Gateway Solution," Product Bulletin SPET011 V2.0R 4/02, 2002.

[53]   Texas Instruments, "Customer Premise Gateway Solutions," Product Bulletin SPET001V1.8R 4/02, 2002.

[54]   Texas Instruments, "TNETV2840 VoIP Gateway Solutions," Product Bulletin SPET012 V3.0 R 4/02, 2002.

[55]   Texas Instruments, "TNETV30xx High Density VoIP Gateway Solutions," Product Bulletin, SPET013 V2.0 R 4/02, 2002.

[56]   Analog Devices, Inc., "ADSP-21mod Family Internet Gateway Processors," H3693b-10-2/00 (rev. B), 2000.

[57]   Analog Devices, Inc., "Multi Port Internet Gateway Processor ADSP-21mod980," C01761-2,5-9/00 (rev.0), 2000.

[58]   ITU-T: "Recommendation G.712, "Transmission Performance Characteristics of Pulse Code Modulation Channels," 11/96, 1996.

# CHAPTER 3

# Switching and Signaling Systems

## 3.1 Evolution of Switching Systems

Subsequent to the invention of the telephone in 1876, by 1884 approximately 350,000 phones were in use, interconnected by a plethora of manual switchboards. Suspicions that the operators at switchboards were diverting the undertaker A. B. Strowger's potential business to competitors led him to develop the first automatic step-by-step exchange in around 1896. This was based on pulse dialing from the rotary phone handset. Variations of this type of switches proliferated and were widely in use up to the mid 1950s. With transistorized circuits in the late 1940s, tone-based signaling became a reality. More advanced control systems and common control of switching arrays progressed with digital control. As a result, crossbar systems and reed relay–based switches proliferated. In all of these systems, the basis of switching was to have a pair of wires selectively coupled to another pair of wires by mechanical means to complete a speech path between subscribers.

As digital hardware and processing techniques matured, engineers saw the advantage of placing a computer subsystem for common control of a switching subsystem. Hence, the stored program control (SPC) of the telephone switching systems evolved. The first SPC exchange trials were by the Bell Labs in around 1958. SPC systems were conducive to providing many special services such as call waiting, call forwarding, and specialized billing. By the early 1970s, there were large exchanges with capacities around 150,000 lines, processing over 750,000 calls per hour during peak periods. While the SPC subsystems provided better and more efficient call processing, the switching within an exchange core and the long-distance transmission systems remained predominantly analog. Crossbar-type exchanges with SPC resulted in reliable switches with lower maintenance. This in turn resulted in services growing rapidly, irrespective of local and national boundaries, via cable, submarine, and satellite transmission.

With semiconductor memories, digital components, and processing techniques becoming cost effective, switching system designers were able to develop the PCM-based fully digital switching systems. In these systems, the processor subsystem was able to tackle the signaling process as well as the switching of voice paths by organized exchanges of PCM voice samples. Such fully digital exchanges came into operation in the late 1970s.

Over the 1980s and 1990s most of the telecommunications systems providers were able to replace older types of exchanges with digital switching systems. This period also saw vast developments related to digital transmission systems. During the same period, various data communication applications made use of a mix of the telephone network's infrastructure as well as dedicated data networks.

Over 800 million subscribers worldwide use the PSTN today. During 1980 to 2000, there were rapid developments in mobile, Internet, multimedia, and broadband, services, which changed the overall scenario of telecommunication systems. This chapter provides an overview of digital switching systems and associated concepts such as signaling as applied to modern telecommunication systems. For more details on the historic development of switching systems, [1–4] are suggested.

## 3.2   Essentials of Switching, Signaling, and Control

Today's PSTN reference model can be simply depicted, as in Figure 3.1(a). Figure 3.1(b) indicates the same situation dividing the network into three layers: access, service, and infrastructure.

In modern switching systems, three basic concepts—signaling, switching, and control—are integrated to provide a physical or a virtual path between the subscribers for a communication session. In simple terms, a subscriber connected to the access network indicates the desire to access the network (for example, a telephone subscriber listens to dial tone and then dials digits) and then gets connected to another subscriber or a location with the aid of signaling. The nearest local switch, which maintains specific information about the particular subscriber, interprets the signaling information and creates a path to another subscriber, distance exchange, or a site using the concepts of switching. Modern switching concepts can be primarily divided into circuit switching and packet switching.

Modern switches are designed around a real-time processor-based control system, operating as the central element that controls the subscriber subsystem (access subsystem), switching subsystem (switch fabric through which the physical or logical connection is made), and the access to transport networks.

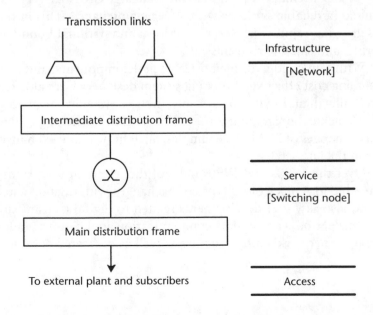

**Figure 3.1**   Three layers of the PSTN.

## 3.3   Principles of Switching

Early manual systems had the operator as its central controller. Manual control gradually evolved into fully automatic SPC systems. The processor system in an SPC exchange controls a complex set of switching matrices of wire pairs. In this process, various elements such as rectangular cross-point arrays and multistage switching modules are used. The simple principle maintained throughout the process was that there was always a physical path for a copper pair from the access network to get connected to another copper pair either within the local access area or to a distance exchange.

The principle of a single rectangular crosspoint array and multistage switching matrices as examples are shown in Figure 3.2. Figure 3.2(a) indicates the case of $N$ inlets and $M$ outlets. Figure 3.2(b) shows a multistage case for $N$ inlets and $N$ outlets. The first stage consists of multiple $n \times k$ matrices, coupled to a second stage of $k$ matrices with $N/n$ inputs and $N/n$ outputs each (called the junctors) and finally connected to the third stage with $k \times n$ matrices. More details of such techniques with important parameters such as blocking, blocking probabilities, paths, and path

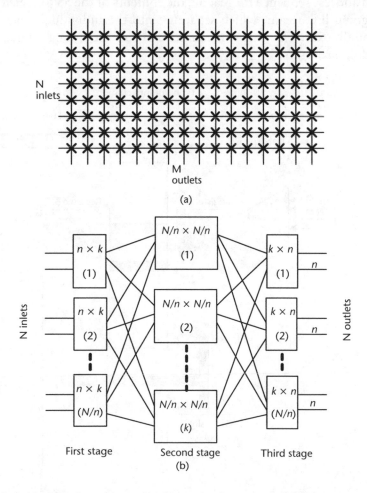

**Figure 3.2**   Switching matrices: (a) rectangular cross point array, and (b) three-stage switching matrix.

finding times are available in [4, 5]. This concept of having a physical path via a switch matrix is called *space division switching.*

The maturing of SPC techniques while semiconductor memories became less expensive allowed designers to adopt PCM to digitize the voice signals and store the samples in a semiconductor memory. With processor blocks controlling the overall operation in real-time, consecutive PCM, coded voice samples from a bank of memories (incoming voice) could be transferred to the output stage with suitable interchange of time slots. Selective switching of voice channels were thus reduced to a write-select-read sequence of PCM samples in a time switch interchange (TSI) module.

### 3.3.1 Time and Space Switching

By the early 1980s, the TSI concept had evolved well into practical switching systems. A situation with the multiplexer, demultiplexer, and associated memories is shown in Figure 3.3. Voice samples in the PCM bit stream from the multiplexer are received into register A and written into the speech store (SS) sequentially. Signaling related to the conversation updates the control store (CS). This in turn provides the read address sequence for reading the contents of the SS to register B, coupled to the outgoing bit stream, which separates into the individual voice channels after the demultiplexer. The time slot counter controls the timing related to the writing and reading processes.

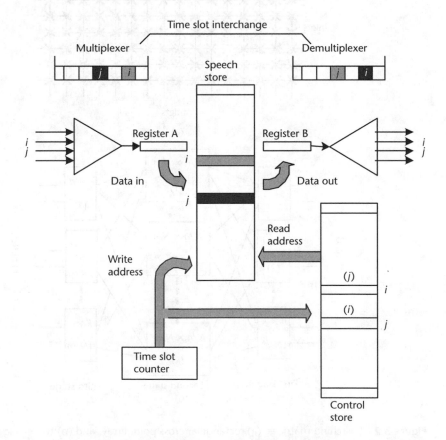

**Figure 3.3**  TSI with multiplexer/demultiplexer and memory switch.

Because a write and a read are required for each channel (sampled every 125 $\mu$s) entering and leaving the memory, the maximum number of channels $N$ that can be supported by a single TSI is $125/2t_c$, where $t_c$ is the memory cycle time in microseconds.

Figure 3.3 clearly indicates that the speed of the memory chips govern the number of channels that can be switched on a nonblocking basis. In general, *time switch* modules can be configured as sequential write/random read or vice versa, as in Figure 3.4(a), showing a case of time slots 5 and 8 getting exchanged in sequential write/random read mode.

Similar to the time switch, a *space switch* can be configured as in Figure 3.4(b) with $n \times n$ switching points designed as electronic gates. Each column in the matrix is controlled by a CS, which is used to indicate the relationship between incoming time slots and the respective outgoing connections that allow the distribution of incoming time slots to various outgoing paths. Similarly, the time slots in incoming and outgoing T1/EI links in transit exchanges can be redistributed. The overall effect of time and space switching is illustrated in Figure 3.4(c). A *group switch* (GS) is a combination of time and space switches.

### 3.3.2  Multidimensional Switching

Large, versatile digital exchanges require combinations of space switch modules (SPM) and time switch modules (TSM). A large exchange could have combinations such as time-space (TS), space-time (ST), time-space-time (TST), and space-time-space (STS) switches. Figure 3.5(a) and (b) indicate TS and STS examples, respectively. It is important to realize that duplex voice requires bidirectional paths with corresponding elements in the SS in each direction.

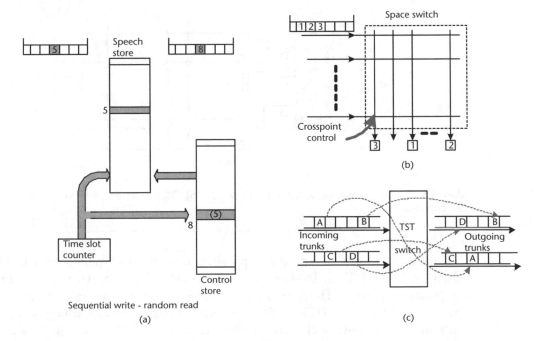

**Figure 3.4**  Switch modules: (a) time switch modules, (b) space switch, and (c) effect on the incoming and outgoing streams.

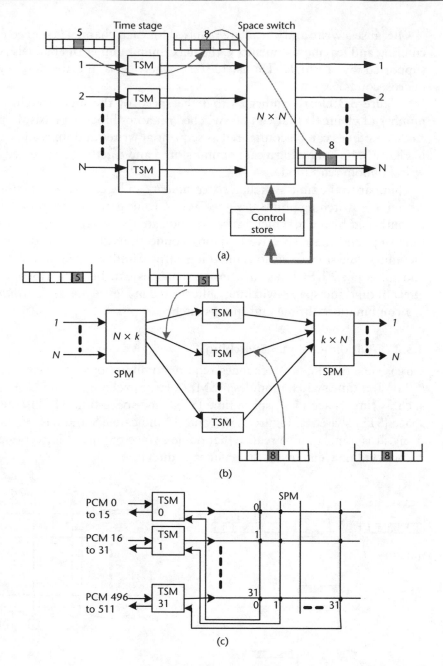

**Figure 3.5** Multidimensional switches: (a) TS switch, (b) STS switch, and (c) TST with multiple PCM connections.

Figure 3.5(c) shows a typical group switch, having a TST switch with 32 TSMs and a 32 × 32 SPM. Each TSM can receive 16 PCM links giving 16 × 32 = 512 multiple positions (MUP). Thirty two such modules connected to an SPM provide a capacity of 512 × 32 = 16,384 MUPs. Because a telephone call always seizes two inputs and two outputs, such a group switch can theoretically switch 8,192 simultaneous calls. In practice this number will be less due to signaling and recorded announcements occupying some time slots. More details are available in [6].

## 3.4   Circuit Switching and Packet Switching

In telecommunications systems, three types of switching are generally possible—circuit switching, message switching, and packet switching.

Circuit switching is the case where a physical channel or its equivalent is established between the two parties, and it remains until the call is completed. Even if there is no conversation present, such as in the case of silent periods in a duplex conversation, an equivalent channel remains end to end until the call is completed. Selection of an appropriate path through the public systems is based on signaling between the two subscribers.

In message switching, a public network carries the entire message from one end to the other. A classic example is telegraphy. Though these systems are nearly extinct in public telecommunications networks, they are used in the aviation industry.

In packet switching, data to be transmitted is broken down into segments, or packets, and each packet is *enveloped* with address and other overhead information to ensure that it travels to the correct destination without errors. The packet travels across the network on a store-and-forward basis until it reaches the final destination. Packet-switched networks use the capacity only when there is something to be sent. This way, all available capacity can be filled with packets. Packet switching is more suited for data communication, where error-free transmission is important and variable delays are tolerable. Circuit switching is traditionally used for voice systems, where presence of errors is acceptable up to a limit but where real-time communication is a must.

### 3.4.1   Basics of Packet Switching

Figure 3.6(a) depicts a packet-switched network. The store-and-forward nodes are interconnected to each other via data links in a mesh. The nodes are processor systems that are capable of many tasks, such as receiving and transmitting files from the information sources, breaking files into smaller units, and forming packets with headers and trailers. This overhead information provides error checking for error-free transmission of packets, routing information regarding the available paths with alternatives, and priority information.

#### 3.4.1.1   Packetizing

The format of a packet varies from one network to another. In the early packet-switched networks, the assumptions were that links were slow, subject to errors, and expensive. In such cases, the subscriber data is supplemented with many additional bits for control and error handling, resulting in increased overhead. In modern systems, where the transmission links are faster and error free, such overhead could be relatively small. Figure 3.6(b) indicates a typical packet format. A header contains destination and source addresses, an operation code (indicating whether the packet is data or control information), sequence information (to help assembly of packets at the destination), and a byte count. The trailer contains an error-detecting code, such as cyclic redundancy check (CRC). The delivery of packets is associated with an acknowledgment procedure, which allows the sender to

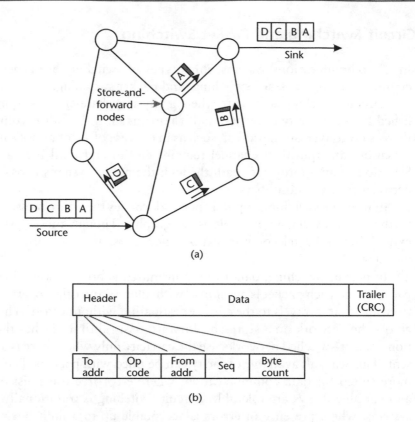

(a)

| Header | Data | Trailer (CRC) |
|--------|------|---------------|

| To addr | Op code | From addr | Seq | Byte count |
|---------|---------|-----------|-----|------------|

(b)

**Figure 3.6**  Packet switching concepts: (a) simplified network, and (b) typical packet format.

determine whether a packet has been delivered to the destination. If a packet has not been received correctly by the destination, a retransmission is initiated following one of several algorithms.

### 3.4.1.2  Virtual Circuits and Datagrams

The switching of packets may be via two basic methods, as *virtual circuits* or as *datagrams*. Virtual circuits adopt basic aspects of both circuit and packet switching. The transmission capacity is dynamically assigned on an "as needed" basis, but all packets of a multipacket information file follow the same route through the network. Before interactive communications begin, a route is established through the network and all participating nodes are informed of the connection path. The path exists only for the duration of the data transfer. These are also referred to as *switched virtual circuits* (SVCs), or *connection-oriented* packet switching. In essence, a virtual circuit is a logical concept involving addresses and pointers in the nodes of the network, but with no dedicated transmission resources.

A *permanent virtual circuit* (PVC) can also be established, where successive sessions between any two end points always use the same path. An attractive feature is the absence of the call-establishment phase of the virtual circuit, while the vulnerability to node or link failures is a disadvantage.

In datagram packet switching, a virtual circuit is not established prior to communication. Packets are switched individually based on their destination address.

Hence, packets corresponding to a single communication may travel via different paths over the network and may arrive out of order. The advantages are the
absence of a call-establishment phase and the ability to route packets, avoiding
network congestion and failures. This technique is called *connectionless* packet
switching.

### 3.4.1.3   Routing

Routing is an important concept used in packet-switching systems. It enables a suitable path to be found for delivery of packets to the destination. The best routes are
determined by switches based on criteria related to factors such as network traffic,
link costs, and delays. All routing techniques allow for a certain amount of adaptation or dynamic routing, circumventing line or node failures and responding quickly
to network failures or congestion.

### 3.4.1.4   Flow Control

Routing and flow control are two closely related operational requirements of any
communications network. The same basic principle for controlling the flow in circuit switched network also applies to packet networks. Whenever the network is
experiencing stress due to loss of capacity or heavy demand for services, new service
requests must be blocked at the periphery before they tie up the common resources
at nodes, the store and forward buffers, and the transmission links. Flow control in
a packet network is primarily concerned with buffer management. Transmitted
packets are held in a buffer until acknowledgments are received. Flow control procedures ensure that these buffers do not overflow in the event of a network problem.
For more details on the basics of packet switching, [7] is suggested.

## 3.5   Signaling Systems

Signaling functions of a communications network refer to the means for transferring network-related control information between various terminals, nodes, and
users. Signaling functions can be broadly categorized into two primary types: supervisory and information bearing. Supervisory signals convey the status or control
information of the network elements. Obvious examples are off-hook, dial tone,
ringing, busy tone, and call terminating. Information bearing signals are call-
ing/called party addresses and toll charges.

### 3.5.1   Subscriber Signaling in the PSTN

A simple example of signaling between the subscriber and the local exchange (LE) is
shown in Figure 3.7(a). The process starts with supplying the dial tone to calling
subscriber ($A$), followed by the number-sending process until the exchange detects
the condition of the called subscriber ($B$). If $B$ is free, the ringing tone and a ringing
signal is sent to each party until $B$ answers. This commences the conversation
process. When one of the parties places the receiver on-hook, a clearing signal is sent
and the call ends. When the calling line identification (CLI) service is present, $A$'s

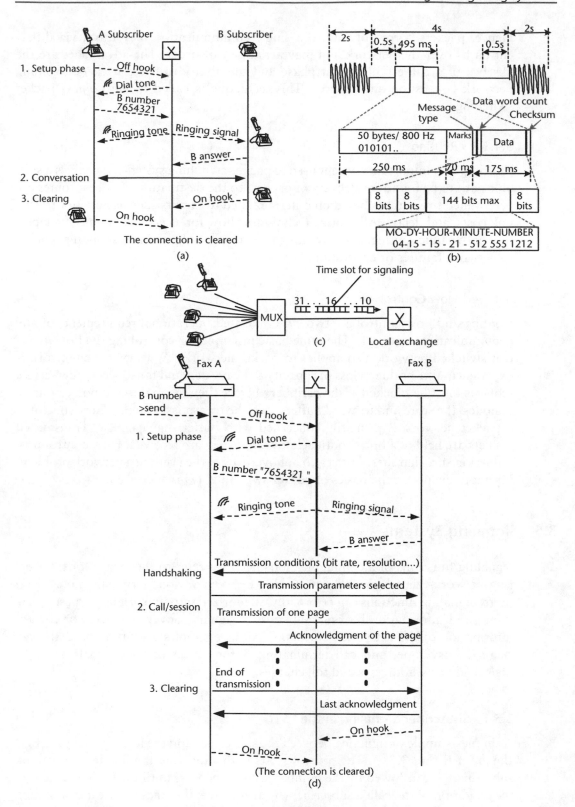

**Figure 3.7** Signaling examples in voice systems: (a) establishing a basic voice connection, (b) a typical CLI signal, (c) signaling between a subscriber multiplexer and an LE, and (d) fax communication.

number and other identifying information is sent typically using DTMF or FSK before or during the ringing period. Figure 3.7(b) shows a typical CLI signal format. When metering pulses are delivered from the exchange to subscribers where call meter services are used, exchange equipment connects an oscillator sending 12- or 16-kHz pulses to the subscriber end. In this case, filters are used to block the signals outside the voiceband from the subscriber and the exchange end voice equipment.

When a T1/E1 link is used between a multiplexer and an LE, a time slot or a bit in the voice time slot of each subscriber is used for signaling [see Figure 3.7(c)]. Chapter 7 provides further details on this.

Figure 3.7(d) indicates the simplified example of fax signaling. After the basic procedure of signaling between *A* and *B* to establish the call, and after *B* answers, the two faxes enter into the process of handshaking. This process exchanges several parameters to determine the transfer rate. A Group 3 fax machine can have a built-in modem with automatic bit-rate adaptation. The transfer rate will depend on the quality of the voice line, as described in [5].

### 3.5.2  Interexchange Signaling in the PSTN

If more than one LE is involved in setting up a connection or in activating a supplementary service, these exchanges must interchange information. Several systems for interexchange signaling are described in [6]. Two basic forms—*in-channel* or *channel associated signaling* (CAS) and *common channel signaling* (CCS)—are present. CAS (sometimes referred to as *per-trunk signaling*) uses the same transmission facilities or channel for signaling and voice. CCS uses one channel for all signaling functions of a group of voice channels. While most older systems use CAS, modern networks use CCS.

CAS systems can be further subdivided into inband and out-of-band techniques. While the former uses voiceband frequencies, the latter uses a range of frequencies outside the voiceband. CAS is further divided into subsets such as line signaling and register signaling [6, 8].

CCS is generally referred to as the *signaling system number 7* (SS7) standardized by ITU-T. CCS requires a separate network as a bearer service for signaling only. The processor systems extract the signaling information relevant to groups of calls together with other network conditions and routes them through a separate data link between switches. The signaling network is an integral part of the overall network. Its task is to support other networks. It can carry traffic transparently in the form of short messages between exchanges.

### 3.5.3  An Overview of ITU-T SS7

Figure 3.8 indicates the way SS7 operates within the modern network, including the PSTN, ISDN, and the PLMN, carrying signaling message packets or message signal units (MSU) between processors in the network.

The SS7 uses the seven-layer open systems interconnect (OSI) concept as depicted in Figure 3.9(a). The protocols for processor communication reside above the OSI layer 3 and are called user parts (UP) and application parts (AP). As shown

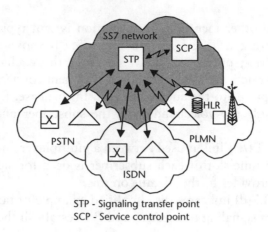

STP - Signaling transfer point
SCP - Service control point

**Figure 3.8** SS7 operation within the overall network.

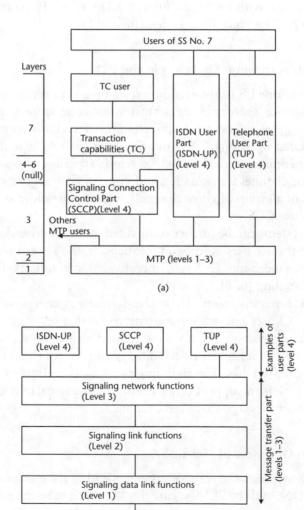

**Figure 3.9** SS7 concepts and message details: (a) architecture, (b) functional levels, and (c) message structure. (*Source:* ITU-T: Q.700.)

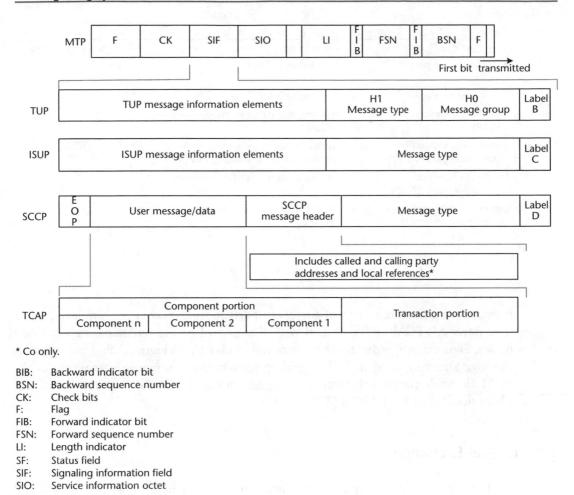

* Co only.

BIB:     Backward indicator bit
BSN:     Backward sequence number
CK:      Check bits
F:       Flag
FIB:     Forward indicator bit
FSN:     Forward sequence number
LI:      Length indicator
SF:      Status field
SIF:     Signaling information field
SIO:     Service information octet

(c)

**Figure 3.9** (continued).

in Figure 3.8, a switching node in the network is called a signal transfer point (STP). It handles traffic between nodes in different types of networks, notably the service control points (SCPs). Nodes connected to the signaling network are called signaling points (SPs). The user services, consisting of UPs and APs of the SS7 network, play an important role as functional building blocks in other networks. UPs and APs handle user information, such as control information for setting up and clearing voice and data connections, and information for centralized supplementary services [intelligent network (IN) services]. Table 3.1 indicates the relationship between UPs and APs with respective applications and associated ITU-T recommendations. They communicate with the same type and levels in the signaling network.

The message transfer part (MTP) and signaling connection control part (SCCP) are at the OSI layers 1 to 3 and form the network service part (NSP). The MTP forwards message signal units (MSUs) between user parts of the same type (e.g., in the case of the PSTN, TUPs between exchanges). TUP messages for the set up and clearing of a telephone connection pass all the exchanges along the traffic path with

**Table 3.1**   The User Services of the SS7 Network

| User Part/Application Part | Application | ITU-T Recommendations |
|---|---|---|
| Telephony user part (TUP) | Signaling in PSTN | Q.721 to Q.725 |
| ISDN user part (ISUP) | Signaling in N-ISDN and PSTN | Q.761 to Q.764 |
| Mobile application part (MAP) | Signaling and database communication in PLMN | |
| Transaction capabilities application part (TCAP) | Support for communication with IN databases and for signaling in PLMN | Q.771 to Q.775 |
| Intelligent network application part (INAP) | Communication with IN databases | |
| Operation and application maintenance part (OMAP) | Communication in management networks | Q.750 to Q.755 |

instructions for switch control for every participating exchange. Similarly, other user parts AP/TCAP or ISUP deal with the control of the participating exchange nodes. Figure 3.9(b) indicates the functional levels of SS7. Figure 3.9(c) indicates the message structure used in SS7. Signaling terminals exchange information via the PCM channels coupled through exchange terminal circuits (ETC) described later. More details are available in [37].

## 3.6   Digital Exchanges

By 1985, most telecom equipment manufacturers were able to supply fully digital exchanges due to the proliferation of inexpensive microprocessors, memories, and other semiconductor components. With modular software being bundled with telecom systems, many operators were able to introduce advanced services such as call forwarding, call waiting, multiparty conference, subscriber barring, and specialized billing services.

In the present-day networks, PSTN nodes can be subdivided into three main categories: LEs, transit exchanges, and international/gateway exchanges. LEs are used for connection of subscribers. Transit exchanges switch traffic within and between different geographical areas. International or gateway switches handle traffic between different operators or national boundaries.

### 3.6.1   Functional Groups of a Digital Exchange

A digital exchange can be logically divided into two separate functional parts as shown in Figure 3.10: the switching system and the processing (or control) system. The processing system, as discussed in Section 3.3, is a complex processor system that handles all control functions, subscriber data, billing information, and routing information, in addition to the control of the switching part.

The switching system contains two switching points: the central group switch and the subscriber stage. The former allows common equipment such as dial tone

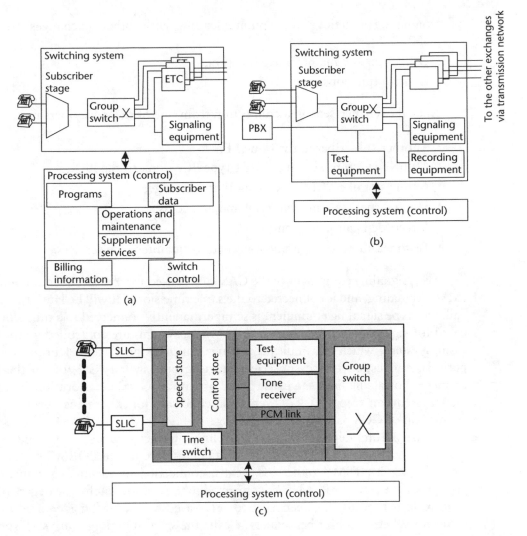

**Figure 3.10** A simplified view of local exchange (a) concept, (b) equipment connected to the group switch, and (c) subscriber stage.

providers and ringers to be connected to subscribers and to concentrate traffic to the central group switch. The subscriber stage, lines to other exchanges (known as ETCs), signaling equipment, recording equipment and test gear are connected to the central group switch, as shown in Figure 3.10(b).

Connection set up is controlled by the processor system by reading a set of test points in the subscriber stage that detect off-hook conditions and provide dial tone through relays in the switching system. A characteristic feature of modern exchanges is their modular design, facilitating capacity expansion and the addition of new functionality. All exchanges include a number of basic functions within the core:

- A group switch for switching functions;
- A trunk stage including ETCs operating as interfaces to the transport network;

- Signaling functions for communicating with other exchanges (such as SS7-CCS);
- Operation and maintenance function;
- Control functions.

A local exchange sets up a connection from one of the subscribers to:

- Another subscriber in the same LE;
- A PBX via a time slot in a T1/E1-type PCM link to the PBX;
- A time slot of a PCM link to another exchange;
- The Internet via a modem pool and an access server;
- A recorded announcement;
- Test/measurement equipment needed for maintenance services.

Also, signaling equipment of the CAS kind is used for multifrequency compelled (MFC) signaling, and is connected to the same time slot that will be later used by the call. SS7-type signaling equipment is semipermanently connected to a time slot dedicated for signaling (see Chapter 7). All of these systems are controlled by the processing system, which is a multiprocessing, multitasking system. During the peak periods, congestion in the subscriber stage, group switch (GS), or on the ETC-connected outgoing lines can cause grade of service (GOS) deterioration. The processor system can keep a track of events related to calls for call tracing purposes or the estimation of GOS parameters and billing.

Figure 3.10(c) depicts the subscriber line interface cirucits (SLIC) and a small TSM and associated SS and CS. The SLIC takes care of the BORSCHT functions (see Chapter 2); TSM handles local calls, connection of various equipment, and grouping time slots for PCM links connected to the group switch. The degree of concentration is typically between 10:1 to 3:1, based on the traffic generated by subscribers. When a subscriber makes a call, the corresponding time slot contains speech samples; otherwise, they are empty. The control system analyzes and determines the order in which the samples are to be read for the desired connections among subscribers, and the corresponding values are written into CS. In this way, subscribers are connected to one another, to the group switch, or to other equipment. The group switch comprises of a set of TSMs and at least one SPM forming a TST stage.

In modern exchanges, the group switch is used both by the PSTN and the N-ISDN. N-ISDN services requiring connection of more than one timeslot are referred to as *wideband* or *n*.64 kbit/s. Another new service is the broadcast function for services such as weather, business, or public announcements. These services call for the connection of an announcement service to many simultaneous listeners.

The control function of an exchange ensures that the data needed in a connection is read and modified as necessary. This includes all data about subscribers or the connected exchanges and the services and facilities they use in addition to the equipment connected to the group switch. Modern exchanges can handle large busy hour call attempts (BHCA). The BHCA indicates the control capacity of the exchange. Most powerful control systems used in modern systems allow over

1 million BHCA [5]. Table 3.2 is a simplified explanation of the setting up of an outgoing call, as depicted by the steps in Figures 3.11(a–d).

### 3.6.2  Processor Systems

Digital exchanges used in the PSTN today handle anything from a few thousand subscribers to nearly 100,000, or equivalent trunk groups in the case of transit or gateway systems. Large subscriber numbers and advanced services can only be handled by extremely complex real-time processor systems and modular software. Generally, duplicated processor systems with parallel or hierarchical structures operate in hot-standby modes for reliability. In addition to essential information processing, these systems handle special services as well as system supervision and fault reporting tasks. Most of the processor systems are coupled to maintenance centers to allow centralized maintenance and supervision. Figure 3.12(a) illustrates this concept, dividing the switching equipment, processing equipment, and the associated software. Switching equipment is coupled to the main processors via fault-tolerant lower level processor subsystems. Real-time software organized in a highly modular structure communicates with the switching equipment by sending orders and receiving data from the switching equipment. This example is for AXE systems where hierarchical processing is used.

**Table 3.2**  Setting Up of an Outgoing Call

| Phase | Tasks |
| --- | --- |
| Phase 1—Figure 3.11(a): Off-hook detection | Check subscriber database for the barring level of the subscriber (the subscriber may be permitted to dial a limited number or all types of outgoing calls), hot line service activation (where no dial tone is sent), and whether any other services such as unconditional call forwarding are activated (where a different dial tone should be sent). Reserve a memory area in the control unit for storing the $B$-subscriber number and associated connection information. Use the connection of a tone receiver for detecting DTMF signaling. |
| Phase 2—Figure 3.11(b): $B$-subscriber number detection | This requires analysis of the $B$-subscriber number for determining $B$-subscriber location, charging function, and length of the number dialed. If the $B$ subscriber is within the local exchange, send a query to the database for the SLIC of the $B$ subscriber, the level of incoming call barring of $B$, and special services allowed for $B$. If $B$ is in another exchange, perform routing analysis involving the subscriber category, alternative routings, and the conditions of the network. Charge for the service. After analysis, prepare set-up connection by reserving a time slot in the GS together with other needs for signaling. |
| Phase 3—Figure 3.11(c): Call set up | While a connection is set up from the far end (possibly via several intermediate exchanges), a signal is sent to $A$ subscriber's exchange indicating that $B$ is free. The GS reserves a path between the subscriber stage and the selected time slot on the outgoing PCM link. Enable appropriate disconnection of the tone receivers, allowing the time slot connection. The $B$ subscriber's exchange sends ringing signals/tones. On lifting of the $B$ handset, the voice samples are switched and activation of charging and call monitoring begins. |
| Phase 4—Figure 3.12(d): Call release | If $A$ concludes the call, the call is released without a noticeable delay and any time supervision (as $A$ pays in general). If B hangs up first, a supervisory period is initiated allowing any other connections $A$ is continuing with. When the call is finally released, the control system disconnects the call charging. Disconnection takes place of all equipment used for the call, release of time slots, and signaling performed to release all other exchange equipment. |

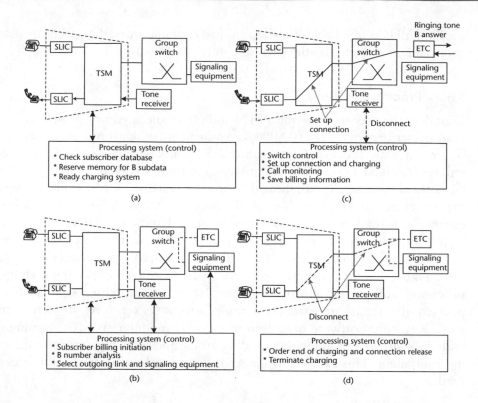

**Figure 3.11**   Process and connection stages of a phone conversation: (a) A-subscriber lifting a hand set, (b), B-number analysis for the outgoing connection, (c) outgoing call set up, and (d) release of the call.

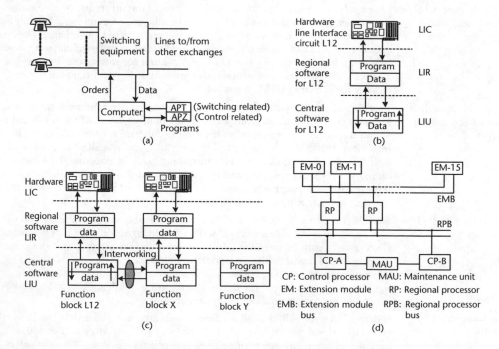

**Figure 3.12**   An example of switching and control systems in a typical exchange—AXE architecture from Ericsson: (a) basic concept, (b) an example of a function block, (c) interworking of function blocks, and (d) processor hierarchy. (*Source:* L. M. Ericsson, Sweden.)

In a typical AXE exchange, the overall activities are divided into many function blocks, which consist of switching hardware, regional software, and central software. This is shown in Figure 3.12(b), which indicates the example of a LIC. As shown in Figure 3.12(c), several function blocks can communicate with each other through the central software using software signaling. Central operating system allows interworking between function blocks by a cyclic system that cycles fast, allowing a meaningful timeshare process.

Figure 3.12(d) indicates the processor interconnections. Duplicated central processors (CPs) communicate via a duplicated regional processor bus (RPB) with many regional processors (RPs). The duplication of buses and the processors maintain reliability. Duplicated RPs control the switching hardware via an extension module bus (EMB). The duplicated extension modules (EMs) and the EMB allows reliable switching system hardware functions. Duplicated CP pairs such as CP-A and CP-B, providing the necessary CP capacity, communicate with each other via a maintenance unit (MAU). This complex architecture divides a given function and its associated tasks into smaller units. The CPs make the major decisions, while more routine tasks are handled by the RPs. RP pairs carry specialized software, dependant on the connected switching equipment such as the group switch, LICs, or ETCs. EMs receive the RP output commands (in bytes), which are decoded and used by the switching hardware. For example, a set of subscribers and their associated LICs are grouped in such a way that when a given set of subscribers go off hook, the conditions are read by the associated RP pair, informing the CP to command the necessary other function blocks to commence the call establishment.

Figure 3.13(a) indicates the distributed control architecture of the Fujitsu FETEX-150 switching system. A main processor (MPR) system communicates with several call processors (CPRs) and the line processors (LPRs). The MPR manages the overall system and performs maintenance and operation functions. The CPR provides call processing functions using the LPR located in the line concentrators or remote switching units. Figure 3.13(b) shows the concept of the *hypothetical independent exchange method* with hypothetical speech path and signaling links. For further details, [9] is suggested.

### 3.6.3  ISDN Systems

Traditional networks developed in separate subsets such as the telephony, telex, and packet switched networks, with basically one network for each service category. Transition from analog to digital systems paved the way for integrated digital networks (IDNs) with digital switching and transmission. Further developments led to ISDNs with advanced signaling for different signal bandwidths. ISDN is a circuit-switched concept, as shown in Figure 3.14.

As discussed in Chapter 6, ISDN makes use of the higher frequency capability of the copper pairs, allowing two basic ($B$) rate (64-kbit/s) channels for voice or data traffic and a 16-kbit/s data ($D$) channel for signaling, forming a basic rate access (BRA) situation as in Figure 3.14. For connection of PBX systems, primary rate access (PRI) is used, where 30 $B$ channels and a $D$ channel are present.

System consequences of an LE with PSTN/ISDN combination and its switch control in the LE will be more complex due to handling of more $n.64$-kbit/s channels. Usually optional packet handlers are installed in the switching part with access

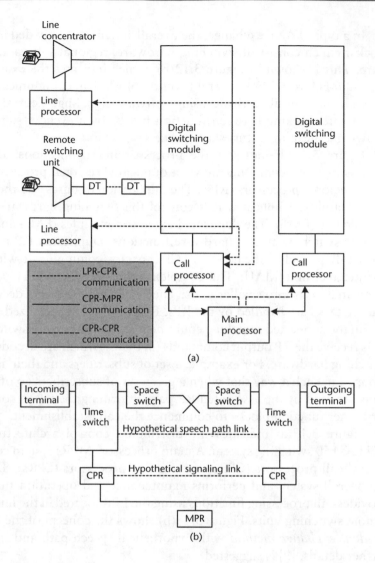

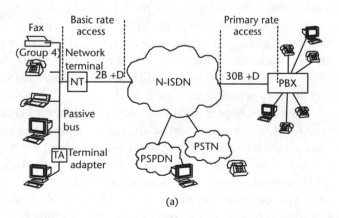

**Figure 3.13**   Processor hierarchy in Fujitsu FETEX-150 system: (a) basic concept, and (b) hypothetical independent exchange method. (*Source:* [9]. ©1990 IEEE.)

**Figure 3.14**   ISDN concept.

servers with statistical multiplexing capability. Narrowband ISDN (N-ISDN) is currently progressing towards broadband ISDN (B-ISDN) systems making use of newer ATM switches. For more details, [5] is suggested.

## 3.7   Private Branch Exchanges and Key Phone Systems

### 3.7.1   Private Branch Exchanges and Centrex

To facilitate intercommunication requirements within large organizations, the same concepts used in digital switches were adapted in private branch exchange (PBXs), dividing them into two clear subsets: the Centrex service and the PBX systems. Centrex is a service to simulate a PBX for an organization by means of software installed in a LE. This service is attractive for medium-sized businesses and eliminates the need for in-house service teams or contracts with PBX system providers. Figure 3.15(a) and (b) distinguish the PBX and the Centrex services.

Both cases can be multilocation systems, where branch offices are connected in such a way that the user dials only extension numbers, even though the sites are connected via T1/E1 trunks via the PSTN. For PBXs coupled via leased lines, capacity can be reserved in the PSTN, allowing permanent connections through respective group switches in participating public exchanges. Computer telephony integration (CTI) is another common service when a LAN is present.

With the Centrex function, the extensions at distant locations are translated to special numbers and sent via intergroup traffic. In the mid 1990s, many wireless PBX (WPBX) systems entered the market, providing mobility in an office environment. Another enhancement to PBX or WPBX systems was ISDN facilities. References [9–16] provide some development-oriented insight on this technology.

### 3.7.2   Key Phone Systems

Key phone systems are smaller cousins of PBX systems, where the cost, implementation, and facilities are not as sophisticated as PBX or Centrex systems. Most low-end systems (from few extensions to around 100 extensions) are designed around a

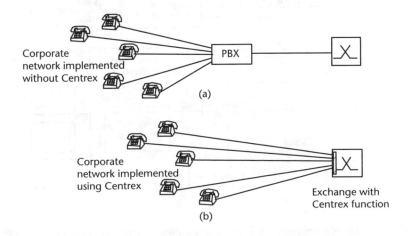

**Figure 3.15**   Comparison of PBX and Centrex systems: (a) PBX, and (b) Centrex.

few microprocessors and some switching hardware, allowing many users to access the PSTN with a few subscriber loops. Designed around simple cross-point switches, key phone extensions typically work with four-wire interfaces. One pair is used for voice, and the other pair is used for signaling (as well as power distribution to key stations). Figure 3.16 depicts elements of a typical key phone system.

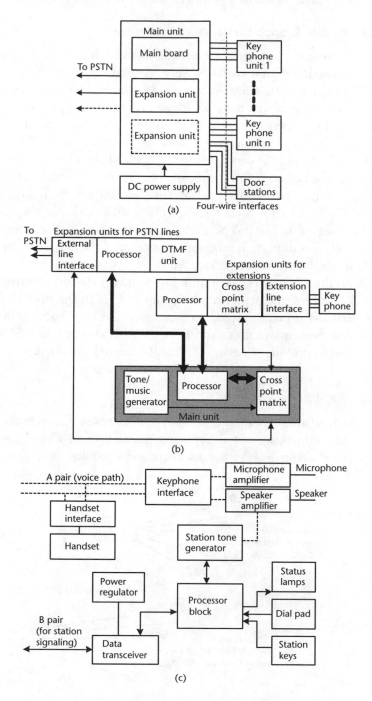

**Figure 3.16**  Components of a typical key phone system: (a) system interconnections, (b) block representation, and (c) key station block diagram.

Figure 3.16(a) indicates the main unit with its main processor board and expansion boards with dedicated processors. Some of the units can have cross-point matrices and external line interfaces as required by the configuration. Figure 3.16(b) indicates a possible configuration with several expansion boards. A typical key station block diagram is shown in Figure 3.16(c).

For small and medium-size office systems, differences between the key phone system architectures and PBX are hardly visible. Many commercial variations with different capacities are available.

## 3.8  Merging Data and Voice

Modern telecommunications systems handle a complex mix of voice, data, video, and multimedia. In such situations, the information flowing through switches is fully digital. This scenario forces service providers to handle the following parameters carefully:

- Bandwidth;
- Burstiness (variation in the bandwidth requirements of a given information set);
- Bit errors and blocking;
- Delay;
- Information security.

Giving special attention to these aspects, modern switch cores are changing fast to handle high-speed digitized signals with appropriate packet-switching techniques. However, this development places special demands on applications needing real-time communications.

## 3.9  Packet Switching Systems

### 3.9.1  X-25 and Other Early Packet Switching Systems

Developing from the early 1970s, packet-switched public data networks (PSPDNs) have been installed all over the world. As summarized in Section 3.4.1, the concept is shown in Figure 3.17(a). As depicted here, the user's data terminal equipment (DTE) is connected to the packet-switched network via the data circuit terminating equipment (DCE). X.25 was the first international standard for wide area packet networks. The name refers to one of many recommendations, as seen in Table 3.3, for packet networks by the then CCITT, for both asynchronous and synchronous access. The application environment for these standards is shown in Figure 3.17(b).

X.25 is essentially a protocol for synchronous transmission across the DTE/DCE interface, as shown in Figure 3.17(b). However, the name X.25 has been applied to a suite of protocols covering three layers of a packet-switched network, as indicated in Table 3.4.

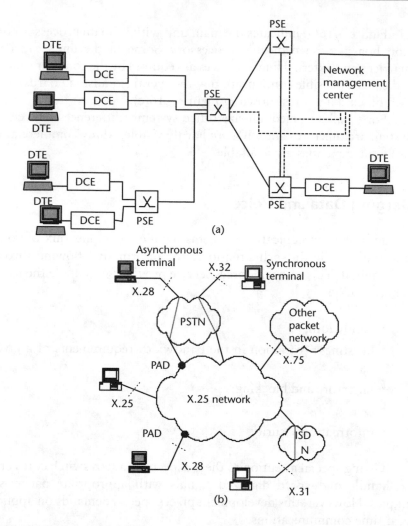

**Figure 3.17** Packet networks: (a) basic structure, (b) early standards.

**Table 3.3** Early Packet Switching Standards

| Standard | Description |
|----------|-------------|
| X.3 | Packet assembly and disassembly (PAD) for asynchronous data equipment |
| X.25 | Interface functions and protocols for the connection of synchronous data equipment in packet-switched networks |
| X.28 | Interface functions and protocols for the connection of asynchronous data equipment to PAD functions in the PSE |
| X.29 | Exchange procedure for control information between a PAD and a synchronous data equipment in PSE |
| X.31 | ISDN terminal connection |
| X.32 | Interface between DTE and circuit-switched equipment such as DCE via PSTN channel |
| X.75 | Interface between X.25 networks |

**Table 3.4** Layers and Details of the X.25 Protocol Suite

| Layer | Function |
|-------|----------|
| Layer 1 | This is the interface to the physical layer, DTE/DCE (X.21 and X.21 bis for digital and analog terminals, respectively, and X.31 for ISDN terminals). |
| Layer 2 | This is the link layer, which defines the data flow over a link. It ensures error-free transport of packets to and from layer 3. |
| Layer 3 | This defines the addressing and packaging (packet assembly). It handles virtual circuits (VCs) and defines packet types for call set up, data phase, and clearing. It also multiplexes logical channels to a single physical channel and addresses flow control and error handling. |

### 3.9.1.1   Packetizing and Signaling in X.25

X.25 is a network that operates on a store-and-forward basis with variable length packets. Layer 2 in X.25, which ensures error-free transport of packets, applies a version of the high-level data link (HDLC) procedure. This is the basis for many packet data networks [7]. The variation used in X.25 is called link access procedure B or D. (LAPB/LAPD). The frame structure [8] formed in HDLC is illustrated in Figure 3.18(a). The information field is used by the data packets from layer 3. Only one packet can be transported in each frame. The frames are separated by one or more eight-bit flags. The signaling process is indicated in Figure 3.18(b).

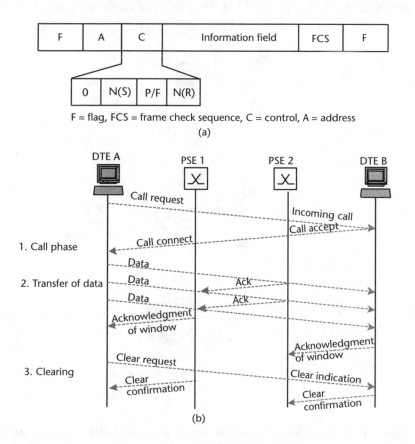

**Figure 3.18**   HDLC: (a) frame format, (b) signaling and data exchange.

### 3.9.2   Frame and Cell Relay

Up to the early 1990s, X.25 was the only technology to offer flexible bandwidth, or bandwidth on demand (BOD), services. Introduction of new data services in the early 1990s showed that the typical X.25 services using mostly 64-kbit/s links were creating bottlenecks. Also X.25 networks were designed with the assumption that BERs were in the order of $10^{-6}$ or worse. With fiber and digital microwave links developing with much lower BERs, in the order of $10^{-9}$ or better, while fast digital switches were developing in the PSTN and PSDPN, frame relay (FR) services were introduced. By reducing much overhead within the error correction and flow control blocks, frame relay was able to transport data much faster over packet networks than X.25.

Protocols for FR are relatively simple and active within the layers 1 and 2 of the OSI model, with only a small amount of overhead. High transmission capacities of 2 Mbps up to 50 Mbps are involved. Services are provided mostly via PVCs, while SVCs are gradually being introduced. LAN-to-LAN connections have resulted in a high rate of FR installations and services.

FR technology was developed using the X.25 experience for the demands of the new data-centric customer bases, and the standardization process commenced in 1990. Figure 3.19(a) indicates how an FR router connects a LAN to the network via an FR switch. In a popular LAN interconnection situation, an FR access device (FRAD) in a LAN is usually combined with a router that has specific protocols for an FR network [Figure 3.19(b)]. The other end of the user network interface (UNI) is

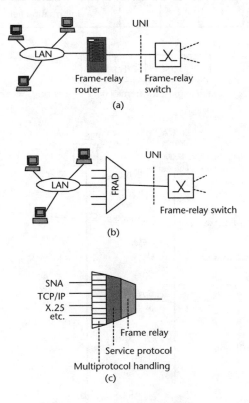

**Figure 3.19**   Frame relay: (a) LAN connection via FR, (b) FRAD, (c) multiprotocol handling via FRAD.

equipped with the FR switch, also known as an FR network device. FRAD is considered the equivalent of the PAD in the X.25, though the FRAD does not assemble data. As seen in Figure 3.19(c), the FRAD has a multiprotocol interface that permits interfacing of protocols other than FR variants. Any commercial FRAD contains three primary components: a multiprotocol handler, service protocol, and the FR component. The FRAD can assemble several multiprotocol information sources to a single logical connection. However each end user sees a transparent connection.

In FR, the physical layer and part of framing functions (as described by standard Q.922) are implemented by the network, while the upper part of the framing functions defined by Q.922 is implemented by the interface. By restricting to a two-layer protocol within the network, the overhead is much removed, compared to X.25. FR uses LAPF at layer 2, an adapted version of LAPD, which was originally designed for N-ISDN.

Figure 3.20(a) indicates the relationship of protocols in an FR network. The physical layer standards involved are derived from ISDN systems. The frame format as per Figure 3.20(b) follows the HDLC format. It consists of two flags delimiting the beginning and end of the frame, an address field [the data link connection identifier (DLCI)], the information field (which is variable in size), and a frame check sequence. The DLCI field is variable from 10 to 23 bits. The fields within these bytes are described in Table 3.5. Note that Figure 3.20(b) indicates only the case of two byte (default) header format. See [6].

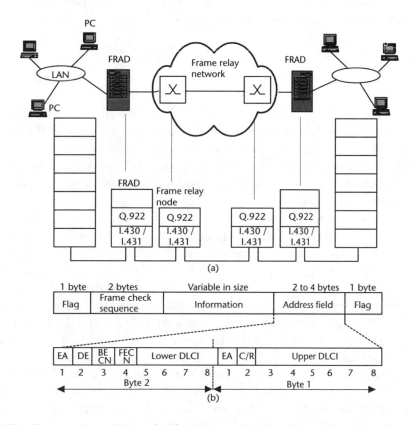

**Figure 3.20**   FR protocols and frame details: (a) protocols in an FR network, and (b) frame formats. (*After:* [6].)

**Table 3.5**   Details of the FR Frame

| Field | Description | Remarks |
|-------|-------------|---------|
| DLCI | DLCI indicates the logical connection between different FR users. | DCLI defines the next destination to which the frame is transported. This identifies the virtual circuit. |
| | The first six bits of the byte 1 and the first four bits of the byte 2 form the DCLI. | There is a total of $2^{10}$ logical connections. Many are reserved for special tasks. By assigning different DLCI values to different frames, several simultaneous logical connections can be made on the same circuit. |
| Command / response (C/R) | CR is not used by the FR protocol. | It is sent transparently through the nodes to users. |
| Extended address (EA) | EA is used to indicate whether the DCLI is over 10 bits. | EA=0 means another byte is to follow. |
| Forward explicit congestion notification (FECN) | FECN indicates the overload in the network. | FECN is sent to the receiving end. |
| Backward explicit congestion notification (BECN) | BECN is the same as FECN, used in the direction of sender. | |
| Discard eligibility (DE) | DE is an indication of frame discard in case of congestion. | DE is used as a priority case for discarding. Frames without a DE can also be discarded |
| Information field | This is user data. | This has a maximum of 1,600 bytes. (The actual size is decided by the service provider.) |
| Frame check sequence (FCS) | FCS checks the frame for errors. | |

Figure 3.21(a) indicates the use of FECN and BECN in the direction of congestion within the FR network, connecting to LANs. FECN and BECN are used in flow control, where the congested node sends the FECN and BECN to receive and transmit nodes, respectively. As shown in Figure 3.21(b), DLCI has a local significance only between two nodes, and each node can change these as needed along the path between two users.

FR networks operate with the assumption of very low BER and thus minimal need for error control. Elements of the end-user systems, rather than the network itself, guarantee error-free transmission. The FR interface specification provides the signaling and data transfer mechanism between the end points and the network based on a set of user parameters, the most important of which are the committed information rate (CIR) and the excess information rate (EIR). The operator can set the CIR value for individual VCs. For a typical case of a 2-Mbit/s network link, the operator may choose any value from 0, 8, 32, or $n.64$ kbit/s. In such a case, 2 Mbit/s is the speed of the physical link, and the CIR is the guaranteed data rate of the VC. The same physical connection can carry several VCs with different CIR values. However, at times during which some VCs are not active, the available capacity can be taken by the active VCs. The EIR is the maximum permissible transmission capacity for a VC during such short periods. For more information on FR, [17] is suggested.

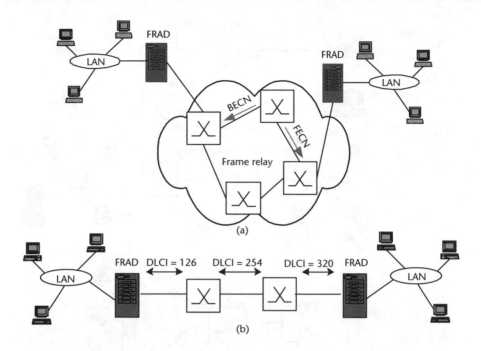

**Figure 3.21**   Use of FECN, BECN, and DLCI in an FR network: (a) congestion notification, and (b) local significance of DLCI in a connection.

### 3.9.3   ATM and Broadband-ISDN

X.25 and FR, both of which are packet-switching techniques, have one important common factor, which is the variable packet size. However in handling connections and traffic, the variable packet length can cause inefficiencies, particularly in situations such as VoIP, where the connection delays are important for QoS. With the networks carrying more data-centric traffic and with the need for wideband connections, fixed-length packet size becomes important. Cell relay techniques use such constant, small, fixed-size frames called *cells*. Hence, the fixed-length cell-relay techniques such as ATM and distributed queue dual bus (DQDB) were introduced. Our focus here will be only ATM. For the interested reader, DQDB is discussed in [6].

Evolution of broadband networks was synonymous with the evolution of ATM and B-ISDN techniques. B-ISDN is a further development of the N-ISDN concept, with the telecommunications community recognizing the need for wider bandwidth networks for handling voice, data, and video in the mid 1980s. While N-ISDN was based on switching bandwidths of $n.64$ kbit/s up to 2 Mbit/s, B-ISDN can switch different bit rates variable from 64 kbit/s to over 100 Mbit/s. At this point, Figure 3.22 provides a clear comparison of switching techniques evolved.

Figure 3.22(a) indicates the different connections in dedicated networks. In ISDN, not only can transmission between nodes be integrated, different transfer modes are also permitted (e.g., circuit mode and packet mode). Figure 3.22(b) illustrates this. When an ISDN subscriber makes a voice call, the circuit-mode switch in ISDN will be used, and the call will be routed via ISDN or PSTN. If the subscriber wants to send packet data, the circuit mode switch has to route the traffic to a

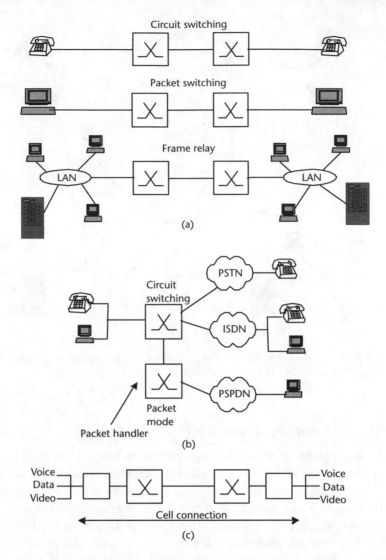

**Figure 3.22** Networks and transfer modes: (a) different connections in dedicated networks, (b) integrated subscriber line and node, and (c) full integration in B-ISDN.

packet handler, whose task is to route the packets to the destination, either locally or a PSPDN.

By contrast, in B-ISDN, full integration is possible at a cell level, as per Figure 3.22(c). In this case voice, data, and (motion) video information is transferred via ATM switches, where an ATM cell is a small packet having a fixed length of 53 bytes, with five of them comprising a header. The 53-byte cell format was standardized as a compromise between the U.S. and European needs. The advantage of a short cell of this kind is for real-time services such as telephony, where there is a strong sensitivity to network delays of over 8 to 10 ms. Filling a cell of this size from an individual call connection takes about 6 ms in today's technology. However, in data communication services, longer cells are more effective due to less total overhead. Table 3.6 compares the characteristics of X.25, FR, and ATM.

**Table 3.6**  Characteristics of X.25, FR, and ATM

| Characteristic | X.25 | FR | ATM |
|---|---|---|---|
| Transfer mode | Connection oriented or connectionless | Connection oriented | Connection oriented or connectionless |
| Packet length | Variable | Variable | Fixed (53 bytes total) |
| Switching | Software oriented, based on the address field of the packet | Software oriented, based on the address field of the packet | Hardware oriented |
| Speed (bandwidth) | 64 kbit/s to 2 Mbit/s | 2–50 Mbit/s | Flexible (1.5 Mbit/s to over 2.5 Gbit/s) |
| Retransmission | Link by link | End to end | |
| Channel performance | High BER (e.g., leased lines) | Low BER (e.g., optical fiber) | Low BER |
| Primary application | Data or still picture exchange | Interconnect of LANs | All applications, including B-ISDN |

### 3.9.3.1  General Features of ATM

ATM is a transfer mode to handle a variety of types of traffic. Data, digitized voice, or video information is broken into fixed-length cells of 53 bytes and transferred over a channel based on a *labeled multiplex* basis, enabling the flexible use of the bandwidth. Video information will occupy more cells, compared to voice information, due to the high information rate required. Figure 3.23(a) indicates the

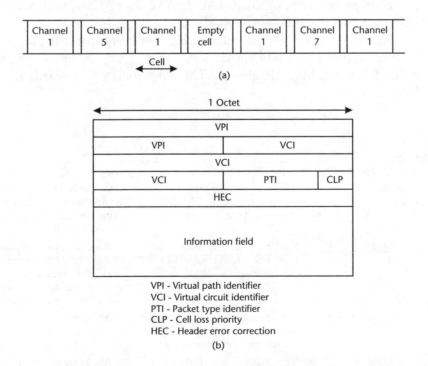

**Figure 3.23**  Labeled multiplexing and the cell format in ATM: (a) labeled multiplexing for flexible bandwidth, and (b) cell format and header details. (*Source:* ITU-T: I.371.)

principle where channel 1 is allowed more cells than the others. Figure 3.23(b) indicates the cell format, with details of the ATM cell header.

As shown in Figure 3.23(b) the five-byte cell header is divided into different fields:

- Address field in the form of virtual path identifier (VPI) and virtual channel identifier (VCI);
- Payload type identifier (PTI), specifying the contents of the cell (whether user information or network information);
- Cell loss priority (CLP) information, in case of congestion;
- Header error control (HEC) with a check value for detecting bit errors;
- Generic flow control (GFC) used for flow control information.

*Classes of Service*
ITU-T has standardized a reference model for ATM with three protocol layers corresponding to the OSI model. These are the physical layer (layer 1), ATM layer (layer 2), and ATM adaptation layer (AAL). In this reference model, five service classes are defined [35] based on the following parameters related to the information processed and sent:

- The need for synchronization between the sender and receiver;
- Constant bit rate (CBR) or VBR;
- Connection-oriented or connectionless transfer.

Four protocols designated AAL 1, AAL 2, AAL3/4, and AAL 5 support the classes, A, B, C, and D, as summarized in Table 3.7.

Figure 3.24 indicates the relationships between processing layers and the relationship between VCs and virtual paths (VPs), via a transmission link such as 155-Mbps SDH link. More details on ATM and associated standards are well described in [7].

**Table 3.7**  Classes of Service in ATM

| Type | AAL Protocol | Type of Services Handled | Example Applications |
|------|--------------|--------------------------|----------------------|
| 1 | AAL 1 | CBR services for connection-oriented transfers with end-to-end synchronization | PCM coded voice, circuit-emulated connections with bit rates of $n.64$ kbit/s or $n.2$ Mbit/s, and video signals coded for CBR |
| 2 | AAL 2 | VBR-based connection-oriented transfers, which need end-to-end synchronization | VBR-coded video signals, compressed using MPEG 2 standard (when the motion elements in the picture is high, the bit rate needs to be increased), and ADPCM-based compressed voice |
| 3/4 | AAL 3/4 or 5 | VBR-based, connection-oriented transfers, which do not need end-to-end synchronization | LAN interconnect, FR, and X.25 |
| 5 | AAL 3/4 or 5 | VBR-based, connectionless transfers without end-to end synchronization | LAN interconnect, TCP/IP, and SMDS |

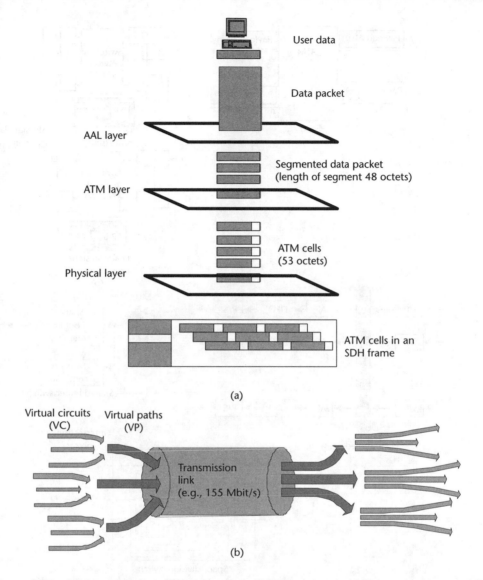

**Figure 3.24**  ATM processing at different layers and VC/VP elements: (a) layer processing, and (b) VC/VP relationships in a 155-Mbit/s link.

### 3.9.3.2  ATM Switching

ATM is a packet-switching technology that directs traffic using a label contained in the packet's header. Using the information within the header, an ATM switch transports ATM cells from an incoming logical channel to one or more outgoing logical channels. A logical channel is identified by a combination of the number of the physical link and the identity of the channel (specified by the VPI and the VCI).

ATM switches operate on a similar principle to the circuit-mode (voice) switches. The logical channels correspond to time slots, and the identity is based on the VPI and VCI in the cell header. The second function is analogous to the space switch in a circuit-mode switch. Figure 3.25(a) illustrates these principles, where the labels indicate the VCI/VPI combinations of the cell headers, and the payload transfer paths are indicated by A and B.

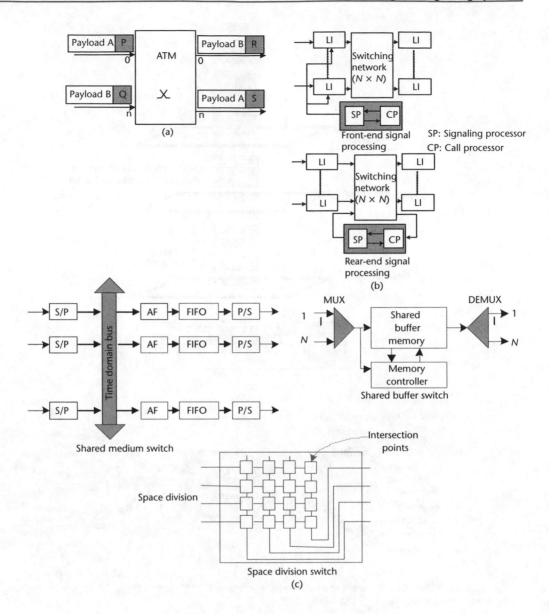

**Figure 3.25** ATM switching: (a) principle, (b) stages of an ATM switch, and (c) different architectures. (*Source:* [36]. © 2002 Artech House. Reprinted with permission.)

When ATM cells reach the switch, the cell header is opened, the VCI/VPI parameters are read, and the cells are switched to the right receivers. To ensure that the cells are not lost or incorrectly routed, a bit error check is performed on the cell header based on the header error correction (HEC) indicated in Figure 3.23(b).

Figure 3.25(b) indicates the stages of an ATM switch. Different architectures of ATM switches are shown in Figure 3.25(c) with shared medium, shared buffer (or shared memory), or space division possibilities. For details, [36] is suggested. Switching time through an ATM node is almost 100 times faster than the normal packet switching, allowing wideband information to be handled easily.

### 3.9.3.3 ATM Standards

During the mid 1990s, the ITU completed a series of standards on ATM for operation and management. These incorporate network resource management, traffic management, congestion control, fault management, and performance management. ITU I series standards for B-ISDN, ITU G-78X series of SDH standards, and many others are included in the overall process, which allows the development of a global ATM network. References [18, 19] are recommended for details.

### 3.9.3.4 ATM-Based Enterprise Networks

ATM was seen in the early 1990s as the technology to meet bandwidth and application requirements across both LANs and WANs. As shown in Figure 3.26(a), various UNI and network node interface (NNI) connections could be carried via different physical media, such as plesiochronous digital hierarchy (PDH) or SDH (see Chapter 7). The structure shown there can be same for LANs, metropolitan area networks (MANs), or WANs. A virtual channel connection (VCC) is set up between any source or destination pair in the ATM network [17], and the network signaling allows setting up a VCC between the two end systems by the use of VCIs and VPIs.

In the latter part of the 1990s, user organizations seriously considered using ATM as a migration path from fiber distributed data interface (FDDI) and Fast Ethernet, encouraging users to couple their LANs and desktop computers to ATM switches. With the release of LAN emulation specification [17] and the availability of suitable software, many LAN vendors were actively developing ATM products. One common way to interconnect desktops [20] is to use an adapter card on the host computer, which allows the protocol conversion process and the electrical to optical (E/O) and optical to electrical (O/E) conversions required to connect to ATM networks. Figure 3.26(a) and (b) illustrate the ATM WAN and LAN network concepts coupling desktops and LANs and the block operation of such a card [20]. However, due to the limitations of many desktop computers with their software bottlenecks, direct coupling may not be the best solution, except for such cases as servers. A better way is to use an ATM switch and LAN emulation [20], as per Figure 3.26(c).

### 3.9.4 IP Switching and Label Switching

The last decade has seen several major changes in the Internet. Not only has there been an explosive growth in terms of size and volume, but there has also been an increase in the number of multimedia and real-time applications, which have put significant pressure on the network to support higher bandwidth and provide QoS guarantees. Though the Internet has been successful in terms of scalability due to its connectionless nature, the hop-by-hop packet forwarding paradigm has turned out to be insufficient in supporting these demands. As an improvement in traditional IP routing, *IP switching* techniques have been developed, which makes use of the label switching technology of ATM for packet forwarding. These techniques are outlined in [21].

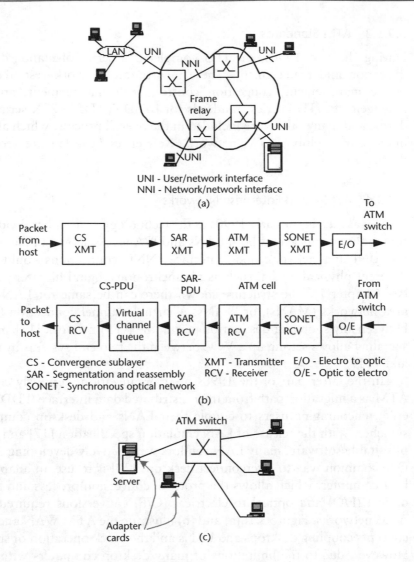

**Figure 3.26** ATM WANs and LANs: (a) ATM network (WAN), (b) ATM adapter card operation (Source: [20], ©1996 IEEE.), and (c) ATM LAN.

### 3.9.4.1  Switching Versus Routing

X.25, FR, and ATM are all based on connection-oriented packet switching, where switching is based on VC/VP identifiers contained in the data-link layer frames. The switch looks up a table for the outgoing VCI/VPI for packets coming on incoming VCI/VPIs.

The IP [7] was the first wide area connectionless protocol used extensively. Each IP packet at layer 3 includes its source and destination and moves independently through the network. The technique by which the packets were forwarded to the destination was termed *routing*. Even though routing is a form of switching, we can distinguish routing by its connectionlessness. The responsibility for maintaining the *state* of an end-to-end connection lies with the end nodes, and the network does not track a circuit, connection, or flow. The network nodes will be indifferent to anything but handing off packets to the next node [22]. On the other hand, in a switching network, the intermediate nodes keep track of the connection—the path and bandwidth (which relates to quality). This distinction applies on all types of data links—SDH/SONET, ATM, and Ethernet.

### 3.9.4.2    IP Switching

The goal of this technology is to make IP faster and offer the QoS support. The approach is to discard the connection-oriented ATM software and implement the connectionless IP routing directly on top of the ATM hardware. An IP switch shown in Figure 3.27(a) uses an ATM switch with all control software above AAL 5 removed and replaced by standard IP routing software. To gain the benefits of switching, a mechanism has been defined to associate *IP flows* (sequences of packets that get the same forwarding treatment at a router) with ATM labels [21]. This approach takes advantage of the robustness of IP routing, as well as speed, capacity, and scalability of ATM switching.

An IP flow is characterized by the fields in IP/TCP/user datagram protocol (UDP) headers such as source and destination addresses and port numbers and IP protocol type. When a packet is assembled and submitted to the controller in an IP switch, apart from forwarding it to the next node, it also does *flow classification*. This is a policy decision local to an IP switch. Depending on the classification, the switch decides to forward (e.g., for short-term flows such as database queries) or switch (e.g., long-term flows such as file transfers) the next packets of that flow. Each packet's next hop and output port are determined by a table look up with the packets destination IP address as the key. Classification is used to derive output port queuing and scheduling rules. The forwarding table is established and updated by the management engine based on the decisions of the active IP routing protocols [23]. This process is shown in Figure 3.28(a).

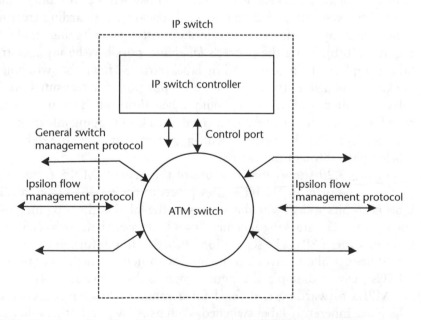

**Figure 3.27**    Structure of an IP switch.

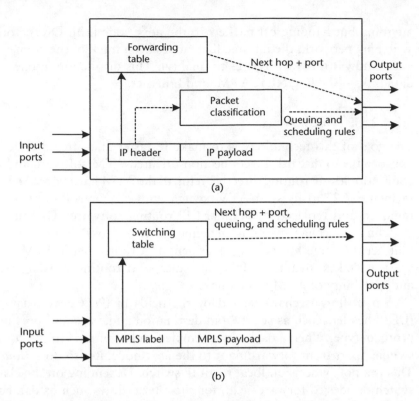

**Figure 3.28** Operation of the forwarding engines in: (a) IP switching, and (b) MPLS. (*Source:* [23]. ©2000 IEEE.)

### 3.9.4.3 Label Switching

Label switching techniques such as multiprotocol label switching (MPLS) add the ability to forward packets over arbitrary nonshortest paths and emulate high-speed tunnels between IP-only domains. Each packet's forwarding treatment is entirely determined by a single index lookup into a switching table, as shown in Figure 3.28(b), using the packet's *label* and possibly the input port. The packet's label is replaced by a new next-hop label retrieved from the switching table, and the packet is queued at the appropriate output port for transmission. The switching table is loaded a priori with a unique next-hop label, output port, queuing, and scheduling rules for all current label values. This mapping information is established and managed by the management engine in response to external requests for a labeled path through the switch.

Figure 3.29 shows the structure of the generic MPLS frame. An MPLS label stack of one or more 32-bit entries precedes the payload (e.g., an IP packet). The label is 20 bits wide. Switching occurs on the label in the top (and possibly the only) stack entry. The stacking scheme allows label-switched paths (LSPs) to be tunneled through other LSPs. The separation of forwarding information from the content of the IP header allows MPLS to be used with devices such as optical cross-connects (OXCs), whose data plane cannot recognize the IP header [24].

MPLS forwarding is defined for a range of link layer technologies, some of which are inherently label switched, such as ATM, and others that are not, such as

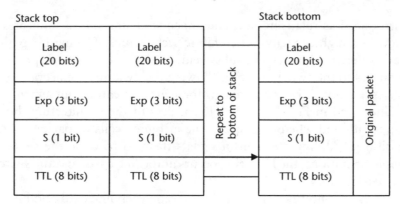

Exp - Additional bits for experimentation
S - Set to 1 to indicate the final stack entry before the
original packet
TTL - Time to live

**Figure 3.29**   The MPLS frame structure. (*Source:* [23]. ©2000 IEEE.)

packet over SONET/SDH. Reference [23] shows how the MPLS frames are placed within the link's native format in each case. ATM and FR switching is based on a link-layer copy of the stack top entry.

### 3.9.5   An Illustrative Modern Data Network

The FR network of David Pieris Information Technologies Ltd., Sri Lanka, provides extensive islandwide coverage. As seen in Figure 3.30, the FR backbone consists of five strategically located multiservice switches. These switches, with a switching capacity of 1.6 Gbit/s, can handle FR, ATM, and transparent HDLC traffic. The backbone operates on a digital microwave media running SDH at the physical layer and is configured as a ring for redundancy. The physical layer terminates in cross connects that interface the switches that provide FR, ATM, and transparent HDLC.

The last-mile connection from the closest network node to the customer premises is also wireless and are point-to-point or point-to-multipoint links connected through digital cross connects. The link speeds can vary from 64 kbit/s to 2 Mbit/s in $n \times 64$ kbit/s increments. At the customer premises, the access network terminates at a router or FRAD. The end-to-end VCs are configured and assigned CIRs in FR switches. Any type of multiprotocol traffic from one end point to the other can be routed through these VCs.

## 3.10   Switch Fabric Interfaces and ICs

Switch fabrics are fundamental building blocks in a wide range of switching systems. While many semiconductor manufacturers introduced basic semiconductors for voice switches in the early 1970s, ATM switch fabric ICs and interfaces were introduced as relatively new components. Figure 3.31(a) indicates a generic overview of a high-capacity switch fabric architecture. The ingress path connects a line card's network processing subsystem consisting of network processors and traffic mangers to the switching core, allowing the dynamic connection of ingress parts to the egress ports. The components for switching can be implemented in many ways, as indicated in Figure 3.31(b). Regardless of the architecture, however, most are input buffered where they queue the incoming cells or packets at the ingress stage until a scheduling mechanism signals them to traverse the switch core. In many implementations, the buffering occurs at the line card, and the switch cards contain limited memory.

A fabric interface chip (FIC) resides on each line card and interfaces to either a network processor or a traffic manager [25].

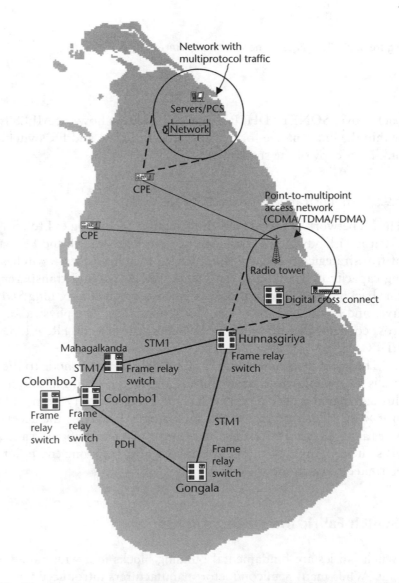

**Figure 3.30** FR network provided by David Peiris Information Technologies Ltd. in Sri Lanka. (Reproduced with permission of David Pieris Information Technologies Ltd.)

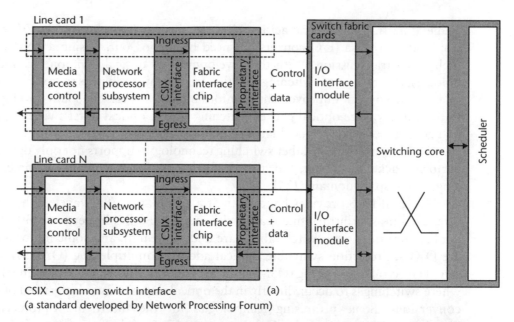

CSIX - Common switch interface (a)
(a standard developed by Network Processing Forum)

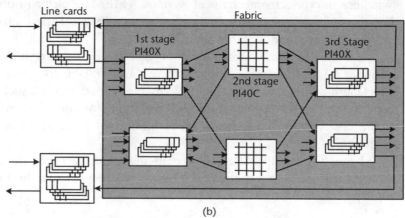

(b)

**Figure 3.31** Switch fabrics: (a) common switch fabric architecture, allowing multiple input ports to multiple output ports. (*Source*: [25], ©2003 IEEE), and (b) a switching systems based on Agere Systems switch fabric P140. (*Source*: [26].)

Semiconductor switch fabric ICs are gradually appearing from the product folios of the telecom component manufacturers, and time-division, space-division, or combination versions are common in the market. Some examples are indicated in [26–28]. Figure 3.31(b) indicates a 40-Gbit/s switch fabric configured using Agere Systems P140 switch fabric with line cards for an ATM or IP switching. Details are beyond the scope of the chapter and [20, 26–29] are recommended.

## 3.11 Optics and the Future

Switching speeds achievable using semiconductors cannot keep up with the transmission capacity offered by optical systems. Optical systems can achieve tremendous transmission capacities. However, in every switching node, O/E conversion and vice versa (E/O conversion) need be carried out, posing a bottleneck to high-speed transmission. With the development of optical components and technologies, the transfer of switching function from semiconductors to optics will result in a massive overall improvement of information transfer. The main developments in this direction are optical packet switching (OPS), optical burst switching (OBS), and generalized multiprotocol label switching (GMPLS).

GMPSL, an emerging label-switching technology, supports not only devices that perform packet switching, but also those that perform switching in time, wavelength, and space domains [24].

While GMPLS provides bandwidth at a granularity of a wavelength, OPS can offer an almost arbitrary fine granularity, while OBS lies in between [30]. Several applications of optical switch fabrics are under research, and some important ones are OXCs, protection switching, optical add/drop multiplexing (OADM), optical signal (spectral) monitoring (OSM), and network provisions. Optical switch fabrics, where switching is to occur directly in the optical domain, avoids the need of O/E/O conversions. Some promising technologies for the future are optomechanical switches, microelectromechanical systems (MEMs), electro-optic switches, and thermo-optic switches. A discussion on these is beyond the limits of this chapter, and [30–33] are suggested for details.

Today's data networks have four layers: IP for carrying applications and services, ATM for traffic engineering, SONET/SDH for transport, and dense wavelength-division multiplexing (DWDM) for capacity. With such multilayer architectures, any one layer can limit the scalability of the network, as well as add to the cost. Figure 3.32 shows the evolution of networks towards all-photonic technologies.

Meanwhile leading telecom system manufacturers are developing futuristic architectures to couple multiservice networks, such as that by Ericsson discussed in [34].

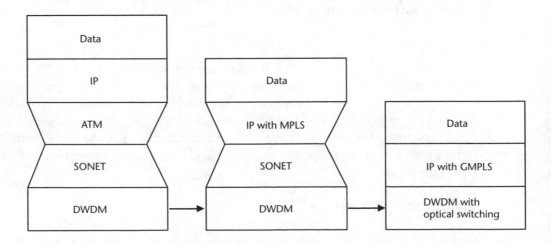

**Figure 3.32**   The evolution towards photonic networking. (*After:* [24].)

# References

[1]   Reeves, A. H., "The 25th Anniversary of PCM," *IEEE Spectrum,* May1965, pp. 53–63.

[2]   Joel, A. E., "The Past 100 years in Telecommunications Switching," *IEEE Communications Magazine,* May 1984, pp. 64–83.

[3]   Joel, A. E., "Digital Switching—How It Has Developed," *IEEE Transactions on Communications,* July 1979, pp. 948–959.

[4]   Thompson, A.R., "Telephone Switching Systems," Artech House Publishers, 2000.

[5]   Ericsson and Telia, Ericsson Telecom AB, Telia AB, Studentlitteratur AB, 1998. *Understanding Telecommunications—Volume 2,* Lund, Sweden: Studentlitteratur, 1998.

[6]   Dixit, S. and Elby, S., "Frame Relay & ATM Networking," *IEEE Communications,* June 1996, pp. 64–82

[7]   Stallings, W., *Data and Computer Communications,* Seventh Edition, Englewood Cliffs, NJ: Prentice Hall, 2003.

[8]   Bellamy, J. C., *Digital Telephony,* Third Edition, New York: John Wiley, 2000.

[9]   Fujiyama,Y., H. Masuda, and K. Mano, "Architecture Enhancement of Digital Switching System," Vol. V, *Proc. XII International Switching Symp.,* Sweden, 1990, pp. 217–220.

[10]  Milstead, R. M. and P. S. Warwick, "The Digital PBX: Current and Future Views of Applications for Information Environment," *Proc. Int'l. Conf. on Private Switching Systems and Networks,* June 1988, pp. 93–98.

[11]  Foote, D. L. and M. W. Medin, "Extending ISDN Functional Layering Concepts in PBX Architectures," *Proc. Int'l. Conf. on Private Switching Systems and Networks,* June 1988, pp. 1–5.

[12]  Kerr, G. W., "A Multisite Videophone PBX System," *Proc. Int'l. Conf. on Private Switching Systems and Networks,* June 1988, pp. 93–98, 215–219.

[13]  Hamada, K. I., et al., "Digital Cordless Telephone Equipment for Wireless PBX Systems," *Proc. 43rd IEE Vehicular Technology Conf.,* May 1993, pp. 963–966.

[14]  Bonzoano, L. and V. Palestini, "DECT Performance in the Wireless PBX Application," *Int'l. Conf. on Communication,* May 1993, Vol. 2, pp. 1269–1272.

[15]  Lin, P., and Y. Lin, " Implementation and Performance Evaluation for Mobility Management of a Wireless PBX Network," *IEEE Journal on Selected Areas in Communications,* Vol. 19, No. 6, June 2001, pp. 1138–1146.

[16]  Lai, W. R., and Y. B. Lin, "Resource Planning for Wireless PBX Systems," *Proc. First IEEE Enterprise Networking Mini Conf.,* June 1997, pp. 63–72.

[17]  Frame Relay Forum, "ATM in Europe: The User Handbook," Ver. 1.0, July 1997.

[18]  Anderson, J., et al., " Operations Standards for Global ATM Networks: Network Element View," *IEEE Communications Magazine,* December 1996, pp. 72–84.

[19]  Jain, R., and G. Babic, "Performance Testing Effort at the ATM Forum: An Overview," *IEEE Communications,* August 1997, pp. 110–116.

[20]  Hou, Y. T., L. Tassiulas, and H. J. Chao, "Overview of Implementing ATM Based Enterprise Local Area Network for Desk Top Multimedia Computing, *IEEE Communications Magazine,* April 1996, pp. 70–76.

[21]  Agrawal, S., "IP Switching," August 1997, http://www.cis.ohio-state.edu/~jain/cis788-97/ip_switching/index.htm.

[22]  "Switching Versus Routing," *Network Magazine,* http://www.networkmagazine.com/shared/article/showArticle.jhtml?articleId=8702661.

[23]  Armitage, G., "MPLS: The Magic Behind the Myths," *IEEE Communications Magazine,* January 2000, pp. 124–131.

[24]  Banerjee, A., et al., "Generalized Multiprotocol Label Switching: An Overview of Routing and Management Enhancements," *IEEE Communications Magazine,* January 2001, pp. 144–150.

[25]  Elhany, I., K. Busch, and D. Chiou, "Switch Fabric Interfaces," *IEEE Computer,* September 2003, pp. 106–108.

[26]  Agere Systems, "Protocol Independent Switch Fabrics," product brief, May 2002.

[27]  Agere Systems, "-X/-C Protocol Independent (ATM/IP) Switch Elements," product brief, May 2001.

[28]  Agere Systems, "Protocol Independent Switch Fabric (P140SAX)," advance product brief, October 2002.

[29]  Goldberg, L., "Fractional Terabit ATM Switch Demonstrated: New Modular Architecture Accommodates User's Growing Bandwidth Needs," *Electronic Design,* February 1997, pp. 31–32.

[30]  Papadimitriou, G. I., C. Papazoglou, and A. S. Pomportsis, "Optical Switching: Switch Fabrics, Techniques and Architectures," *IEEE Journal of Lightwave Technology,* Vol. 21, No. 2, February 2003, pp. 384–405.

[31]  Massetti, F., et al., "High Speed, High Capacity ATM Optical Switches for Future Telecommunication Transport Networks," *IEEE Journal on Selected Areas in Communications,* Vol. 14, No. 5, June 1996, pp. 979–998.

[32]  Forbes, M., J. Gourlay, and M. Desmulliez, "Optically Interconnected Electronic Chips: A Tutorial and Review of Technology," *Electronics and Communication Engineering Journal,* October 2001, pp. 221–232.

[33]  Neukermans, A., and R. Ramasawamy, "MEMS Technology for Optical Networking Applications," *IEEE Communications Magazine,* January 2001, pp. 62–69.

[34]  Hallenstal, M., U. Thune, and G. Oster, "ENGINE Server Network," *Ericsson Review,* No. 3, 2000, pp. 126–149.

[35]  European Market Awareness Committee: "ATM in Europe: The Use Handbook," The ATM Forum, version 1.0, July 1997.

[36]  Lee, B. G., and Kim, W., *Integrated Broadband Networks: TCP/IP, ATM, SDH/SONET, and WDM/Optics,* Norwood, MA: Artech House Inc., 2000.

[37]  ITU-T: "Q.7000: Specifications of Signalling System No. 7" Revision 03/93.

# CHAPTER 4
# Cellular Systems

## 4.1 Introduction

While switching and transmission at the core of telecommunications networks have evolved in wondrous ways, the facilities by which we gain access to the network, the local loop, or the last mile, remained fundamentally unchanged from the era of Alexander Graham Bell until the 1980s. Within the two decades 1980–2000, developments in semiconductors and radio techniques spurred enormous technological developments to overhaul the antiquated last mile with both wired and wireless access technologies.

By avoiding the use of a physical medium altogether, wireless access breaks out of the linear, one-dimensional problem of space for copper and expands into two-dimensional area coverage. Wireless access techniques come basically in two types: fixed and mobile. Cellular systems providing mobile wireless access to the global telecommunications network is the subject of this chapter.

## 4.2 Advantages and Constraints of Wireless Access

Summarized next are some of the most important developments to follow from wireless access [1]:

- The cost structure of the copper access plant-specifically, the use of price averaging and the negative relationship between subscriber density and cost per subscriber-places rural areas at a disadvantage. Wireless access makes the cost per connection more uniform, varying less with distance and subscriber density.
- Many new services are often impeded from being delivered to customers by the performance limitations of the copper access plant. Wireless technology aids the early deployment of these services.
- A subscriber can obtain service in much less time with wireless access compared to wireline access. If a subscriber relocates, reconfiguration of the wireless access plant is more convenient. Where true mobility is required, it is the only option.

However, compared to copper, wireless systems require advanced hardware, elaborate planning, field measurements, and surveys. The situation is further complicated by spectrum issues, radio propagation problems, and the demand for Internet and multimedia services requiring higher bandwidth than traditional voice services.

## 4.3   Basic Components of Wireless Access

Figure 4.1(a) shows a simple model of a wireless access system with its three basic components.

The subscriber station is a fixed or mobile radio transceiver. This provides a single circuit to the network. Multiple circuits are also possible in present applications. The functions of the subscriber station are to format the user's messages (voice, images, or data) for transmission over the radio channel and do the reverse for received signals, establish wireless circuits to the network through base stations, and

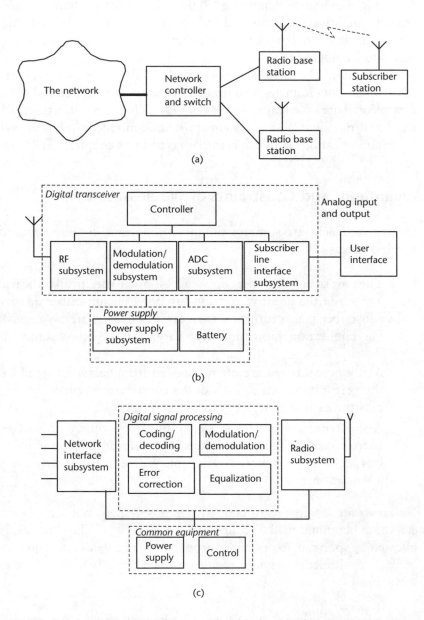

**Figure 4.1**   Wireless access system: (a) basic building blocks, (b) a functional block diagram of a subscriber station, and (c) a functional block diagram of a radio base station.

transfer signals between the base station and itself. A functional block diagram of a subscriber terminal is shown in Figure 4.1(b).

The base station or the cell site shown in Figure 4.1(c) is a centrally located, multicircuit transceiver that radiates over its coverage area. The size of the area covered may range from a few hundred meters to tens of kilometers and may radiate omnidirectionally or in sectors. Modern wireless access systems use sophisticated adaptive beam steering techniques to operate in highly congested radio environments. Base stations are linked to the mobile switching center (MSC)/network controller via high capacity links.

The network controller manages the operation of the entire wireless access system, providing the interconnection between itself and external networks. It determines the assignment of circuits to individual subscribers, monitors system performance, and may support the provision of advanced services through the maintenance of subscriber databases. The external network to which the wireless access system connects may be the PSTN, ISDN, or a private LAN or WAN.

## 4.4   Cellular Systems as an Access Technology

### 4.4.1   The Evolution of Cellular Systems

The earliest significant use of mobile radio was in 1921 by the Detroit police department in the 2-MHz band. Around 1940, further frequencies around 30 to 40 MHz were made available. This experience allowed the development of private mobile radio (PMR) systems not generally connected directly to the telephone network. After World War II, systems developed significantly with additional spectrum, first around 150 MHz and then around 450 MHz. Details of these are found in [2, 3]. In the 1960s, automatic channel selection was introduced in the 450-MHz band in the improved mobile telephone system (IMTS). IMTS systems required a single base station positioned at a high elevation with a wide coverage area of radius 20 to 50 miles [2]. For each channel, the base station transmit power was typically 200 to 250W. The equivalent isotropically radiated power (EIRP) was as high as 500W.

As the demand for mobile telephony grew, a totally different concept of small cells with low-power base stations and spectrum reuse was verbalized in 1947 by D. H. Ring of Bell Laboratories in an unpublished work [2]. Two other wireless technologies being developed at the time, citizen's band (CB) radio and cordless telephone systems, contributed significantly to the development of what has come to be known as cellular technology today. The cellular concept was developed by the Bell Laboratories during1970s, leading to the successful initial implementation of the advanced mobile phone system (AMPS) in 1983.

### 4.4.2   The Cellular Concept

A cellular system is a high-capacity land mobile system in which the available frequency spectrum is partitioned into discrete channels and assigned in groups to cells covering the geographic service area (GSA). The channel groups are capable of being reused in different cells within the service area. All mobiles in a cell communicate with the base station. Transmitters in adjacent cells operate on different frequencies to avoid interference. With the transmit power and antenna height in each cell being relatively low, cells that are sufficiently far apart can reuse the same frequencies without causing undue cochannel interference, as shown in Figure 4.2(a). Here, the cells containing the same letter use the same frequencies and are called *cochannel* cells. A group of cells using all the available frequencies is called a *cluster*. Through frequency reuse, a cellular system in one service area can handle a number of simultaneous calls greatly exceeding the total number of allocated channels [4].

Growth in traffic within a cell will require a revision of cell boundaries so that the area formerly regarded as a single cell can now contain several cells. The process called cell splitting fills this need. Figure 4.3 shows different stages of the cell splitting.

*Handoff* is the act of transferring a call from one cell to another without interruption while a call is in progress. It is necessitated by the use of small cells. This involves monitoring of signal strength and smooth transfer to another channel in a different cell.

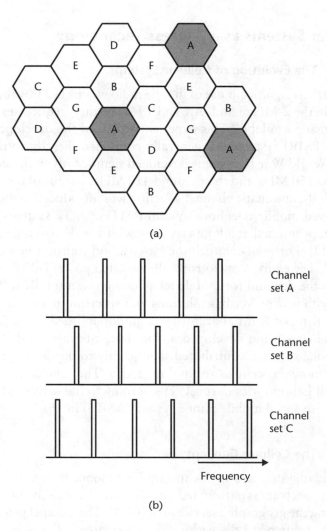

**Figure 4.2** Frequency reuse in cellular systems: (a) example of a reuse pattern, and (b) channel subsets used in different cells.

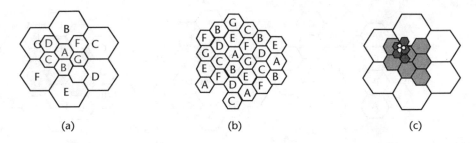

**Figure 4.3**  Cell splitting: (a) early stages, (b) after completion, and (c) splitting into progressively smaller cells to match traffic.

### 4.4.3  The Cellular Geometry

The main purpose of defining cells in a mobile communications system is to delineate areas in which specific channels or a specific base station will be used. A reasonable amount of geographical confinement of channel usage is necessary to prevent interference.

#### 4.4.3.1  Cell Coverage

In practice, cell coverage is distinctly misshapen due to terrain features and obstructions, as shown in Figure 4.4(a). However, for planning purposes, the assumption of a regular cell structure is necessary. While a simple circular cell shape results in ambiguous coverage, any regular polygon can cover an area with no gaps or overlaps. Hence, early system designers from Bell Laboratories adopted the regular hexagonal shape shown in Figure 4.4(c) that has become the standard in cellular system design and marketing.

#### 4.4.3.2  Frequency Reuse Patterns

Figure 4.5(a) shows clusters or frequency reuse patterns of size 3, 4, and 7. In the reuse pattern, all cochannel cells are equidistant from each other. To achieve this, the reuse pattern should satisfy:

$$N = i^2 + j^2 + ij \tag{4.1}$$

where $i$ and $j$ are integers such that $i, j \geq 0$. These are called *shift parameters*. Figure 4.5(b) illustrates how to lay out a reuse pattern of 19 using $i = 3$ and $j = 2$.

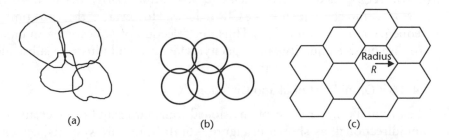

**Figure 4.4**  Cell coverage: (a) irregular, (b) circular, and (c) hexagonal.

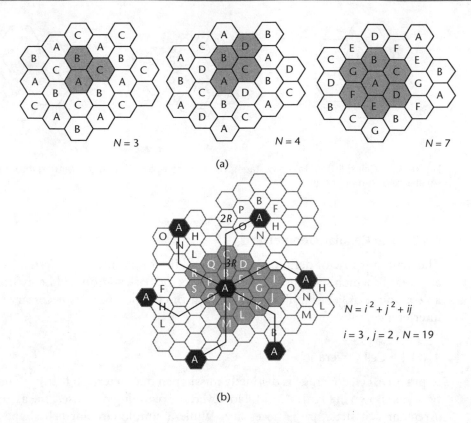

**Figure 4.5** (a) Frequency reuse patterns of 3, 4, and 7, and (b) generation of frequency reuse patterns.

Some geometrically realizable reuse patterns as stipulated by (4.1) are given in Table 4.1.

The ratio of $D$, the distance between the centers of nearest neighboring cochannel cells, to $R$, the cell radius, is called the *cochannel reuse ratio* and is related to the reuse pattern as:

$$\frac{D}{R} = \sqrt{3N} \tag{4.2}$$

In practical systems, the choice of $N$ is governed by cochannel interference considerations. The common measure of interference is the *carrier-to-interference ratio* (CIR). As $N$ increases, the relative separation (D/R) between cochannel cells increases, and consequently the CIR reduces. However, at the same time, geographic frequency reuse becomes less. Thus, the selection of $N$ becomes an important compromise between quality and capacity, which we will illustrate subsequently.

### 4.4.3.3    Omnidirectional and Directional Cells

The original cellular concept envisioned transmitter and receiver antennas that are omnidirectional, as shown in Figure 4.6(a). In mature systems, cells are *sectored*, where each channel is transmitted and received over one of three 120-degree

**Table 4.1**   Reuse
Patterns

| Shift Parameters | | Reuse Pattern |
|---|---|---|
| i | j | N |
| 0 | 1 | 1 |
| 1 | 1 | 3 |
| 0 | 2 | 4 |
| 1 | 2 | 7 |
| 0 | 3 | 9 |
| 2 | 2 | 12 |
| 1 | 3 | 13 |
| 2 | 3 | 19 |

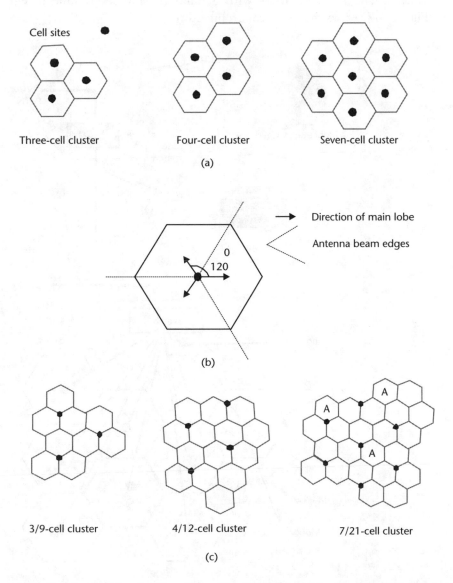

**Figure 4.6**   (a) Omnidirectional cells patterns, (b) sectorization of a cell, and (c) sectorized cell patterns.

sector antennas as shown in Figure 4.6(b). This is practically implemented as in Figure 4.6(c), where the cell sites are at the same locations as in Figure 4.6(a) but use sectorized transmission. Reuse patterns in Figure 4.6(c) are referred to as 3/9, 4/12, and 7/21. Sixty-degree sectorization is also used. The role of sectorization in reducing interference will be illustrated later.

### 4.4.4   The Physical Architecture of Cellular Systems

A cellular system infrastructure comprises base stations connected through fixed links to an MSC, also called a mobile telecommunications switching office (MTSO). This is a local switching exchange with additional features to handle the dynamic location information, subscription data, and management functions of a cellular system. The MSC interacts with a database and interconnects with the PSTN. Figure 4.7 shows this physical architecture.

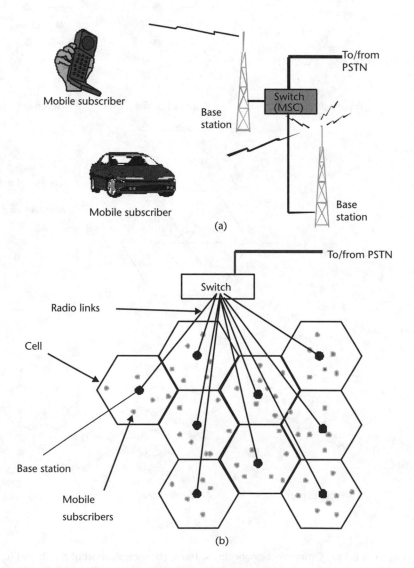

**Figure 4.7**   The physical architecture of a cellular system.

### 4.4.5   The RF Channel Structure

In a cellular system, two primary operations happen: control/monitoring and voice/data transfer. The system thus has two types of RF channels, as shown in Figure 4.8. These are *traffic (or voice) channels* and *control channels*. Forward (base to mobile) and reverse (mobile to base) channels are needed for duplex communications. The control (signaling or set-up) channels are used for call set up and other signaling and control functions, while the traffic channels are for subscriber traffic.

Signaling in a cellular system is *common channel*. However, because a cellular transceiver can operate on only one channel at a time, for control functions during a call, the voice (traffic) channels are used to send control commands between the mobile and the base through *blank and burst* techniques or *inband signaling*.

The actual implementation of RF channels is different, depending on the implementation of the air interface (the link between the mobiles and the base). A channel may be a pair of frequency bands or a pair of time slots within a pair of frequency bands. The usage of these channels is described next.

## 4.5   Cellular System Operation

The AMPS, the earliest cellular standard, is taken as an example to illustrate the basic processing steps. References [3, 5, 6] give further information on system operation.

### 4.5.1   Identification Numbers and Codes

Each mobile unit has two unique numbers: the *mobile identification number* (MIN) and the *electronic serial number* (ESN). The former is the mobile's phone number, also called the mobile station ISDN number (MSISDN), and is programmable. The latter is assigned at the time of manufacture of the mobile unit. The combination of the two allows the system to carry out proper billing and discourage fraudulent use. The mobile phone also has a *station class mark* (SCM), which identifies its maximum transmit power.

The cellular network has a *system identification number*. This enables a mobile phone to identify the system to which it has subscribed. Each cell site has a *digital color code* (DCC) and a *supervisory audio tone* (SAT), which are transmitted on all

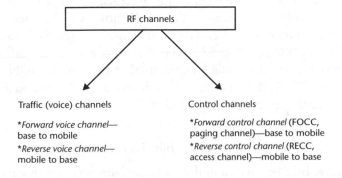

**Figure 4.8**   RF channel classification in cellular systems.

control and voice channels, respectively, simultaneously with control signals or traffic. These are used by mobiles and base stations to identify a valid signal from cochannel interference, as explained later. The *mobile attenuation code* (MAC) specifies the power level to be used.

### 4.5.2  Mobile Unit Initialization

When a user activates a mobile unit, the receiver scans the designated set-up channels, selects the strongest, and locks on. As each base station is assigned a different set of set-up channels, locking onto the strongest usually means selecting the nearest cell site. The mobile receives the system identification number from this channel, which continuously broadcasts it. The mobile then transmits its MIN and ESN to the base station on the corresponding reverse control channel. The base station in turn assigns a pair of control channels to the mobile. As there are only a few control channels in each cell, these channels are shared by all mobiles in the cell. The mobile then tunes to the assigned forward control channel (paging channel), which continuously broadcasts system information and incoming call alerts. It monitors these broadcasts and updates the operating parameters as necessary. This *self-location scheme* is used in the idle stage and has the advantage of eliminating the load on the system for locating mobiles. Initialization is repeated at regular intervals.

After self location, the base station can send the mobile's location information to the MSC, where it is entered into a database of all active mobiles. To deliver an incoming call to a mobile, the MSC and the associated location databases must be updated as the mobile moves through the coverage area. After initialization, by listening to a paging channel, idle mobiles can detect if they move to another location area or system or if they receive calls. In the former case, they repeat the self-location procedure immediately. In the latter, the process for call receiving is started.

### 4.5.3  Mobile Originated Call

The user places the destination number into an originating register in the mobile unit and pushes the *send* button. A request for service is sent on the reverse control channel (access channel) obtained during initialization. The request includes the MIN, ESN, and the destination number. The cell site receives it and passes it on to the MTSO. Once authorization is complete, the MTSO selects an available voice channel for the call and informs the mobile of the voice channel, the DCC, and the MAC for the call. This is done through the assigned paging channel. The MTSO simultaneously establishes a circuit to the destination over the telephone network.

Both the base station and the mobile now switch to the assigned voice channel. The base station sends its SAT to the mobile, and the mobile replies with the same SAT, the reception of which confirms that the correct mobile and the cell site are connected. The call then begins. This procedure is illustrated in Figure 4.9, assuming the destination to be a PSTN number.

### 4.5.4  Network Originated (Mobile Terminated) Call

When a subscriber dials a mobile number, the network to which it belongs is identified and the call is routed to the appropriate MTSO. The MTSO sends a message to

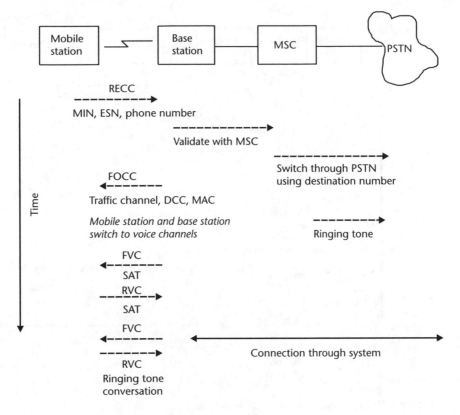

**Figure 4.9**   Call origination by a mobile phone.

the most recent cell to which the mobile has identified itself. The cell site transmits a message on its paging channels, which includes the MIN. The mobile unit, which is monitoring a paging channel, recognizes its own identification and responds to the cell site with its MIN, ESN, and SCM on the access channel. After authentication, the base station informs the mobile about the paging channel, the channel assigned for the call, MAC, and SAT. The mobile tunes to the voice channel, responds with the SAT to the base station, and initiates ringing to alert the user. The call then commences. This procedure is illustrated in Figure 4.10 for an incoming call from the PSTN.

### 4.5.5   Handoff (Handover) and Power Control

During a call, the serving base station monitors the signal strength and quality from the mobile through the CIR, aided by the SAT. If the CIR falls below a predesignated threshold, it requests the MTSO for a handoff for the call. The MTSO requests the neighboring base stations to monitor the CIR for the mobile. If another base station reports a better CIR than the serving base station, the MTSO finds a free channel for the call to be transferred to on in this cell and informs this to the base station requesting the handoff. A signaling message is sent to the mobile on the voice channel (using a blank and burst procedure) from the current base station asking it to retune to the new voice channel. The mobile tunes to the new channel while the MTSO switches the call to the new base station. The interruption and the

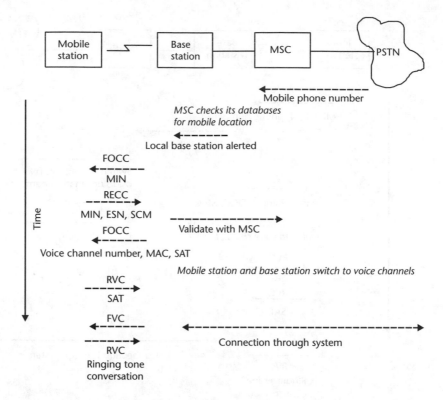

**Figure 4.10**   Call reception by a mobile phone.

change of speech channels are imperceptible to the user. In newer digital cellular systems, mobile-assisted handoff is also possible, where the mobile is able to measure signal levels of channels in adjacent cells and report back to the network. Effective and reliable handoff is an essential feature of a cellular system.

When the base station detects the signal strength from a mobile decreasing progressively, it will request the mobile to increase its transmit power level by sending the appropriate MAC by a blank and burst procedure on the voice channel. This ensures that all transmissions within a cell use the minimum necessary power levels at a given time, which greatly helps in reducing the average cochannel interference. Handoff may be initiated only when the signal strength becomes lower than the threshold while the mobile is transmitting at its maximum level.

### 4.5.6   Location Tracking and Roaming

Roaming allows mobile subscribers to originate and receive calls using their own telephone number when they are in an area served by a different operator than the one to which they have subscribed. Location tracking is a key requirement for roaming. The two operators must also have a roaming agreement between them, through which their subscriber information is exchanged.

A mobile service provider can have within its service area, either *home subscribers* (the operator's own) or *visiting subscribers* (subscribers of other operators). The MSC maintains subscriber databases called the *home location register* (HLR) for its own subscribers and the *visitor location register* (VLR) for visiting subscribers. The

HLR and VLR contain up-to-date location information and the service profile of the respective type of subscriber.

When a mobile initializes in a service area of a different operator, the MSC recognizes it as a visitor and assigns a temporary *mobile station roaming number* (MSRN) to it. It then records the subscriber's location information and the MSRN in its VLR. Through internetwork signaling, the VLR sends this information to the subscriber's HLR in the home MSC.

A call to a mobile is always received first by its home MSC, as call routing is based on the MIN. The home MSC queries the HLR. If the called subscriber is roaming, the HLR will contain the visiting network identification. The call will then be routed to the visiting MSC, which will query the VLR and route it to the subscriber, as illustrated in Figure 4.11.

## 4.6   The Cellular Radio Environment

In urban environments where subscriber density is high, line of sight between the mobile and the base station is a luxury the designer cannot afford to assume. In such situations where many man-made structures and other large reflecting and obstructing bodies are present, the radio signal is subjected to many effects such as:

• Multipath propagation effects and fading;
• Path loss;
• Interference.

What the mobile receives is the result of many signals traveling via many paths having different lengths and arriving at different phases, as shown in Figure 4.12. As

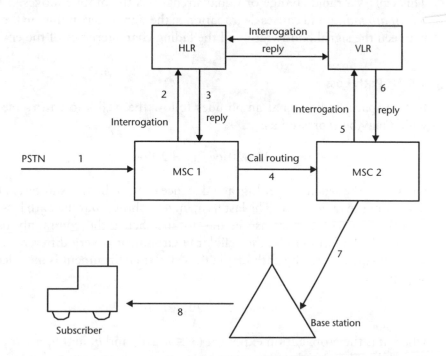

**Figure 4.11**   Implementation of roaming: Example of routing an incoming call.

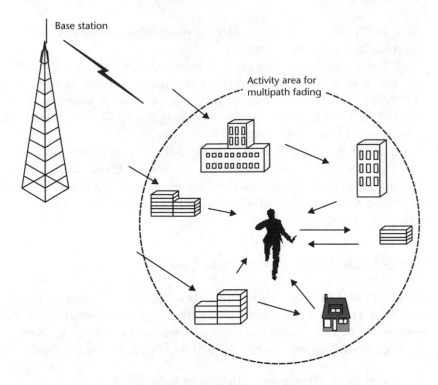

**Figure 4.12** Multipath propagation in the mobile environment.

a result, the signals may add constructively at one position and destructively at just another half a wavelength away. (In the 900-MHz frequency, this is about 15 cm). This causes a rapid change of signal strength as the mobile moves. This is termed *multipath fading* and causes degradation in the signal, depending on the relationship between the signal bandwidth and the fading characteristics of the environment.

### 4.6.1 Path Loss

In free space, radio signal amplitudes follow the well-known inverse-square law, which may be expressed as:

$$L = 32 + 20\log f_{MHz} + 20\log d_{km} \tag{4.3}$$

where $L$ is the signal power loss at a distance $d$ from the transmitter, and $f$ is the frequency of the radio wave. The last term in (4.3) shows that the path loss increases by 20 dB for a tenfold increase in the distance-hence the commonly used term 20 dB/decade. Unfortunately, the cellular environment is very different from the free-space environment. The path loss in the cellular environment is modeled commonly as [6]:

$$L = n\log d_m - 20\log h_T h_R + \beta \tag{4.4}$$

where $n$ is the propagation exponent ($3 \le n \le 6$) and $h_T$ and $h_R$ are the base station and the mobile height. $\beta$ accounts for effects of frequency dependency, surface irregularities, line-of-sight obstacles, manmade structures, vegetation, and terrain features. The propagation exponent, usually determined through measurements, gives rise to a path loss of 30 to 60 dB/decade. A commonly used empirical model derived by the ITU for cellular planning purposes in urban areas is shown in Figure 4.13. $n = 4$ has been assumed.

This model predicts the mean signal level as a function of distance and is therefore useful in determining the radio range of a base station. Figure 4.13 shows the possible cell radii at different frequencies of operation, assuming a receiver threshold of –107 dBm ($\mu$V/m).

### 4.6.2   Fading

Figure 4.14 depicts the many manifestations of fading in the mobile environment. The two principle effects are *large-scale (slow)* and *small-scale (fast)* fading. Figure 4.15 shows the signal power $r(t)$ received by a mobile receiver, where $r(t)$ is partitioned into two components as:

$$r(t) = m(t)\, r_0(t) \tag{4.5}$$

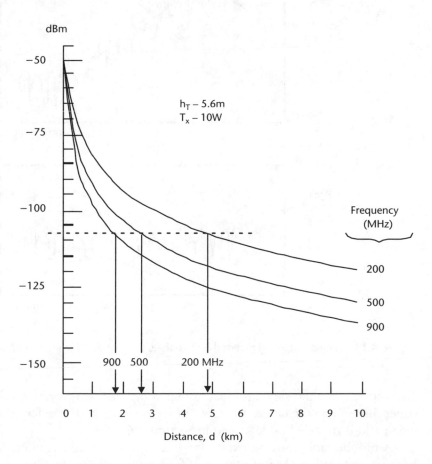

**Figure 4.13**   Signal strength versus distance according to the ITU model (base station height 5.6m, transmit power 10W, receiver height 1.5m). (*Source:* [6]. ©1993, Artech House, Inc. Reprinted with permission.)

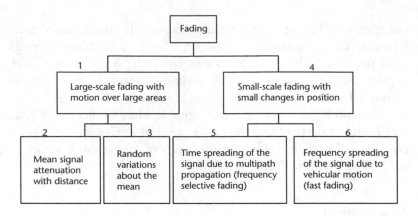

**Figure 4.14**   Manifestations of fading.

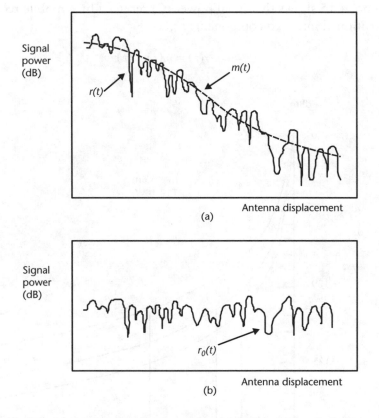

**Figure 4.15**   Fading: (a) signal received by a mobile, and (b) the small-scale fading component.

$m(t)$ and $r_0(t)$ are the slow and fast fading components, respectively. The former is superimposed on the latter, as shown in Figure 4.15. $m(t)$ is referred to as the local mean and is modeled as described in Section 4.6.1.

A mobile radio moving over a large area must process signals that experience both types of fading. Manifestation of fading depends on the physical nature of the environment, the characteristics of the signal being transmitted, and the user's mobility.

### 4.6.2.1   Large-Scale Fading

This represents the average signal power variation due to motion over large areas. The effects due to large scale fading are depicted in blocks 1, 2, and 3 of Figure 4.14. As discussed in Section 4.6.1, propagation models predict the mean signal strength at a given distance. However, it has been observed that the mean also has a random component. This is also observed in Figure 4.15(a), where the dotted line representing the average signal strength exhibits a slow fluctuation with distance instead of monotonically decreasing, as in Figure 4.13. Details of such random fluctuations, their statistical characterization, and their field measurements are presented in [5] and [7].

### 4.6.2.2   Small-Scale Fading

This refers to the rapid changes in signal amplitude and phase that can be experienced as a result of small changes (as small as half a wavelength) in the mobile's position. This is also referred to as *Rayleigh fading*, as the fading signal amplitude in this environment has been found to follow a Rayleigh distribution [7]. Small-scale fading manifests itself in two mechanisms, *frequency selective fading* and *fast fading*, as shown in blocks 5 and 6 in Figure 4.14. Frequency selective fading and fast fading cause distortion of the signal spectrum and signaling pulse shape, respectively. In the case that a particular mobile channel is not frequency selective or fast, the Rayleigh fading is said to be *flat* or *slow* and causes only random fluctuations of signal strength.

For details on fading, the reader is referred to [8, 9]. Reference [7] describes how cellular system design parameters such as cell size and transmit power are determined in link budget preparation by taking different aspects of fading into account.

### 4.6.2.3   Effects of Fading and Their Mitigation

Typical BER curves for different communications environments are illustrated in Figure 4.16. Table 4.2 summarizes the three major performance categories observed in this figure.

Mitigation of fading depends on whether the distortion is caused by frequency selective fading or by fast fading. Error control coding and interleaving are

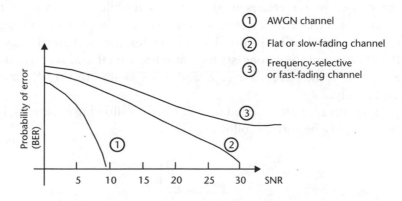

**Figure 4.16**   BER curves in different communications environments.

**Table 4.2**  Performance Categories

| Category | Description |
|---|---|
| AWGN channel | This is the performance that can be expected when using any nominal modulation type in AWGN. |
| Flat or slow-fading channel (the Rayleigh limit) | The performance is approximately a function of the reciprocal of the $C/N$. For reasonable SNR, performance will be significantly worse than in an AWGN environment. However, by increasing the $C/N$ or $C/I$, the performance may be made satisfactory. |
| Frequency selective or fast-fading channel | Irreducible *error floors* are caused by severe distortion. The desired performance cannot be achieved by increasing $C/N$ or $C/I$. |

commonly used to mitigate fast fading. Adaptive equalization, spread spectrum techniques, and orthogonal frequency division multiplexing (OFDM) are some of the techniques used or proposed to combat frequency selective fading. With these techniques, performance should move towards the Rayleigh limit. Next, further improvements can be effected by using some form of diversity reception to provide the receiver with a collection of uncorrelated samples of the signal and by using further error correction methods. This should bring the performance curve towards the AWGN curve. More information is found in [7, 9].

### 4.6.3  Interference

Interference in cellular systems are of two types: cochannel interference (CCI) and adjacent channel interference (ACI).

### 4.6.3.1  CCI

Frequency reuse inevitably causes interference in cellular systems. We have shown that the tighter the frequency reuse (i.e., the smaller the reuse pattern $N$), the more the effective the number of channels available in a given geographical area. However, with decreasing $N$, CCI will become stronger, due to the decreasing reuse distance. The challenge is therefore to obtain the smallest possible $N$ for the required system performance.

The quantity determining the performance of a cellular system is the CIR. The carrier power relates to the signal a mobile receives from the base station in its own cell. The interference consists of the sum of all signals it receives from cochannel cells. Figure 4.17 illustrates the cochannel cells in the first tier (the closest and the strongest) for omnidirectional and sectorized cells. The numbers of such interferers are 6, 2, and 1 for omnidirectional, 120°, and 60° cells, respectively.

The average CIR, referred to as $C/I$ in the following equations, in a cellular environment may be derived following [5] as:

$$\frac{C}{I} = \frac{\left(\dfrac{D}{R}\right)^{n}}{N_I} \tag{4.6a}$$

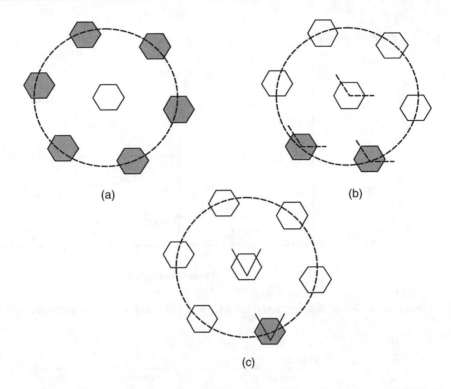

**Figure 4.17**  Cochannel cells of the first tier for: (a) omnidirectional cells, (b) 120° sectored cells, and (c) 60° sectored cells.

where $D/R$ is the reuse ratio, $n$ is the propagation exponent, and $N_I$ is the number of significant cochannel interferers. Following (4.2), we can rewrite (4.6a) as:

$$\frac{C}{I} = \frac{\left(\sqrt{3N}\right)^n}{N_I} \tag{4.6b}$$

Hence,

$$N = \frac{1}{3}\left\{\left[N_I\left(\frac{C}{I}\right)\right]^{1/n}\right\}^2 \tag{4.6c}$$

This is the important relationship between signal quality $C/I$ and reuse pattern $N$ used in ascertaining the smallest possible reuse pattern for a required signal quality. However, $N$ must be the smallest integer satisfying (4.1) as well. Figure 4.18 highlights an acceptable $C/I$ of 18 dB. A total of 100 channels are assumed for the system. The number of channels per cell for each $N$ is therefore $100/N$. It is seen that the smallest possible reuse pattern is 7.

The advantage of sectorization can be understood by using $N_I = 2$ in (4.6c) corresponding to 120° sectorization, which results in a reuse pattern of 4 for the same $C/I$. Table 4.3 summarizes the $C/I$ versus $N$ relationship for omnidirectional and 120° sector cells. For a $C/I$ of 18 dB, $N$ can be reduced to 7 after sectorization.

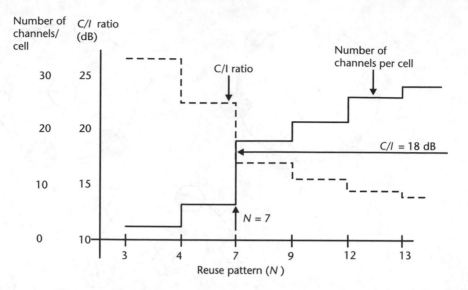

**Figure 4.18**   Relationship between C/I and the reuse pattern N for omnidirectional cells.

**Table 4.3**   *C/I Versus N*

| N | C/I (dB) Omni Cells | C/I (dB) 120° Cells | Number of Channels Per Cell (Total of Hundred) |
|---|---|---|---|
| 3 | 11.3 | 16.1 | 33 |
| 4 | 13.8 | 18.6 | 25 |
| 7 | 18.7 | 23.4 | 14 |
| 9 | 20.8 | 25.6 | 11 |
| 12 | 23.3 | 28.1 | 8 |
| 13 | 24.0 | 28.8 | 7 |
| 16 | 25.8 | 30.6 | 6 |
| 19 | 27.3 | 32.1 | 5 |
| 21 | 28.2 | 32.9 | 4 |

### 4.6.3.2   ACI

Although intermediate frequency (IF) sections in radio transceivers significantly attenuate signals from adjacent channels, additional precautions are necessary to avoid ACI. Fortunately, because only a fraction of the allocated channels is used in one cell, it is possible to avoid the use of adjacent channels in the same cell. In determining channel sets for frequency reuse, the largest possible frequency separation is maintained between adjacent members of the same set. If $N$ channel sets are required (i.e., a reuse pattern of $N$), the $n$th set would contain channels $n, n + N, n + 2N,...$ [4].

### 4.6.3.3   Use of the SAT for Interference Detection

The SAT has been used from early cellular systems to identify the desired signal from interference. These are three tones in the 5- to 6-kHz range. All base stations in a cluster are assigned one SAT, with the neighboring clusters assigned the other two SATs, as illustrated in Figure 4.19. The cells that have the same SAT and the same channel set are separated by $\sqrt{3}D$, where $D$ is the cochannel reuse distance.

Each base station adds its SAT to each forward voice channel (FVC). The mobile detects this and retransmits it on the reverse voice channel (RVC). Hence, during a call, the SAT is always present on the voice channels. The reception of a wrong SAT indicates a cochannel interfering signal. Interruption of the tone indicates a lost connection and, if the tone is not received within a predetermined short interval, the call is terminated. The DCC serves the same purpose on the control channels.

## 4.7   Frequency Assignment and Multiple Access Schemes for Cellular Systems

Frequency assignments for cellular systems are usually in two bands: one for the base-to-mobile (forward) direction and one for the mobile-to-base (reverse) direction. This is referred to as frequency division duplexing (FDD).

Cellular systems are designed to carry many simultaneous voice channels in the allocated frequency band. All subscribers in a cell need to communicate with the base station simultaneously without interfering with each other. The multiple access scheme determines how each of these frequency bands are shared by subscribers to maximize the number of simultaneous calls. The basic multiple access techniques are frequency division multiple access (FDMA), time division multiple access (TDMA), and code division multiple access (CDMA), with practical implementations usually being hybrids.

### 4.7.1   FDMA

In FDMA, the spectrum allocated for the service in each direction is subdivided into a set of contiguous subbands, as shown in Figure 4.20. These subbands are allocated to cells according to the CCI and ACI considerations, as discussed in Section 4.6.

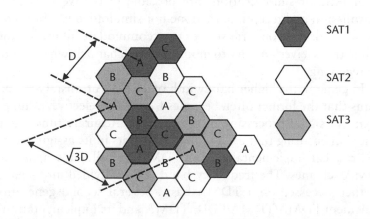

**Figure 4.19**   The use of SATs.

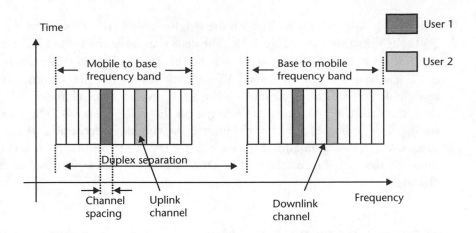

**Figure 4.20** FDMA.

Each mobile station is assigned a pair of channels, one in each direction, for the duration of a call. All first generation cellular systems use this technique, known as FDMA/FDD. A serious disadvantage of FDMA is that a separate transceiver is required at the base station for each mobile station in its coverage area. A significant advantage of using FDMA is that each user's transmissions are over a narrow channel (e.g., 30-kHz bandwidth) and hence is not vulnerable to frequency selective fading.

### 4.7.2 TDMA

In TDMA, each user is allowed the entire available bandwidth, but only for a fraction of the time and on a periodic basis. A framing structure, as shown in Figure 4.21(a), is used, with a user given one time slot in each frame. If the user generates continuous data, it is buffered and transmitted in a burst during the assigned time slot. Usually, transmit and receive time slots for a communication are displaced from each other, as shown in Figure 4.21(b), where user 1 transmits on time slot 1 on frequency $f_1$ and receives on time slot 3 on frequency $f_2$ (TDMA/FDD).

Significant simplifications are possible in receiver design in TDMA because transmission and reception are done not simultaneously but sequentially. The base station needs only one transceiver to accommodate all users. Another feature is the mobile transceiver's ability to monitor other channels when it is neither transmitting nor receiving.

In general, the higher bandwidth of a TDMA transmission compared to FDMA means that the former often becomes frequency selective, adding considerable complexity in signal processing at the receiver. Accurate time synchronization is also critical in ensuring that each mobile transmits in its assigned time slot.

Practical implementations of TDMA in cellular schemes are hybrid FDMA/TDMA schemes. The frequency band is first divided into smaller channels, which are then accessed on a TDMA basis. Several second generation cellular systems implement FDMA/TDMA/FDD. TDMA and its implementation in cellular systems are discussed in [10].

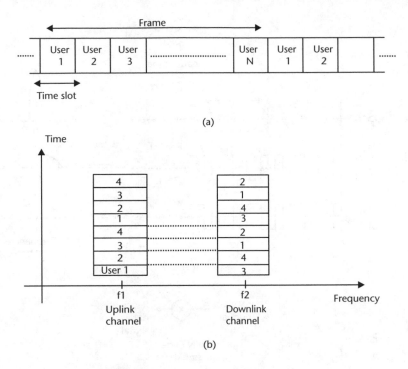

**Figure 4.21**    TDMA: (a) framing in TDMA, and (b) TDMA/FDD.

### 4.7.3    CDMA

CDMA is a direct-sequence spread-spectrum technique. Each user is allowed to use the entire bandwidth (like in TDMA) for the complete duration of the call (like in FDMA). Hence, all users' signals overlap each other in frequency and time. However, the use of spread-spectrum techniques allows each signal to be recovered in the presence of interference. CDMA is used in one second generation cellular system and is proposed as the major multiple-access technique for third generation systems.

In CDMA, multiple access is through a code that is unique to a particular user. The user generates data (e.g., the output of the speech coder) at a rate $R_b$. The code assigned to the user is a pseudorandom data stream at a rate $R_c >>> R_b$. The user's data stream is multiplied (or *chipped*) by this code in a process known as *spreading*, as illustrated in Figure 4.22(a). The frequency spreading of user data depends on the ratio $\dfrac{R_c}{R_b}$, known as the *processing gain*, $G$, of the system. At the receiver, *despread-ing* is carried out to recover the data. This involves the multiplication of the received signal with the same spreading code, as shown in Figure 4.22(b). Signals spread by other codes (other users' signals) are not despread and appear as wideband noise to the despread signal.

Figure 4.23(a) shows this spread-spectrum technique being used for multiple access. Users 1, ... $N$ in a cell use different spreading codes $c_1$, ... $c_N$. Even though the spectra of all users overlap each other at the base station, as shown in Figure 4.23(b), the $k^{th}$ user's signal can be recovered by despreading the composite received signal by $c_k$. After despreading, all other users' signals appear as noise to user $k$'s signal, as

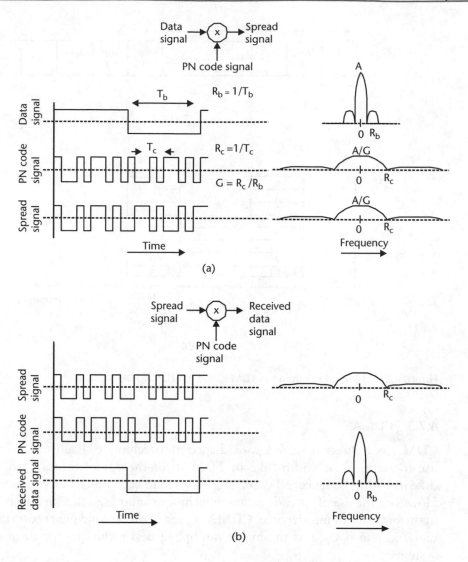

**Figure 4.22** Direct-sequence spread spectrum: (a) spreading process, and (b) dispreading.

shown in Figure 4.23(c). This noise is known as *multiple access interference* (MAI) or *intracell interference*.

The choice of spreading codes is critically important in CDMA. The code should have good autocorrection properties, where the autocorrelation of a sequence with a delayed version of itself is very low. The codes of different users must be orthogonal or nearly so, so that one user's code should not be able to despread another user's signal. If this happens, there will be interference between the two. A number of orthogonal or nearly orthogonal code families with the required features are used in cellular systems. The reader is referred to [11] for more information on this topic.

Based on this discussion, it is apparent that the capacity of a CDMA system is limited by the amount of interference generated by other users employing the same frequency band. It is seen from Figure 4.23 that the higher the $G$, the more the individual signals are spread, and hence the lower the MAI. This indicates a

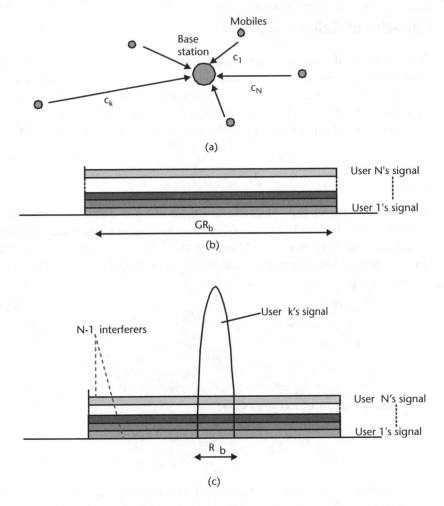

(a)

(b)

(c)

**Figure 4.23** CDMA: (a) multiple access, (b) received signal spectrum at the base station, and (c) received signal spectrum after dispreading one user's signal.

qualitative relationship between the number of simultaneous users in a cell and $G$. As $G$ increases, more subscribers can be accommodated in a cell. This relationship is discussed quantitatively in [12, 13].

Power control is needed in CDMA to ensure that signals from all mobiles reach the base station at approximately the same power level irrespective of their distance from the base station. Without power control, signals reaching the base station from nearby mobiles will drown signals coming from mobiles that are further away. This is known as the *near-far* problem.

In CDMA, MAI increases as the number of users increase. This causes a graceful degradation of system performance, rather than an abrupt degradation as in the case of TDMA or FDMA. Also with CDMA, the possibilities exist of trading off capacity for cell coverage and capacity for quality or data rate. The soft-handoff possible with CDMA is generally considered to be superior to the hard-handoff technique carried out with other multiple access schemes.

## 4.8  Capacity of Cellular Systems

The capacity of a cellular system may be defined as the amount of traffic carried per megahertz of bandwidth per square kilometer (i.e., the spatial traffic density per unit bandwidth) [7]. This is expressed in units of Erlangs/megahertz/square kilometers. It is a three-dimensional function of time, frequency, and space.

The overall capacity of a cellular system, $C_T$, can be expressed as:

$$C_r = \frac{n_{cell} a_c}{WS} \tag{4.7}$$

where $n_{cell}$ is the number of channels per cell, $a_c$ is the traffic carried per channel (in Erlang/channel), $W$ is the total system bandwidth (in megahertz), and $S$ is the area of a cell (in square kilometers). Assuming that there is a total of $n_T$ channels assigned for the system, the frequency reuse pattern is $N$, and the bandwidth allowed for each channel is $f_{ch}$,

$$n_{cell} = \frac{n_T}{N}$$
$$n_T = \frac{W}{f_{ch}} \tag{4.8}$$

Hence, $C_T$ can be expressed as,

$$C_T = \frac{a_c}{f_{ch} NS} = \frac{1}{NS} \cdot \frac{1}{f_{ch}} \cdot a_c \tag{4.9}$$

Equation (4.9) can be decomposed as a component with respect to space $C_s$, a component with respect to frequency (bandwidth) $C_f$, and a component with respect to time $C_t$, where

$$C_s = \frac{1}{NS} \tag{4.10a}$$

$$C_f = \frac{1}{f_{ch}} \tag{4.10b}$$

$$C_t = a_c \tag{4.10c}$$

The number of subscribers in an area of 1 km$^2$ can be expressed as

$$N_{sub} = \frac{a_{cell}}{a_{sub}} \cdot \frac{1}{S} \tag{4.11}$$

where $a_{sub}$ is the offered traffic per subscriber in the busy hour and $a_{cell}$ is the offered traffic per cell. If the grade of service (blocking probability) is $B$,

$$a_{cell} = \frac{n_{cell} \cdot a_c}{1 - B} \tag{4.12}$$

Hence,

$$N_{sub} = \frac{W}{a_{sub}(1-B)} \cdot C_T \qquad (4.13)$$

Because $W$, $a_{sub}$, and $B$ are already defined system parameters, maximizing system capacity becomes a problem of maximizing $C_T$. This reduces to the optimization of utilization of space, frequency, and time, as implied in (4.9) and (4.10). Table 4.4 summarizes the techniques for the improvement of capacity with respect to its three components. Subsequent parts of this chapter illustrate how cellular systems implement these techniques practically and the different technology options that are available.

## 4.9 First Generation Cellular Systems and Standards

First generation cellular systems use FM for speech transmission, FSK for signaling, and FDMA for multiple access. FDMA cellular systems design was pioneered during the 1970s by the AT&T Bell Labs. The initial implementation was AMPS [2]. Subsequently, several other similar systems were deployed, including the total access communication system (TACS), the Nordic Mobile Telephone (NMT) in Europe, the Nippon Telephone and Telegraph (NTT), and Japanese Total Access Communication System (JTACS) systems. Table 4.5 summarizes their main characteristics.

### 4.9.1 AMPS

AMPS is the most widely used analog cellular system. Figure 4.24(a) shows the frequency bands occupied by the AMPS system. The uplink and downlink are separated by 45 MHz, and the channel spacing is 30 kHz. The frequency allocation is divided into A and B bands, with the former allocated to a nonwireline operator and the latter allocated to a wireline operator, allowing two operators to coexist in the same area. The regions A', A", and B' are referred to as extended AMPS (EAMPS), as they were added later after initial commercial deployment [6]. In addition, 10-kbit/s FSK is used on the 21 control channels available. Mobile stations in AMPS are categorized according to their maximum allowed effective radiated power (ERP). Within this maximum, several power levels are defined. The operating power level for a mobile will be the lowest required for satisfactory $\frac{C}{N}$ or $\frac{C}{I}$. To meet

**Table 4.4** Techniques for Improvement of Capacity in Cellular Systems

| Component of Capacity | Techniques for Improvement |
|---|---|
| $C_s$ (utilization of space) | Smaller cells and smaller reuse pattern through sectorization, diversity reception, and forward error correction |
| $C_f$ (utilization of bandwidth) | Reduce the bandwidth requirement for each channel through low-bit-rate voice coding, high frequency stability, and agility |
| $C_t$ (utilization of time) | Efficient multiple access technologies for channel sharing |

**Table 4.5**  First Generation Cellular Standards

| Parameter | AMPS | TACS | NMT900 | NTT |
|---|---|---|---|---|
| Frequency (MHz) | | | | |
| Reverse | 824–849 | 890–905 | 890–905 | 860–885 |
| | | | | 843–846 |
| Forward | 869–894 | 935–960 | 935–960 | 915–940 |
| | | | | 898–901 |
| Duplex separation (MHz) | 45 | 45 | 45 | 55 |
| Channel spacing (kHz) | 30 | 25 | 25/12.5 | 25/12.5/6.25 |
| Number of full-duplex channels | 832 | 600 | 1,999 | 600–2,400 |
| Voice transmission | FM with ±8-kHz deviation | FM with ±9.5-kHz deviation | PM with ±5-kHz deviation | FM with ±5-kHz deviation |
| Data transmission | FSK with ±8-kHz deviation, 10 kbit/s | FSK with ±6.4-kHz deviation, 8 kbit/s | FFSK with ±3.5-kHz deviation, 1.2 kbit/s | FFSK with ±4.5-kHz deviation, 0.3 kbit/s |
| Mobile Tx. power (W) | 3 | 7 | 6 | 5 |
| Base station ERP (W/channel) max | 100 | 100 | 100 | 100 |

(*Source*: [14].)

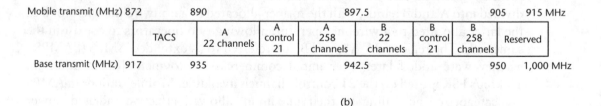

(a)

(b)

**Figure 4.24**  (a) AMPS frequency allocation, and (b) TACS frequency allocation.

the CCI objectives, the frequency reuse typically employed in the AMPS systems is either 12 or 7/21.

### 4.9.2  TACS

TACS is an adaptation of the AMPS system to suit European frequency allocations. Frequency assignment is shown in Figure 4.24(b). TACS provides 1,000 duplex channels. However, only the first 600 are allocated (as systems A and B), while the remaining 400 were held in reserve for GSM, which is now operational. Later frequency allocations are known as extended TACS (ETACS). JTACS is a variant of the TACS system.

### 4.9.3   Data Transmission over First Generation Cellular Systems

Analog cellular systems are circuit switched and designed for voice communications. Consequently, the most straightforward way to carry data over analog cellular systems is via modems at each end. The cellular environment, which is very harsh due to noise, interference, and fading, present very adverse conditions for data communications. Actual speeds achievable through cellular modems are 9,600 b/s or less. Performance is improved by operating in a stationary environment. An error-correcting protocol called MNP-10 is generally used with cellular modems. It allows a start from a low speed of operation and packet size and increases up to a level that is suitable for the current wireless link.

A method of sending packet-switched data, particularly over the AMPS system, called cellular digital packet data (CDPD) is available. This is implemented as an overlay to the existing AMPS system by making use of idle cellular channels, and by moving to another channel if the current channel is needed for voice service. Sometimes a small number of channels may be dedicated to the CDPD service.

Figure 4.25 shows the structure of a CDPD network. Each mobile end system (M-ES) communicates with a mobile database system (MDBS) colocated with the

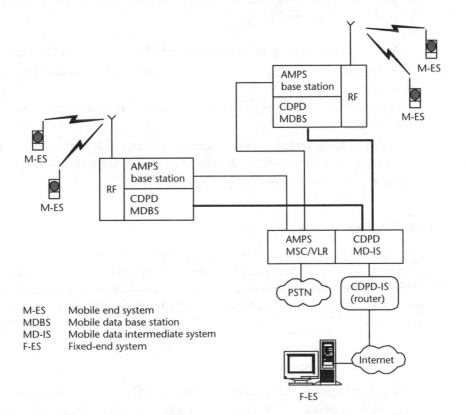

**Figure 4.25**   CDPD network architecture.

base station providing normal voice service. All the MDBSs in a service area are linked to a mobile data intermediate system (MD-IS), which provides a function analogous to that of the MSC. The MD-IS may be linked to other MD-ISs and to external networks. More information is available in [14, 15].

Each 30-kHz CDPD channel will support transmission rates up to 19.2 kbit/s using Gaussian minimum shift keying (GMSK) modulation. However, adverse channel conditions will limit the actual data throughput, due to the need for error detection and retransmission protocols. CDPD can provide an overall increase in user capacity compared to cellular modems due to the inherent efficiency of a connectionless packet data service.

## 4.10   Second Generation Digital Cellular Systems

Second generation mobile cellular systems use digital voice transmission in the air interface aided by advances in voice coding, as illustrated in Chapter 2. Advances in large-scale IC technology and digital signal processing hardware have helped complex, yet compact, energy-efficient systems to be built at low cost.

The primary motivation towards the development of second generation cellular systems was the improvement of capacity to handle this explosive demand for mobile services. Digital systems can support more users per base station per megahertz of spectrum through TDMA and CDMA, allowing cellular operators to provide service in high-density areas more economically. Digital architectures also offer additional advantages including:

- A more natural integration with the evolving digital wireline network;
- Flexibility for mixed voice-data communications and support for new services;
- A potential for further capacity increases as reduced-rate speech coders are introduced;
- Reduced RF transmit power (increasing battery life) through the use of error correction, interference cancellation, and equalization techniques;
- Encryption for security and privacy.

### 4.10.1   Digital Cellular Standards

There are four main standards for second generation cellular systems: GSM and its derivatives, digital AMPS (D-AMPS), personal digital cellular (PDC) and cdma One (IS-95). The first three are based on TDMA, and the fourth is based on CDMA. Table 4.6 summarizes the characteristics of the three TDMA-based second generation systems.

### 4.10.2   The GSM System

In 1982, Conference Europeene des Postes et Telecommunication (CEPT), the main governing body of the European posts and telecommunication authorities, created the Group Special Mobile committee for the task of specifying a common

**Table 4.6**  Characteristics of TDMA-Based Digital Cellular Systems

| Standard | Mobile Tx/Base Tx (MHz) | Access Method | Carrier Spacing (kHz) | Modulation | Channel Bit Rate kbit/s | Full-Rate Speech Coding kbit/s | Channels per Carrier (fr/hr) |
|---|---|---|---|---|---|---|---|
| IS-54 | 824–849/ 869–894 | FDMA/TDMA/ FDD | 30 | $\pi/4$- differential quadrature phase shift keying (DQPSK) | 48.6 | 7.95 (13 w/FEC) | 3 /6 |
| PDC | 810–915/ 940–960 | FDMA/TDMA/ FDD | 25 | $\pi/4$-DQPSK | 42.0 | 6.7 (11.2 w/FEC) | 3/6 |
| GSM | 890–915/ 935–960 | FDMA/TDMA/ FDD | 200 | GMSK | 270.8 | 13 (22.8 w/FEC) | 8/16 |

pan-European cellular system in the 900-MHz band. The system thus developed became so widespread all over the world that later it was renamed the Global System for Mobile. The basic GSM system operates in the 900-MHz band, with two frequency-upshifted derivatives, the Digital Cellular System-1800 (DCS-1800) and Personal Communication System-1900 (PCS-1900). The prime reason for the higher frequency band was the lack of capacity in the 900-MHz band.

The specific parameters chosen for the GSM system were shown in Table 4.6. The speech coding and modulation process in the GSM uplink is shown in Figure 4.26(a). A combined FDMA/TDMA access scheme has been selected with eight channels to be accommodated on a single RF carrier. Encoding and error protection are done on 20-ms segments (frames) of speech. A 13-kbit/s regular pulse excited linear predictive (RPE-LPE) codec is used for full-rate speech. The specifications allow the use of 16 channels per RF carrier with a future half-rate codec. The encoded speech is error protected by a block/convolution coding scheme, after which the channel bit rate is 22.8 kbit/s. The 20-ms frame, now containing 456 bits (260 bits of speech and 196 bits of error protection), is then interleaved to combat the effects of burst errors. Each channel transmits its data in bursts at a rate of 270.8 kbit/s, in order to accommodate eight voice channels/carrier. The 456 bits are partitioned into eight 57-bit subblocks and positioned in the time slot, as shown in Figure 4.26(b), distributing a speech segment across four frames. Finally, a 26-bit equalizer training segment is included in the center of the traffic burst. Details of channel equalization, which becomes necessary due to the high data rate, are found in [12]. The guard bits and tail bits are provided to assist the transceiver to power up and down smoothly.

The resulting 156.25-bit duration burst, called a *normal burst* and shown in Figure 4.26(b), constitutes the basic timeslot in GSM. The resulting 270.8-kbit/s data stream is GMSK modulated, passed through a duplexer, and transmitted in the assigned time slot. Figure 4.26(b) shows the multiplexing hierarchy in GSM, indicating how traffic and control channels are formed. A detailed description of these channels and their usage is found in [12, 14].

In addition to interleaving and channel equalization, GSM uses frequency hopping to combat fading. In the particular technique adopted, each TDMA burst is transmitted via a different RF channel. If the present TDMA burst encounters severe fading, the next will not, provided that the two channels are sufficiently apart such

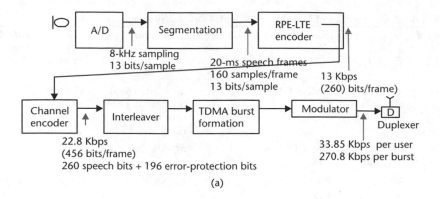

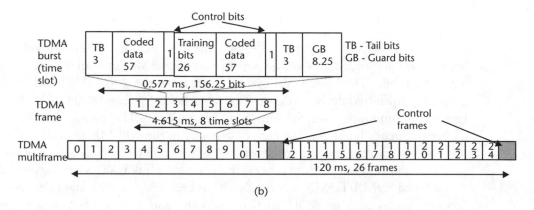

**Figure 4.26** (a) Speech encoding and modulation in GSM uplink (After: [14]), and (b) TDMA frame structure in GSM (*After:* [12]).

that their fading characteristics are independent. The frequency-hopping algorithm is described in [12].

### 4.10.3   IS-54/IS-136 (D-AMPS)

D-AMPS was developed to improve the capacity of the AMPS system. These standards require that the existing AMPS and the new digital system coexist using dual-mode terminals and a common frequency band, with the digital systems to gradually replace the analog systems. Thus, D-AMPS is designed to coexist with the analog AMPS system using dual-mode terminals. The IS-54B standard for D-AMPS was completed in 1992. In late 1994, its capabilities were significantly enhanced with the IS-136 standard.

D-AMPS utilizes a total of 50 MHz of spectrum allocated to the AMPS system, as shown in Figure 4.24(a). Each 30-kHz channel is time division multiplexed with frames of duration 40 ms having six time slots. For mobile-to-base communications in a full-rate D-AMPS system, three mobiles transmit to a single base station on a TDMA basis. When the half-rate codec is introduced, six mobiles will be able to share an RF channel.

The speech encoding technique used is 7.95-kbit/s VSELP (refer to Table 2.3). A channel coder and interleaver results in a channel data rate of 13 kbit/s, which includes error detection bits, control channel data, training sequence bits, and guard bits. The bit stream is then input to the modulator, which transmits in periodic bursts of 6.67 ms in the assigned time slot. Figure 4.27(a) illustrates this. Each time slot encompasses traffic and control channels in the format shown in Figure 4.27(b). The control channels available and their uses are described in [14]. The modulation scheme is $\frac{\pi}{4} - DQPSK$, which modulates the 48.6-kbit/s data bursts onto a 30-kHz RF channel.

The Japanese digital cellular system, PDC, has significant similarities to D-AMPS. Some of the key differences are the use of diversity reception in the mobile stations to simplify the equalization functions, simpler signaling protocols, and a lower data rate of 42 kbit/s in the three-user TDM RF channel.

### 4.10.4  IS-95

Though the primary objective of introducing digital cellular systems in North America was to increase the traffic capacity, D-AMPS was able to provide only a three-fold increase over the AMPS system. In an attempt to improve on this, the IS-95 standard was developed and adopted in 1995. This is designed to coexist with AMPS and D-AMPS and uses a hybrid FDMA/CDMA/FDD scheme. Frequency division is achieved by dividing the available spectrum into 1.23-MHz channels by combining 41 AMPS channels of 30 kHz each.

In the forward link, these 1.23-MHz channels implement 1 to 64 channels including a pilot channel, a synchronization channel, seven paging channels, and 55 traffic channels, as shown in Figure 4.28(a). The channels are assigned unique Walsh codes [11] for spreading, as shown in Figure 4.28(b). The pilot, sync, and paging channels are common control channels shared by all users in the cell. The traffic channels are dedicated to individual communications. The use of these channels is detailed in [14, 16]. The modulation process in the base station (forward link) is shown in Figure 4.28(b). Voice coding is by a QCELP (refer to Table 2.3) variable rate codec. The digitized voice is protected by convolution coding and interleaving. The data stream is then scrambled by the long pseudonoise (PN) sequence, and

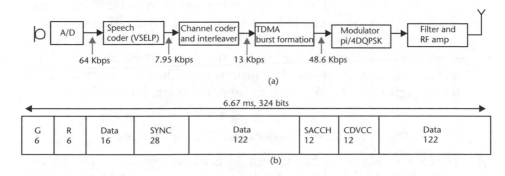

**Figure 4.27**   (a) Voice coding and modulation in the D-AMPS uplink, and (b) time slot format in D-AMPS. (*After:* [14].)

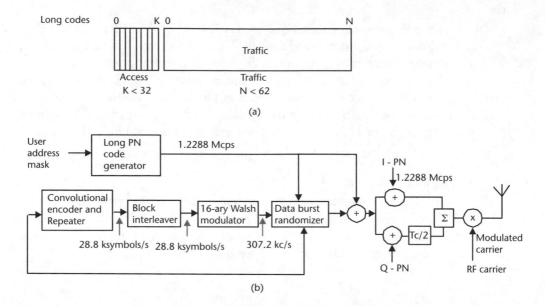

**Figure 4.28**  IS-95: (a) forward link channel structure, and (b) processing in the forward link. (*After:* [14].)

power control information is added. The resulting data is spread using an orthogonal Walsh code and randomized by the short I- and Q-PN sequences. The signal then BPSK modulates the RF carrier and is transmitted.

The long PN code provides privacy by scrambling the voice data. The short PN sequences distribute the energy of the transmit signal so that it appears noiselike. The Walsh code spreads the signal over the 1.23-MHz bandwidth with a processing gain of 64.

The reverse link allows up to 62 different traffic channels and 32 access channels as shown in Figure 4.29(a). Processing in the reverse link is shown in Figure 4.29(b). The digital data to be transmitted is encoded and interleaved for error protection. Then the data is Walsh modulated. This is a 64-ary modulation method that encodes six bits into each modulation state. The long code, which is masked by the user's ESN, is used to distinguish between CDMA users on the reverse link. The I and Q spreading is for a similar purpose as in the forward link.

Unlike the forward link, it is nearly impossible to establish truly orthogonal traffic channels on the reverse link because mobiles are scattered over the cell at different distances to the base station. Hence, synchronization is not possible and spreading codes become less effective. Mobile radios are constrained in power, size, and cost. Hence, the reverse link processing at the mobile is relatively simple. Further aspects of the IS-95 system are summarized in Table 4.7.

### 4.10.5  Data over Second Generation Cellular Systems

The basic GSM system, having been designed for voice, could provide only 9.6-kbit/s circuit-switched data, which was later enhanced to 14.4 kbit/s. A novel feature of GSM was the short message service (SMS). Generation 2.5 (2.5G or 2G+) is a

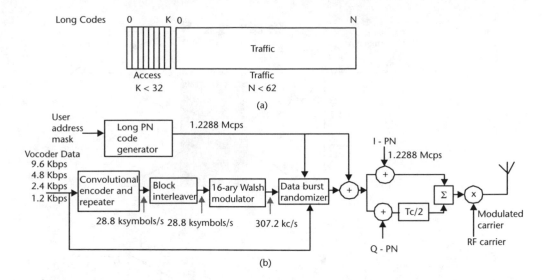

**Figure 4.29** (a) Reverse link channel structure in IS-95, and (b) processing in the reverse link. (*After:* [14].)

**Table 4.7** Key Features of IS-95

| Feature | Description |
|---------|-------------|
| Diversity | This is for protection against multipath fading. The types of diversity offered are through error correction coding and interleaving (time diversity), frequency diversity through spectrum spreading, and space diversity through RAKE receivers [17]. |
| Power control | This is to overcome the near-far problem (see Section 4.7.3) in the reverse link. Accurate power control techniques in CDMA systems enable transmissions from all mobiles in the cell to be received within 1 dB of each other. Power control accuracy determines the number of users sharing a cell [14, 16, 17]. |
| Soft handoff | This is a make-before-break handoff procedure. In the transition region between two cells, the call is delivered simultaneously via the two base stations, reducing the probability of dropped calls during handoff (see Figure 4.30). FDMA and TDMA systems employ break-before-make handoff. |
| Soft capacity | CDMA system capacity is interference limited, while that of FDMA and TDMA is limited by the number of noninterfering frequency bands or time slots. CDMA has the inherent ability to tolerate interference. The processing gain determines the interference each user causes to the others [13, 14]. The operator has the flexibility to admit more calls during peak periods at a somewhat lower QoS (lower processing gain, increased bit error rates). |

designation that broadly includes several enhancements for the second generation systems, supporting, in addition to voice, fax and enhanced data rates.

Generation 2.5 includes techniques called high-speed circuit-switched data (HSCSD), general packet radio service (GPRS), and enhanced data rates for GSM evolution (EDGE). Though most of these technologies were developed as enhancements to GSM, they are also applicable to other systems. These upgrades provide

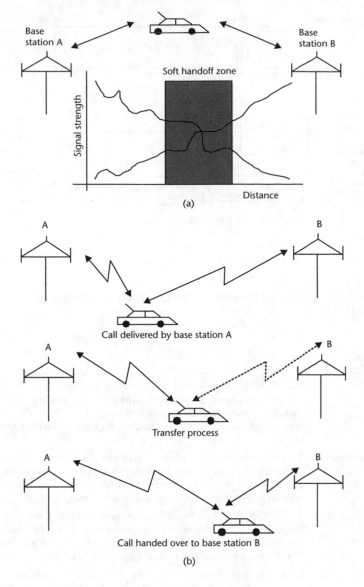

**Figure 4.30**  Soft handoff.

evolutionary paths for the second generation systems to the third generation, and they may sometimes provide almost the same capabilities as the third generation.

### 4.10.5.1  HSCSD

HSCSD is based on the simple concept of allocating multiple time slots per user and does not require any major changes to the network infrastructure. One time slot can carry either 9.6 kbit/s or 14.4 kbit/s, with the total rate being this multiplied by the number of time slots. Data rates up to 57.6 kbit/s are achievable. Though this is relatively inexpensive to implement, it suffers from the disadvantages associated with circuit switching. The use of several radio channels continuously will result in more congestion. Thus, only a few operators around the world have introduced this.

### 4.10.5.2 GPRS

A significant portion of today's mobile data communications needs are based on bursty traffic, such as e-mail and Web browsing, for which a circuit-switched connection is wasteful. With GPRS, packet-switched data up to 115 kbit/s can be achieved [18].

GPRS retains the GSM radio modulation, frequency bands, and frame structure. However, significant functional and operational changes are required to support packet-based data delivery. The conventional GSM structure is extended by a new class of logical network entities, the GPRS support nodes (GSNs), as shown in Figure 4.31(a). Their functions are summarized in Table 4.8.

Figure 4.31(b) illustrates the operation of GPRS. In the case of a mobile-originated transmission, the SGSN encapsulates the incoming packets and routes them to the appropriate GGSN, where they are forwarded to the correct public data network (PDN). Packets coming from a PDN are routed to the GGSN based on the

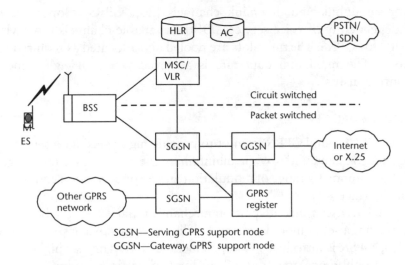

(a)

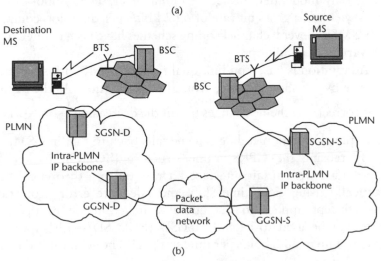

(b)

**Figure 4.31** (a) GPRS network architecture (After: [14]), and (b) packet transfer in GPRS (Source: [19]. ©1997 IEEE).

**Table 4.8**   GSNs

| Node | Function |
| --- | --- |
| Serving GSN (SGSN) | This is responsible for communication between the mobile station and the GPRS network, detection and registration of new GPRS mobiles, delivery of data, and tracking of mobiles. |
| Gateway GSN (GGSN) | This acts as the GPRS gateway to external networks and provides interworking functions, such as routing, translation of data formats, protocols, and addresses. |

destination address. The GPRS register (GR), which may be integrated with the GSM-HLR, maintains the GPRS subscriber data and routing information. The GPRS protocol architecture is described in detail in [18, 19].

The guiding principles around which GPRS was designed are the always on-line facility, billing based on volume, bandwidths in the range of those achievable with wireline modems, improved usage of radio resources, and separate allocation of uplink and downlink channels [20]. GPRS is operated in parallel but independently of voice services. Of the available channels in GSM, a subset is set aside for GPRS. The time slots are pooled and allocated as required for data connections. The maximum data rate is achieved when all eight time slots are used continuously.

### 4.10.5.3   EDGE

Both HSCSD and GPRS attempt to provide high-speed data services by implementing add-on capabilities while maintaining the air interface unchanged. By contrast, EDGE supports a range of modulation and error correction coding techniques in the air interface.

EDGE uses a link adaptation mechanism that selects the best possible combination from a set of modulation and coding schemes to suit the time-varying channel. Also, 8-PSK is introduced to supplement GMSK. This has higher spectral efficiency and is only moderately difficult to implement. The symbol rate remains at 270.8 kbit/s, yielding a gross bit rate of 69.2 kbit/s per time slot, compared to 22.8 kbit/s with GMSK. Several channel coding schemes have been defined to ensure robustness in a variety of conditions.

Adaptation is based on link quality, and 8-PSK with simple error correction is effective over a short distance from the base station. As the distance increases, the error correction scheme changes to suit the decreasing $\frac{C}{I}$ or $\frac{C}{N}$. Towards the edge of a cell, the modulation/coding scheme falls back to that in GSM.

By reusing the GPRS frame structure (EGPRS), packet data services can be provided with an air interface bit rate up to a theoretical peak of 554 kbit/s, practically using eight time slots and adequate error protection to 384 kbit/s. This concept applies to the circuit-switched mode of operation as well and hence can be used to extend HSCSD (EHSCSD) with support for air interface data rates up to 28.8 kbit/s per time slot [21]. The achievable data rates as a function of $\frac{C}{I}$ or $\frac{C}{N}$ with link adaptation and the resulting improvements are shown in Figure 4.32.

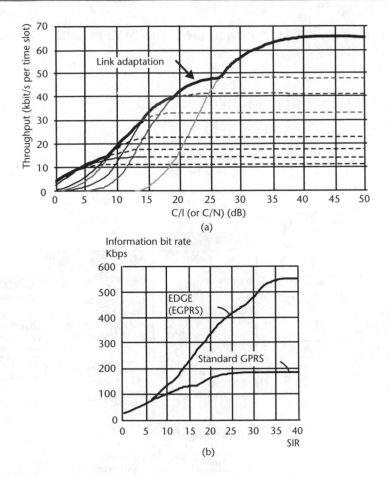

**Figure 4.32** (a) Data rates achievable with EDGE due to link adaptation (*Source:* [21]), and (b) extension of GPRS with EGPRS (*Source:* [22]; Courtesy of Telefonkatiebolaget LM Ericsson).

EDGE can also be used with other cellular systems as exemplified by its implementation on IS-136. Hence, this can be regarded as a generic air interface facilitating the evolution of cellular systems toward third generation capabilities [21].

### 4.10.5.4   Packet Data Service in the PDC System (PDC-P, i-mode)

A proprietary packet data service similar to GPRS and supporting TCP/IP has been developed for the Japanese personal digital cellular system. This service, known as *i-mode*, has been very successful in subscriber acceptance. It utilizes a new physical channel structure and link control procedures to support packet transport at high data rates. By using all three time slots of the PDC carrier, PDC-P can deliver a maximum transmission rate of 28.8 kbit/s. PDC-P utilizes a newly defined user packet channel, which is shared among multiple users for transfer of packet data.

### 4.10.5.5   Enhancements to IS-95

As with GSM and IS-136, cdmaOne is also enhanced to provide improved services, capacity, quality, and data rates. The original data rate of 14.4 kbit/s has been enhanced to provide packet data rates up to 64 kbit/s in IS-95 B, and up to 144 kbit/s in IS-95C.

## 4.11   Third Generation Mobile Systems

Figure 4.33 depicts the envisaged objectives of third generation wireless systems compared to the previous generations.

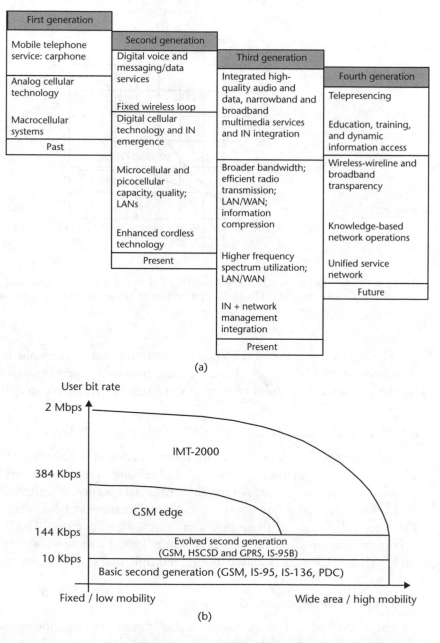

(a)

(b)

**Figure 4.33**   Third generation mobile systems: (a) applications, and (b) capabilities.

### 4.11.1   Requirements of the Third Generation

The major requirements of the third generation are identified as:

- Greater capacity and improved spectrum efficiency;
- Seamless global mobility;
- Integration of wireline and wireless networks;
- Support for both packet-switched and circuit-switched services;
- High data rates (up to at least 144 kbit/s, preferably 384 kbit/s) in all radio environments and up to 2 Mbit/s in low-mobility and indoor environments;
- Symmetrical and asymmetrical data transmission;
- Flexibility to operate in any environment such as indoor, outdoor, and mobile;
- Smooth evolution of current national/regional systems;
- Economics of scale and an open global standard that meets the needs of the mass market.

### 4.11.2   Standardization of Third Generation Systems

The ITU has been working since 1986 toward developing an international standard for wireless access, IMT-2000. This is intended to form the basis for third generation wireless systems. Regional/national fora such as ETSI in Europe, TIA and TIA North America (particularly TIA's committees TR-45 and TR-46), the Association of Radio Industries and Businesses (ARIB) in Japan, and Telecommunications Technology Association (TTA), in S. Korea complement and provide input and direction to IMT-2000 activities.

#### 4.11.2.1   Radio Spectrum Allocation

To meet the objectives of the third generation, availability of a common worldwide frequency spectrum was essential. Based on detailed studies within the ITU, the World Administrative Radio Conference (WARC) identified 230 MHz of global frequency spectrum for IMT-2000 at its 1992 meeting (WARC-92). Based on revised forecasts, additional spectrum was allocated at WARC-2000, adding a further 519 MHz of spectrum. The IMT-2000 spectrum allocation is shown in Figure 4.34.

Some notable features of the spectrum allocation are that both FDD (paired bands) and TDD (unpaired bands) operation are supported and that the IMT-2000 spectrum overlaps existing second generation wireless services. The latter feature allows operators to migrate existing second generation services, adding support for third generation services. Only one frequency band is needed to carry both uplink and downlink traffic in TDD. This is useful, as globally available unpaired spectrum is easier to find than paired spectrum. FDD and TDD are compared in Figure 4.35.

#### 4.11.2.2   Radio Transmission Technologies

In 1988, a total of 15 proposals for third generation radio transmission technologies were submitted to the ITU by the regional/national standardization bodies named

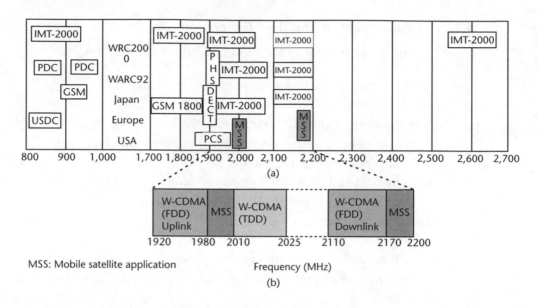

(a)

MSS: Mobile satellite application          Frequency (MHz)

(b)

**Figure 4.34** (a) Spectrum allocation for IMT-2000 and other wireless service (*Source:* [17]; reproduced with permission of Artech House), and (b) detailed IMT-2000 spectrum allocation (©2001 Artech House, Inc.).

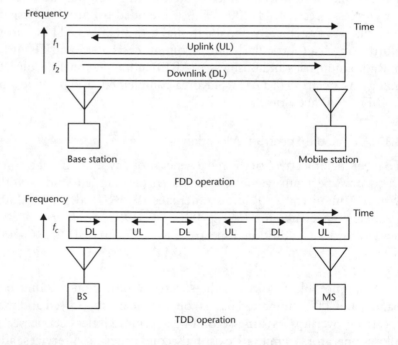

**Figure 4.35** TDD versus FDD.

previously. These are described in [12, 14, 23]. Among the proposals, several candidates had similar characteristics. ETSI's Universal Mobile Telecommunications System (UMTS) terrestrial radio access (UTRA) interface, proposals from ARIB, and the TIA were based on variants of wideband CDMA (W-CDMA). One TDMA specification was proposed by the TIA.

In an effort to harmonize the proposals, the ITU, other standards bodies, and the industry have arrived at a small set of third generation standards based on common features. ETSI, ARIB, T1P1, and TTA have joined forces in the third generation partnership project known as 3GPP, resulting in one W-CDMA based system [24]. Another organization that promotes the cdma2000 system (multicarrier CDMA) is 3GPP2.

Resulting from these efforts, a single third generation CDMA standard and one TDMA standard has evolved. The CDMA standard has three operating modes: direct sequence (DS-CDMA), a multicarrier mode based on cdma2000 (MC-CDMA), and a TDD mode (TD-SCDMA). The TDMA standard selected is UWC-136. Even though the original objective of the third generation standardization effort was to develop one worldwide standard, today we have a what is known as a "harmonized family of four" [22], as summarized in Table 4.9. Figure 4.36 shows the spectra of DS-WCDMA and multicarrier CDMA.

New spectrum allocation at WARC-2000, as illustrated in Figure 4.34, was in keeping with the definition of this family of standards. The requirements in spectrum are identified in [22]. The evolutionary paths provided for second generation cellular system and their functionalities are illustrated in Figure 4.37. A detailed description of third generation standards is beyond the scope of this chapter. References [12, 14, 17, 20, 22–28] are recommended.

**Table 4.9**   Harmonized Family of Four Third Generation Standards

| *W-CDMA* *DS-CDMA* | *W-CDMA* *Multicarrier* | *W-CDMA* *TDD* | *TDMA* *EDGE/UWC-136* |
|---|---|---|---|
| WCDMA as per 3GPP | WCDMA as per 3GPP2 | As per 3GPP | As per ETSI/UWC |
| New spectrum | IS-95 spectrum overlay | Unpaired spectrum | Existing spectrum, 200-kHz TDMA |
| FDD | FDD | TDD | High-level modulation with link adaptation |
| Chip rate 3.84 Mc/s | Chip rate 3.6864 Mc/s | Chip rate 3.84 Mc/s | |
| Asynchronous (synchronous operation supported) | Synchronous | | |

(*Source:* [22].)

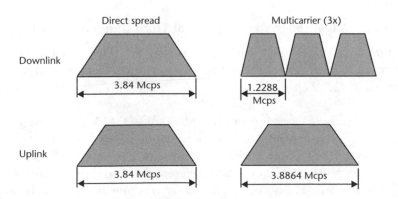

**Figure 4.36**   Comparison of WCDMA and multicarrier CDMA.

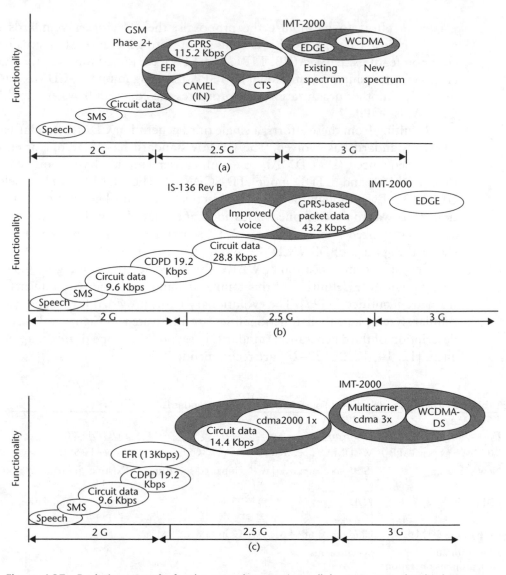

**Figure 4.37** Evolutionary paths for the second generation cellular systems to the third generation: (a) GSM, (b) IS-136, and (c) IS-95.

### 4.11.3  Advanced System Aspects of Third Generation Systems

Compared to second generation systems, the third generation utilizes many new technologies to deliver the required capabilities. Detailed technical descriptions are found in [26, 27].

- *Coherent detection in the uplink.* W-CDMA uses coherent detection in the uplink, a feature that has not previously been implemented in cellular CDMA systems. This improves the performance by up to 3 dB compared to noncoherent detection. To facilitate coherent detection, time-multiplexed pilot symbols are used. The actual performance improvement depends on the pilot to data signal power ratio.

- *Multiuser detection in the uplink.* MAI, the interference between users in an individual cell, is a factor that limits the capacity and performance of CDMA systems. The conventional detector follows a single-user detection strategy, in which each user is detected separately at the base station without regard for others. However, a better detection strategy is one of multiuser detection, where information about multiple users is used jointly to better detect each individual user, by interference cancellation. This technique has the potential to provide significant additional benefits for CDMA systems.

- *Support for adaptive antenna arrays in the downlink.* Diversity concepts are used to reduce the impact of fading. Multiple antenna concepts are a further extension to this, gaining from uncorrelated multipath transmission channels between the different antenna elements. Adaptive antennas improve link quality further by reducing the cochannel interference from different directions and in the more advanced space division multiple access (SDMA) concept, by reusing the same frequency channels simultaneously for different users in different directions [28]. A pilot channel facilitates the deployment of adaptive antennas for beamforming, with a pilot signal being used for channel estimation.

- *Hierarchical cell structures (HCSs).* Hierarchical cell structures consisting of overlaid microcells and picocells on larger macrocells have been proposed to achieve high capacity. With this technique, a cellular system can provide very high system capacity through the microcell layer, at the same time offering wide coverage and high mobility through the macrocell layer. Hotspot microcells are also implemented in high-traffic areas. Cells belonging to different layers will be in different frequency bands, requiring smooth interfrequency handovers. These concepts are illustrated in Figure 4.38.

- *Transmit diversity in the downlink.* This is another technique that improves downlink performance. For DS-CDMA, this can be performed by splitting the data stream and spreading the two streams. For multicarrier CDMA, the different carriers can be mapped into different antennas.

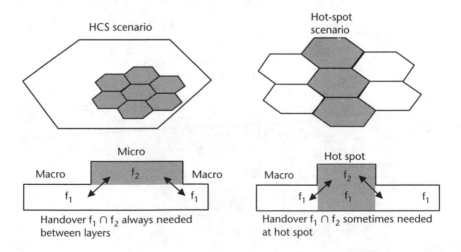

**Figure 4.38**   Hierarchical cell structures and interfrequency handover. (*Source:* [27].©1998 IEEE.)

## 4.12  Mobile Device Hardware, Architecture, and Implementation

The first mobile terminals were single mode and single band. Second generation terminals are implemented for several bands (e.g., combinations of 900; 1,800; and 1,900 MHz). The implementation of dual-mode terminals is proceeding to support the evolution of second generation systems. In the third generation, multimedia, multimode terminals will have to be supported to achieve worldwide roaming and true personal communications. Due to the multiplicity of available standards in all types of fixed and wireless systems, adaptive terminals will have to be introduced.

A considerable amount of research and standardization activities are ongoing, where a key issue addressed is the seamless interworking of different access systems on a common platform. Other challenges are improved display techniques, user interface performance, and battery technology, whose discussion is beyond the scope of this text.

### 4.12.1  Second Generation Mobile Terminals

A typical second generation wireless transceiver is shown in Figure 4.39. Due to the nature of radio communications, physical layer design considerations fall into one of two categories, RF/IF susbsystems and baseband subsystems. There exist sufficient differences between analog RF/IF and baseband processing that the integration of the two in devices is not easily achievable. While RF/IF design is beyond the scope of this text, an overview is found in [29]. However, the RF/IF structure in mobile transceivers is the familiar superheterodyne structure depicted in [30].

In designing the baseband subsystem, engineers can mix and match microcontrollers, DSPs, and ASICs. The choice of which to use is driven by such issues as development time, cost, size, and functionality. As processors optimized for high-speed arithmetic, DSPs are technically capable of doing their function on their own. However, in terms of today's device technology, it is more economical for them to share their work with faster, more flexible ASICs that use hard-coded logic to perform arithmetic and   field programmable gate arrays (FPGAs) whose programmable

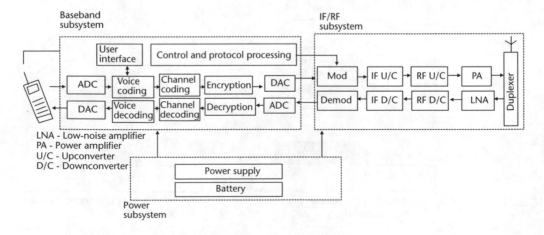

**Figure 4.39**  A typical second generation wireless transceiver.

interconnect and logic functions can be redefined after manufacture [31]. Critical issues in the design of firmware architectures for DSPs, ASICs, FPGAs, and micro-controllers are discussed in [30–32].

Figure 4.40 illustrates the partitioning of baseband functions between a DSP and a microcontroller. Microcontrollers are more suitable than DSPs for branching functions, which are dominant in system-control software [30]. The DSP is thus controlled by the microcontroller and runs physical layer functions in parallel with the microcontroller running system control functions. The physical layer algorithms that the DSP needs to support in a typical wireless transceiver are outlined in Figure 4.40 and further discussed in [30].

The AD20msp430 SoftFone® chipset from Analog Devices is a good example of a practical implementation related to this discussion. This consists of two chips: a digital basbeband controller and an analog/mixed-signal interface. The digital base-band controllers in the AD20msp430 family include the AD6522, AD6525, and the AD6526. The analog/mixed-signal interface chips include the AD6521 and AD6535.

Figure 4.41 shows a functional block diagram of the AD6526 digital baseband processor. This is designed to be used with GSM900, DCS1800, and PCS1900 handsets. The device is divided in to three main subsystems: the control processor subsystem, the DSP subsystem, and the peripheral subsystem. This integrates full-rate and half-rate speech codecs, as well as a full range of data services, including circuit-switched 14.4 kbit/s, GPRS, and HSCSD.

### 4.12.2  Dual-Mode Terminals

A typical second generation dual-mode cell phone is shown in Figure 4.42. It would likely work on one first generation analog system and one second generation digital system. For this, the phone has two pairs of transmitters and receivers. In the analog mode, outgoing signals emerge from the (analog) signal processor and are fed to a chain of IF/RF functional blocks. In a reverse process, the analog receiver selects the

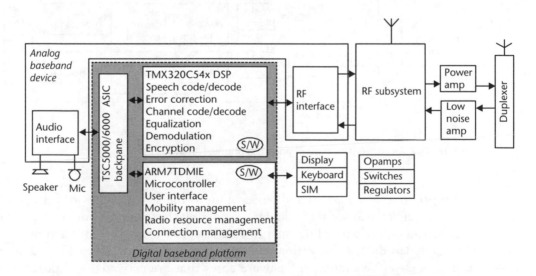

**Figure 4.40**  Partitioning of baseband subsystem functions in a mobile terminal.

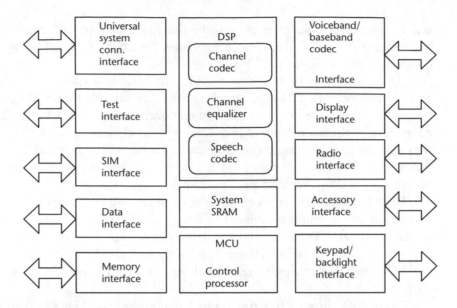

**Figure 4.41** Functional block diagram of the AD6526 GSM/GPRS digital baseband processor. (*Source:* [33]. Courtesy of Analog Devices, Inc.)

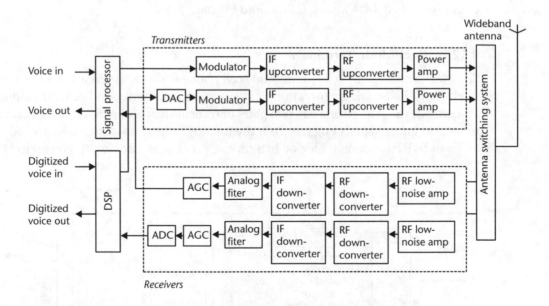

**Figure 4.42** A conventional dual-mode cell phone. (*After:* [31].)

appropriate received signal and passes it on to the signal processor for demodulation. The digital transceiver is similar, except that the operations on the transmit and receive signals are carried out in a DSP. These may include such functions as voice encoding/decoding, equalization, filtering, and encryption. In essence, a dual-mode cell phone employs multiple firmware cores that are activated and deactivated as needed.

### 4.12.3  Software-Defined Radios

Figure 4.43 shows different ways in which multimode mobile terminals can be implemented. Figure 4.43(a) depicts the concept in Figure 4.42 of having several fixed modes in parallel. This concept is inflexible for evolution, as a variety of different modes are introduced. In the concept shown in Figure 4.43(b), parameters of different access systems are stored in memory and used to implement different modes as appropriate in a single signal processing unit. In Figure 4.43(c), a programmable signal processing unit is adapted to the actual access system. The signal processor first downloads the parameter set of the particular system and then runs the appropriate set of algorithms. This is the most flexible of the three design concepts. The two approaches in Figure 4.43(b) and (c) are called software-defined radios (SDR).

The first step in transforming a conventional mobile terminal into an SDR is to make as much of the circuitry digital as possible. On the transmit side, this means digitizing the input voice as close to the microphone as possible, so that all subsequent signal processing functions (e.g., encoding, filtering, and modulation) can be done digitally and hence can be made programmable. The received signal must be digitized similarly as close to the antenna as possible. An SDR architecture that is digital up to the IF stage is shown in Figure 4.44(a).

The ultimate approach to SDR is the direct conversion architecture shown in Figure 4.44(b). Here, the IF stage is completely implemented in software, with direct conversion between the baseband and the RF stages.

The components that form the backbone of SDR systems and set their performance limits are analog-to-digital converters (ADCs), DSPs, filters, and RF amplifiers. The ADC is the most critical element, as its speed determines how close to the RF subsystem the A/D/A conversion can be done. As more powerful ADCs and digital-to-analog converter (DACs) become available, programmability at higher and higher frequencies can be achieved.

Filters and mixers are also hard to implement digitally. The IF section of a GSM phone, for example, runs at several hundred megahertz, which is no problem for any modern silicon IC process. However, to implement this digitally in the same processor with the rest of the phone's functions requires the ability to execute around 100 billion instructions per second. To put the number in perspective, the

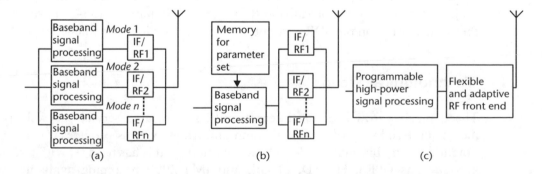

**Figure 4.43**  Different implementation of multimode transceivers: (a) parallel modes, (b) software-defined signal processing, and (c) fully adaptive SDR.

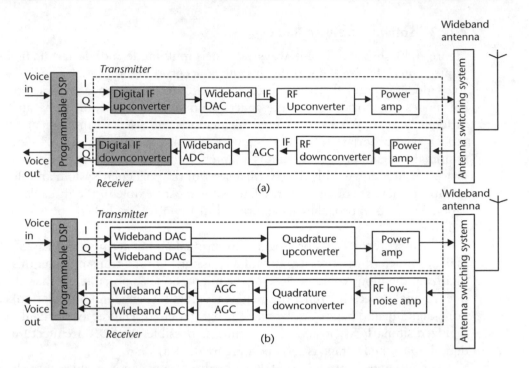

**Figure 4.44** SDR architectures to handle different standards with a single transmitter and receiver: (a) digital architecture up to the IF stage, and (b) direct conversion architecture. (*Source:* [31]. ©2001 IEEE.)

single-purpose silicon chips in present-day cell phones execute approximately 10 to 100 million instructions per second [31].

Upgrading application features by software download over the Internet is well established. This concept can be extended to wireless transceivers by having subscribers download complete radio standards over the air interface as appropriate. To facilitate this, an open application programming interface (API) needs to be defined, where different vendors can develop compatible software and hardware interfaces. Object-oriented technologies such as common object request broker architecture (CORBA) and Java support this facility.

Figure 4.45 shows a conceptual SDR terminal. The key functional units are the application program, radio function library, software specification language, digital radio processor, and the broadband RF stage [34]. References [35–38] provide further information on the SDR concept, enabling technologies, and challenges.

## 4.13 Technological Trends

The explosive growth of the Internet and its associated services as well as the increasing trend towards personal communications systems provides a big market potential for mobile multimedia. The foundation for this has been set with technologies such as GPRS, HSCSD, EDGE, and IMT-2000 becoming available. With increasing data rates and improved displays in mobile devices, more advanced content will be available for mobile platforms. The growing data and Internet traffic

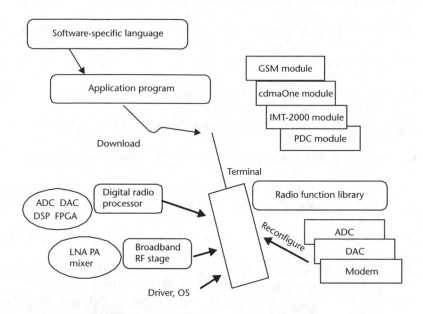

**Figure 4.45**   A conceptual SDR terminal. (*Source:* [34]. ©1999 IEEE.)

results in packet-oriented, asymmetric traffic dominating over circuit-switched traffic in mobile systems. Mobile terminals will become the major man-machine interface in the future instead of personal computers [28].

## References

[1]   Calhoun, G., *Wireless Access and the Local Telephone Network*, Norwood, MA: Artech House, 1992.

[2]   Young, W. R., "Advanced Mobile Phone Service: Introduction, Background and Objectives," *Bell System Technical Journal*, Vol. 58, No. 1, January 1997, pp. 1–14.

[3]   Blake, R., *Wireless Communication Technology*, Clifton Park, NY: Delmar Thomson Learning, 2001.

[4]   MacDonald, H. V., "Advanced Mobile Phone Service: The Cellular Concept," *Bell System Technical Journal*, Vol. 58, No. 1, January 1979, pp. 15–41.

[5]   Lee, W. C. Y., *Mobile Cellular Telephony*, New York: McGraw Hill, 1995.

[6]   Balston, D. M., and R. C. V. Macario, *Cellular Radio Systems*, Artech House, 1993.

[7]   Feher, K., *Advanced Digital Communications*, Englewood Cliffs, NJ: Prentice Hall, 1987.

[8]   Sklar, B., "Raleigh Fading Channels in Mobile Digital Communication Systems Part I: Characterization," *IEEE Communications Magazine*, September, 1997, pp. 136–46.

[9]   Sklar, B., "Raleigh Fading Channels in Mobile Digital Communication Systems Part II: Mitigation," *IEEE Communications Magazine*, September, 1997, pp. 148–155.

[10]  Falconer, D. D., and F. Adachi, "Time Division Multiple Access Methods for Wireless Personal Communications," *IEEE Communications Magazine*, January 1995, pp. 50–57.

[11]  Dinan, E. H., and B. Jabbari, "Spreading Codes for Direct Sequence CDMA and Wideband CDMA Cellular Networks," *IEEE Communications Magazine*, September 1998, pp. 48–54.

[12]  Steele, R., and L. Hanzo, *Mobile Radio Communications*, Second Edition, New York: John Wiley and Sons Ltd., 1999.

[13] Kohno, R., R. Meidan, and L. B. Milstein, "Spread Spectrum Access Methods for Wireless Communications," *IEEE Communications Magazine*, January 1995, pp. 58–67.

[14] Pandya, R., *Mobile and Personal Communications Systems and Services*, New Jersey: IEEE Press, 2000.

[15] Pahlavan, K., and A. H. Levesque, "Wireless Data Communications," *Proceedings of the IEEE*, September 1994, pp. 1398–1430.

[16] Groe, J. B., and L. E. Larson, *CDMA Mobile Radio Design*, Norwood, MA: Artech House, 2001.

[17] Korhonen, J., *Introduction to 3G Mobile Communications*, Norwood, MA: Artech House, 2001.

[18] Brasche, G., and B. Walke, "Concepts, Services, and Protocols of the New GSM Phase 2+ General Packet Radio System," *IEEE Communications Magazine*, August 1997, pp. 94–104.

[19] Cai, J., and D. J. Goodman, "General Packet Radio Service in GSM," IEEE Communications Magazine, October 1997, pp. 122–131.

[20] De Vriendtd, J., et al., "Mobile Network Evolution: A Revolution on the Move," *IEEE Communications Magazine*, April 2002, pp.104–110.

[21] Furuskar, A., J. Nasland, and H. Olofsson, "Edge: Enhanced Data Rates for GSM and TDMA/136 Evolution," *Ericsson Review*, No. 1, 1999, pp. 28–37.

[22] Nilsson, M., "Third Generation Radio Access Standards," *Ericsson Review*, No. 3, 1999, pp. 110–121.

[23] Zeng, M., A. Annamalai, and V. K. Bhargava, "Recent Advances in Cellular Wireless Communications," *IEEE Communications Magazine*, September 1999, pp. 128–138.

[24] Chaudhury, P., W. Mohr, and O. Seizo, "The 3GPP Proposal for IM-2000," *IEEE Communications Magazine*, December 1999, pp. 72–81.

[25] Nilson, T., "Toward Third Generation Mobile Multimedia Communication," *Ericsson Review*, No. 3, 1999, pp. 122–131.

[26] Ojanpera, T., and R. Prasad, "An Overview of Air Interface Multiple Access of IMT-2000/UMTS," *IEEE Communications Magazine*, September 1998, pp. 82–95.

[27] Dahlman, E., et al., "UMTS/IMT-2000 Based on Wideband CDMA," *IEEE Communications Magazine*, September 1998, pp. 70–80.

[28] Mohr. W., and W. Konhauser, "Access Network Evolution Beyond Third Generation Mobile Communications," *IEEE Communications Magazine*, December 2000, pp. 122–133.

[29] Razavi, B., "Challenges in Portable Transceiver Design," *IEEE Circuits and Devices*, September 1996, pp. 12-25.

[30] Kostic, Z., and S. Seetharaman, "Digital Signal Processors in Cellular Radio Communications," *IEEE Communications Magazine*, December 1997, pp. 22–35.

[31] Bing, B., and N. Jayant, "A Cell Phone for All Standards," *IEEE Spectrum*, May 2001.

[32] Gatherer, A., et al., "DSP-Based Architectures for Mobile Communications: Past, Present and Future," *IEEE Communications Magazine*, January 2000, pp. 84–89.

[33] Analog Devices, Inc., AD6526 GSM/GPRS "Digital Baseband Processor," Preliminary Data Sheet, 2003.

[34] Tsurumi, T., and Y. Suzuki, "Broadband RF Stage Architecture for Software-Defined Radio in Handheld Terminal Applications," *IEEE Communications Magazine*, February 1999, pp. 90–95.

[35] Efstathiou, D., J. Fridman, and Z. Zvonar, "Recent Developments in Enabling Technologies for Software-Defined Radio," *IEEE Communications Magazine*, August 1999, pp. 112–117.

[36] Tuttlebee, W. H. W., "Software Radio Technology: A European Perspective," *IEEE Communications Magazine*, February 1999, 118–123.

[37]   Mitola III, J., "Technical Challenges in the Globalization of Software Radio," *IEEE Communications Magazine*, February 1999, pp. 84–89.

[38]   Dick, C., and F. J. Harris, "Configurable Logic For Digital Communications: Some Signal Processing Perspective," *IEEE Communications Magazine*, August 1999, pp. 107–111.

## CHAPTER 5

# Fixed Wireless Access

## 5.1 Introduction

WLL or FWA are generic terms for an access system that uses a wireless link to connect subscribers and can be both a substitute and a complement to copper wire in the local loop. Initial stages of WLL system development (during early 1990s) were for voice services and progressed onto data and broadband services towards the latter part of the 1990s.

The fixed radio access network architecture is generally similar to cellular systems, with a radio base station providing service to an area around it. Subscribers receive service through a radio unit linked to the PSTN via the local base station. As depicted in Figure 5.1(a), a basic WLL system's customer end equipment, the fixed radio access unit (FAU), consists of an antenna, a transceiver unit, and a processor subsystem. The latter converts signals between the form suitable for radio transmission and that for the customer premises equipment. The signal between the transceiver unit and the telephone handset is an analog signal carried via a copper pair. The unit is powered by a battery and a charger system. The system configuration usually results in creating a 48-V dc rail-based analog connection to the customer premises equipment.

Fixed access units communicate with the nearest base station. Sometimes, in order to place base stations in a more flexible manner to provide good coverage, the base station functions are separated into a radio node controller (RNC) and several transceiver units (TRXs). At the nearest local exchange, a transceiver pair between the exchange and the base station or RNC forms a duplex connection via possible options, such as a T1/E1/SDH link or a fiber-optic link. This completes the last mile in fixed wireless access systems. The WLL network manager carries out maintenance and subscriber management functions. The complete system is illustrated simply in Figure 5.1(b).

WLL systems require minimal planning and can be deployed quickly. Construction costs are minimal, and there is no need for rights of way for buried cable. WLL systems can help eliminate the backlog of orders for telephone service, which is estimated at over 50 million lines worldwide [1]. Being principally a fixed service, the location of the residential subscribers is known. Hence, a WLL system provides user coverage at a lesser cost and is likely to support higher transmission rates, support broader bandwidth services, and promise a wider range of future services, including both entertainment and packetized data services, than cellular systems [2]. Key factors preventing the rapid growth of WLL systems include the lack of common worldwide frequency allocations and technical standards.

This chapter describes the types of fixed wireless access technologies available, their capabilities, standardization efforts, and evolving trends, covering

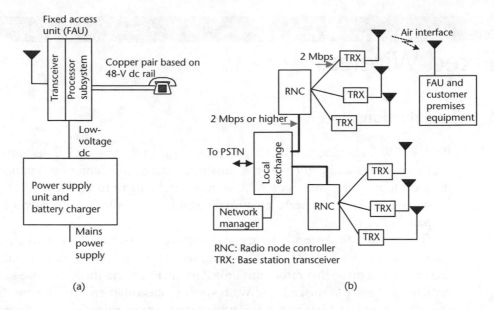

**Figure 5.1** (a) Typical WLL system components at the customer premises, and (b) overview of a WLL system.

voice-centric as well as broadband services. While most technical aspects of WLL are the same as those for cellular systems described in Chapter 4, a few essential new topics are discussed, with an emphasis on systems and standards.

## 5.2 Access Technologies

### 5.2.1 Competing Technologies for Access

With the evolution of basic voice telephony towards broadband services, the following types of access techniques have developed:

- DSL;
- Third generation mobile;
- Fixed/mobile integration;
- Microwave video distribution;
- Fiber-optic distribution;
- Internet telephony;
- Digital broadcasting;
- Satellite systems;
- Power-line communications.

The challenge is to work out the least expensive and most efficient solution for a given set of boundary conditions.

Figure 5.2 shows the main contending access technologies. A point of interest in this figure is that in the evolutionary process, traditional technologies are moving

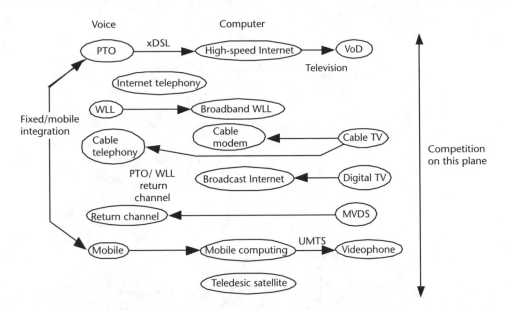

**Figure 5.2** Competing technologies for access. (*Source:* [3]. Reproduced with permission of the IEE.)

both from right to left as well as left to right. The concept of convergence predicts that a range of different media, including voice, data, and video, will be transmitted along the same path, possibly in an integrated form. Details are found in [3–5].

### 5.2.2 Alternatives for Wireless Access

While DSL systems, for example, rest heavily on already existing infrastructure, radio transmission allows rapid installation and is particularly attractive for new carriers entering the market and for thinly populated areas. Expected features and capabilities of such systems are described in [6]. This evolution along several wireless technologies is illustrated in Figure 5.3. Satellite-based wireless access is intended to serve sparsely populated remote regions or for specific services, such as banking or retail networks. Wireless LANs and personal area networks (PANs) are indoor short-range wireless access schemes. Along all these axes, the evolution is towards a unified body of broadband access systems generally envisaged as the *fourth generation* wireless systems.

When examining the position of WLL in this multidimensional competing scenario, it is seen to have broadly, two different roles:

- *As a technology for basic voice provision.* This application scenario may be the only practical means of connections for reducing waiting lists or as an alternative or supplement to basic voice provision (POTS).
- *As an advanced access method providing broadband access.*

While the first item is mostly relevant to developing countries and high-capacity *hot spot* applications, the second is becoming increasingly important due to emerging multimedia communications needs as outlined in other chapters of this text.

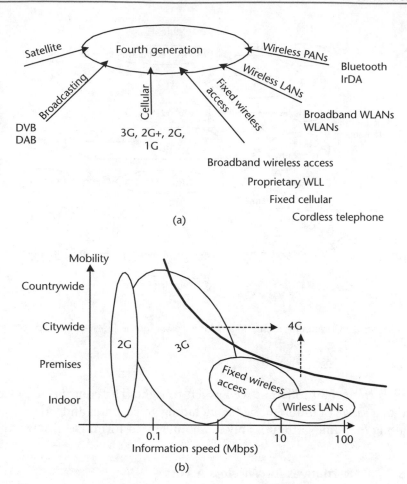

**Figure 5.3**  Different wireless access alternatives: (a) convergence, and (b) capabilities.

## 5.3  WLL Technologies Primarily for Voice Provision

### 5.3.1  Fixed Cellular Technologies in WLL Applications

Because the basic architecture of cellular and WLL systems are the same, it is convenient to adapt cellular technologies for WLL applications. However, key differences that exist between the two environments are summarized in Table 5.1.

First generation WLLs are based on analog cellular technologies, which provide good access techniques for medium- to low-density fixed applications. These operate in the same 900-MHz frequency band and are based on FDMA/FDD. However, the key drawback is that these are optimized for mobility rather than local loop service with low-bit-rate voice coding. Hence, quality equivalent to wireline service is hard to achieve.

The next generation of this type of WLL is based on the digital cellular standards that use TDMA and CDMA. CDMA-based WLLs have gained wide popularity. In addition to providing higher voice quality than analog systems, digital WLL systems are able to support higher speed data services. However, digital cellular technology provides coverage areas that are usually smaller than with analog.

**Table 5.1**    Comparison of Cellular and WLL Environments

| Cellular Environment | WLL Environment |
|---|---|
| Mobility (subscriber location, hand off, roaming) must be supported | Mobility is not an essential requirement |
| Line of sight between the mobile and the base station does not exist | Line of sight can be achieved by suitable base station and antenna spacing |
| Poorer quality than wireline service is acceptable | Quality must be equivalent to wireline services |

### 5.3.2  PCS Technologies in WLL Applications

Many WLL systems make use of the standardized cordless telecommunications systems such as cordless telephone-2 (CT2), digital European cordless telephony (DECT), personal wireless telecommunications (PWT), personal access communications system (PACS), and personal handyphone system (PHS). These low-mobility, low-power wireless communications systems are generally referred to as personal communications systems (PCS). Developed for microcellular environments, their coverage is typically several hundred meters. However, with fixed elevated antennas and other enhancements, their range can be extended to several kilometers for WLL systems. These systems operate in the 1,800- to 1,900-MHz frequency bands. Their suitability for WLL applications is examined in [7–9].

A general comparison of these low-power systems with cellular systems for WLL applications shows the following advantages of the former [9]:

- Superior voice quality with ADPCM used in PCS systems than with the lower bit rate encoding techniques used in cellular systems;
- Ability to provide data rates in multiples of 32 kbit/s, whereas cellular systems provide only 9.6 or 14.4 kbit/s in basic systems;
- Ability to provide much higher capacity in dense urban areas;
- Operation in the 1,800- to 1,900-MHz bands means they avoid interference from the more crowded 900-MHz cellular band;
- Simplicity and low cost. These are results of a simpler infrastructure that does not need support for mobility and frequency planning.

#### 5.3.2.1  The DECT System

Sponsored by the European Union and developed originally as a standard for cordless domestic and business systems and for limited mobility in the public market, DECT was intended to supersede the cordless telephony (CT) series of standards. The DECT subscriber units were low cost and low complexity in comparison with cellular systems such as GSM, with system performance optimized for confined indoor use. PWT is a DECT-based system developed in the United States for unlicensed PCS applications. PWT-E (PWT-enhanced) is a version for licensed PCS. These systems have been widely used in WLL applications around the world.

Figure 5.4(a) shows the architecture of the DECT system. PWT is similar. DECT is an interface between a fixed part (FP) and a portable part (PP). The FP has

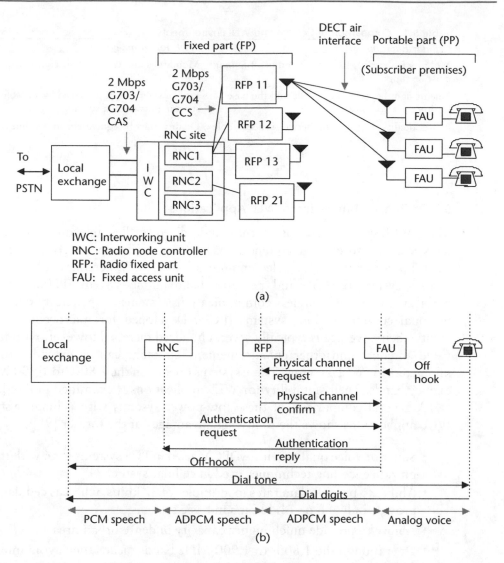

**Figure 5.4**   (a) DECT system architecture, and (b) call processing example (outgoing call).

three major components. The radio fixed part (RFP) terminates the air interface. The central system, the radio node controller (RNC), provides a cluster controller functionality, managing a number of RFPs. The DECT radio system is expected to work in conjunction with a wireline network. The interworking unit (IWU) provides all of the necessary functions for the PWT/DECT radio system to interwork with the attached network (e.g., the PSTN, ISDN, PLMN or a packet-switched network). The PP interfaces the subscriber premises equipment to the DECT air interface.

Figure 5.4(b) illustrates outgoing call processing in DECT. When the phone goes off hook, the subscriber premises FAU transmits a physical channel request to the RFP, and the allocation is confirmed. An authentication request is sent from the RNC, and when acknowledged by the FAU, the off-hook signal is sent to the local exchange. Dial tone is then received at the subscriber, followed by the dialed digits from the subscriber, via the channel formed. Subsequent processing is similar to the PSTN.

DECT is designed to operate in the 1,800- to 1,900-MHz frequency band, with flexibility to use other bands close by. It is based on TDMA/TDD. There are 10 carriers spaced at 1,728 kHz. Each carries 12 duplex TDMA channels at an aggregate rate of 1,152 kbit/s. Therefore, a total of 120 channels are available. The modulation method is Gaussian FSK (GFSK), and 32- kbit/s ADPCM is used for voice coding. Table 5.2 summarizes the parameters of the DECT system and compares the important parameters of additional systems with those of DECT. The normal cell radius for DECT is several hundred meters.

As depicted in Figure 5.5(a), the time slot structure of a DECT RF channel consists of 24 time slots in 10 ms. Twelve slots are defined for base-to-subscriber transmission, and twelve are defined for subscriber-to-base transmission. The full-duplex channel between the FP and the PP consists of a pair of time slots on a single RF channel. Each TDMA time slot (burst) consists of fields for synchronization, signaling, speech data, and error checking. Figure 5.5(b) shows the 10 DECT carriers and the formation of 120 channels.

Channel assignment in DECT is dynamic and does not involve detailed frequency planning as in cellular systems. While a pair of time slots are used for transmission and reception, the other 22 time slots are used by the subscriber unit to scan and evaluate other channels to hand over to a better channel when available. Further information on DECT is found in [11].

**Table 5.2**  Parameters of DECT, PACS, and PHS

| Parameter | DECT | PACS | PHS |
|---|---|---|---|
| Frequency band (MHz) | 1,880–1,990 | Licensed: 1,850–1,910 (uplink); 1,930–1,990 (downlink) Unlicensed: 1,920–1,930 | 1,895–1,918 |
| Duplexing | TDD | Licensed: FDD Unlicensed: TDD | TDD |
| Access | TDMA | TDMA | TDMA |
| Number of carriers | 10 | 200 | 77 |
| Carrier spacing (kHz) | 1,728 | 300 | 300 |
| Modulation | GFSK | $\pi/4$QPSK | $\pi/4$QPSK |
| Channel bit rate (kbit/s) | 1,152 | 384 | 384 |
| Maximum rate (kbit/s) | $11 \times 32$ | $7 \times 32$ | $2 \times 32$ |
| Number of slots/frame | 12 + 12 | 8 | 4 + 4 |
| Speech coding | ADPCM | ADPCM | ADPCM |
| Mobility (km/h) | 20 | 70 | 10 |
| Standardizing body | ETSI | American National Standards Institute (ANSI) | RCR[1] |
| Channel assignment | Dynamic channel allocation (DCA) | Quasi-static automatic autonomous frequency assignment (QSAFA) [10] | DCA |
| Cell size | Small | Largest | Large |
| Diversity | Optional | Yes | No |
| Spectral efficiency (b/s/Hz) | 0.66 | 1.28 | 1.28 |

(*Source:* [9].)

1. Research and Development Center for Radio Systems, currently known as ARIB.

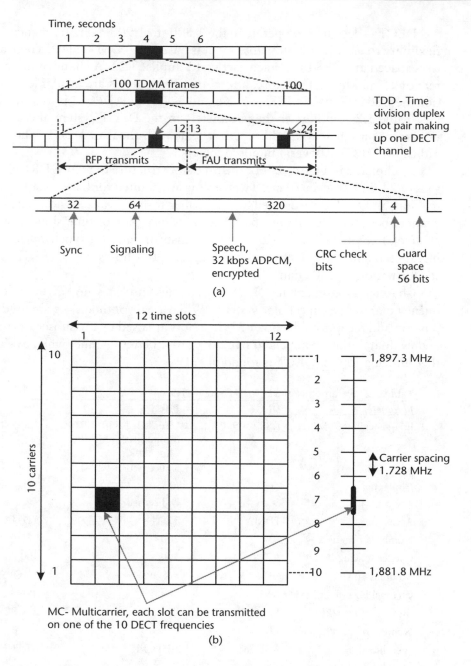

**Figure 5.5**  DECT channels: (a) time slot structure within an RF channel, and (b) formation of RF channels.

## 5.3.2.2  Other Cordless/PCS Systems

While DECT was well deployed in Europe and some parts of Asia, PHS (a digital system operating in the 1.9-GHz PCS band) has been successful in Japan. PACS, also operating in the 1.9-GHz band, is based on Bellcore's wireless access communications system (WACS) and on Japan's PHS. Detailed information and a comparison of these technologies are found in [7–10]. General comparison of the systems, through underlying technologies and implementation instances, can be summarized in Table 5.3.

In summary, no single low-power wireless system is ideally suited for WLL applications. The most suitable system can be chosen on the specific conditions prevailing in the environment. For low-traffic environments, PACS performs best due to its larger cells. The number of base stations per square kilometer in PACS would be significantly lower than other systems. In suburban areas, where capacity is an issue in addition to coverage, DECT has better performance. In urban areas, all three systems have similar performance, and the designer can increase capacity by reducing the cell size. The performance of PACS and PHS is better in terms of the ability to do this. However, DECT can be enhanced with interference-reduction techniques. Its superior dynamic channel allocation algorithm is an advantage in

**Table 5.3**   Summary and Comparison of DECT, PACS, and PHS

| Technical Feature | Comparison |
| --- | --- |
| TDD versus FDD | TDD results in simpler implementation than FDD. It has the advantage of needing only one RF channel (and equipment) for each call. However, TDD requires stringent time synchronization between all transmitters and receivers, with imperfections leading to severe loss of capacity. Maintaining time synchronization becomes more difficult as the coverage area becomes larger. FDD has the advantage of having lower bit rates in each of the two directions and thus needing less equalization to combat multipath fading. |
| Delay spread | Delay spread in multipath environments places an upper limit on transmission rate. DECT is a high-bit-rate system designed for indoor and microcellular environments and can tolerate delay spreads up to about 90 ns. For WLL applications, the operation can be extended up to 300 ns with diversity. PACS and PHS use lower bit rates and can operate in delay spread environments up to 260 ns and 1,000 ns without and with diversity, respectively. PACS has diversity built into the standard. As a result, the possible cell size is smallest in DECT. Some DECT implementations adopt larger cell sizes at the cost of deactivating every other time slot. |
| Frame size | PACS has the smallest frame size. This results in fewer delays in error correction in high bit-error-rate environments compared to DECT and PHS. Requirements for echo cancellation increase with frame length. |
| Channel selection | Channel selection and handoff is done almost exclusively by the subscriber units in DECT and PACS. In PHS, the base stations are also involved, and handoff occurs when performance becomes unacceptable. In the case of DECT and PACS, handoff is based on using the best channel available, which may lead to excessively frequent handoffs unless performance thresholds are also incorporated. The DCA technique used in DECT and PHS is superior to the QSAFA technique in PACS. |
| Blocking | In PHS, a subscriber in a particular cell can communicate only with its base station. In DECT and PACS, the subscriber can access neighboring base stations. Therefore when all time slots in a particular cell are occupied, transmission can continue with neighboring cells. This reduces the blocking probability. |
| Control channels | In PACS and PHS, one time slot in each frame is used as a control channel. In DECT, control information is embedded into the traffic channels. Therefore, DECT has a better channel efficiency. |
| Direct communication | Two PHS portable units in close proximity can carry direct two-way communications. This reduces the load on the central switch and is particularly advantageous in indoor environments. |
| Dual-mode operation | PACS is the only system that allows dual mode FDD/TDD operation in licensed/unlicensed bands. This makes it operable via dual-mode terminals in both unlicensed, private, indoor environments (e.g., wireless PABX) and public licensed wireless access systems. |

these environments. The same conclusions apply for both DECT and its derivatives, PWT and PWT-E.

One major technical problem with WLL systems based on these standards was the effect of multipath propagation, which makes radio planning extremely difficult. In a WLL application with common cell sizes of over 5 km, delay spreads of a few hundreds of nanoseconds are present. The technology should cater to such delay spreads. With the absence from the key PCS systems of technologies such as equalization, their ability to provide optimum performance in a complex radio environment is poor. Efforts to realize better WLL systems led to the implementation of proprietary systems.

### 5.3.3  Proprietary WLL Systems

Proprietary systems such as Nortel's Proximity I series are designed from the start as alternatives to the copper-based local loop. They operate at frequencies around 3.5 GHz, where the spectrum is less crowded. They provide high-quality voice at 32 kbit/s via ADPCM and support voiceband modem and fax transmission at up to 28.8 kbit/s.

The Proximity I system has nearly similar characteristics as DECT, but with different radio systems. This system has two main elements—a base station and a residential service system—linked by a 3.5-GHz digital microwave link. Base stations contain RF equipment for the microwave link to the subscribers' premises, along with subsystems for call and signal processing. A choice of base station systems is available for different environments. An omnidirectional system can support 600 or more customers in rural areas, whereas a trisectored system can serve over 2,000 in more highly urbanized regions. Base stations can be sited up to 40 km from the subscriber's premises. Table 5.4 summarizes the specifications for Proximity I.

**Table 5.4   Specifications of the Proximity I system**

| | | |
|---|---|---|
| Frequency Ranges | 3,425–3,442 MHz | RSS to Base |
| | 3,475–3,492 MHz | Base to RSS |
| Number of RF channels | 54 | |
| Channel spacing | 307.2 kHz | |
| TDMA structure | 10 time slots/frame | |
| Frame duration | 5 ms | |
| Slot duration | 500 $\mu$s | |
| Modulation | $\pi$/4-DQPSK | Root raised cosine filtering with $\alpha = 0.4$ |
| Symbol rate | 256 ksymbols/s | Gross bit rate = 512 kbit/s |
| Net user bit rate/timeslot | 32 kbit/s | A user can access up to three timeslots per frame |
| RSS antenna 3-dB beamwidth | 20° | |
| Subscriber interface | 2 PSTN lines; one voice + one voiceband data calls or two voice calls | |
| Speech coding | 32 kbit/s ADPCM | ITU-T Recommendation G. 721 |
| Maximum PSTN line length | 1 km | RSS to telephone |

(*Source:* [12].)

In the evolution of proprietary FWA systems, choices revolve around the current set of mobile/wireless standards—TDMA (IS-136), CDMA (IS-95), DECT, and GSM. As an example, Nortel has broadened its portfolio with the Proximity T (TDMA), Proximity C (CDMA), Reunion (broadband), and GALA (GSM) family of FWA systems. Figure 5.6 shows the operating frequency ranges of these systems.

## 5.4   Broadband Fixed Wireless Access Systems

Cellular systems and WLL systems described here reside at typical data rates of tens of kilobits per second in wide area coverage, served by microcells and macrocells. Bearer services, which qualify as broadband access (i.e., hundreds of kilobits per second up to 2 Mbit/s), are specified in third generation IMT-2000 systems for both fixed and mobile systems. There are two classes of broadband fixed wireless access (BWA) systems. One is the broadcast of television, with range of tens to hundreds of kilometers. The other is the wireless LANs (WLANs). Wireless LANs with 2- to over 50-Mbit/s transmission rates have coverage range of tens of meters.

The broadcast category is evolving towards multichannel, two-way fixed wireless communications, while the WLANs are evolving towards broadband outdoor systems with connectivity to ATM and TCP/IP transport networks. In the United States, the Federal Communications Commission (FCC) has set aside 15 bands for commercial BWA systems. Frequency allocations in other countries are very similar. Table 5.5 shows the frequency allocations for BWA systems.

Known as the ISM band, the 2.4000- to 2.4835-GHz band is popular with operators because it is unlicensed, and used with equipment manufacturers worldwide. Until recently, there was no question about using this band for communications, as the technology to overcome interference from ISM uses was not available. The arrival of spread spectrum technology for commercial communications opened up this band. Two more unlicensed bands span the frequency ranges 5.725 to 5.875 and 24.0 to 24.25 GHz. The former is known as the UNII band.

### 5.4.1   MMDS and LMDS

Out of the first three licensed bands shown in Table 5.5, the first two were licensed in the 1970s, when they were called MDSs to broadcast 6-MHz television channels. In 1996, the band expanded to allow for multichannel services called MMDS.

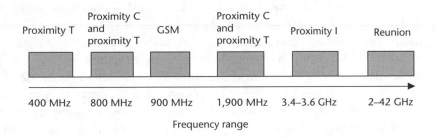

**Figure 5.6**   Nortel's fixed wireless access portfolio. (*Source*: [12]. Reproduced with permission of the IEE.)

**Table 5.5**  Broadband Fixed Wireless Communication Frequencies

| Frequency (GHz) | Usage |
|---|---|
| 2.1500–2.1620 | Licensed multichannel distribution service (MDS) and multichannel multipoint distribution service (MMDS), two bands of 6 MHz each |
| 2.4000–2.4835 | Unlicensed industrial, scientific, and medical (ISM) band |
| 2.5960–2.6440 | Licensed MMDS, eight bands of 6 MHz each |
| 2.6500–2.6560 | Licensed MMDS |
| 2.6620–2.6680 | Licensed MMDS |
| 2.6740–2.6800 | Licensed MMDS |
| 5.7250–5.8750 | Unlicensed ISM band, unlicensed national information infrastructure (UNII) |
| 24.000–24.250 | Unlicensed ISM band |
| 24.250–25.250 | Licensed |
| 27.500–28.350 | Licensed local multipoint distribution service (LMDS) (block A) |
| 29.100–29.250 | Licensed LMDS (block A) |
| 31.000–31.075 | Licensed LMDS (block B) |
| 31.075–31.225 | Licensed LMDS (block A) |
| 31.225–31.300 | Licensed LMDS (block B) |
| 38.600–40.00 | Licensed |

(*Source:* [13].)

Other licensed frequency allocations are for the LMDS in two blocks, A and B. This is also known as local multipoint communication service (LMCS).

Figure 5.7 shows the architecture of MMDS and LMDS systems. These systems employ a point-to-multipoint broadcast downlink with possibilities of either integrated or independent point-to-point uplink. Operation of MMDS/LMDS in an area will normally require a cluster of cells with separate base stations for colocated transmitter/receiver sites. Interference in adjacent cells would be avoided by using differently polarized antennas in adjacent cells. The service provider beams signals to a single point in multiple dwelling units or commercial buildings, and the signals are then distributed to individual tenants. One of the base station sites will serve as the coordination center and connect the cells to external networks. Intercell networking may be implemented using fiber or short hop radio relay connections. Colocation with mobile base stations allows for infrastructure sharing [14].

The MMDS service, which has existed for some time, offers a maximum of 33 analog video channels in a total bandwidth of 500 MHz. These services had a competitive advantage in providing television to rural populations out of reach of cable and ordinary broadcast services. For this reason, MMDS is also referred to as *wireless cable*. The transmit power allowed for MMDS services allowed signals to be carried as far as 70 km from the transmitter to receivers within line of sight [13]. MMDS requires terrestrial wired networks to communicate back to the headend (e.g., to select programming or use VCR-type controls on video-on-demand programming).

Current MMDS operators are looking to use digital compression techniques to increase the number of channels to around 200, making it competitive with presently available wired cable systems and satellite TV systems. With the evolution of digital technologies, typically with MPEG-2 encoding and complex modulation

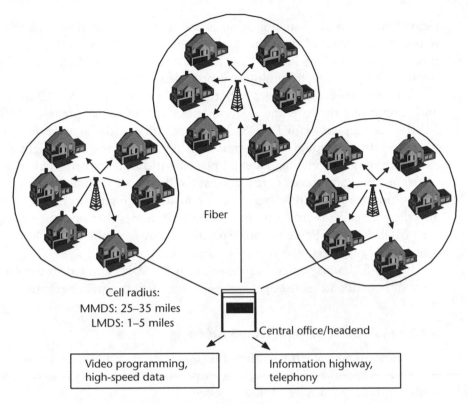

Cell radius:
MMDS: 25–35 miles
LMDS: 1–5 miles

Fiber

Central office/headend

Video programming,
high-speed data

Information highway,
telephony

LMDS - Local multipoint distribution service
MMDS - Multipoint multichannel distribution service

**Figure 5.7** Architecture of LMDS and MMDS systems. (*Source*: [2]. ©1997 IEEE.)

schemes, these systems may also provide two-way connectivity and transport to the transmitter from the headend using ATM or SONET networks, perhaps using TCP/IP protocols.

Whereas MMDS was developed for analog TV distribution, the first digital systems evolved during the late 1990s, leading to LMDS. It originally supported Motion Picture Expert Group (MPEG) video transmission on several carriers with an 8- to 27-MHz spacing, each transporting about 40 Mbit/s. LMDS services operate in the 27.5- to 31.3-GHz band. Typical LMDS applications now include all kinds of interactive services, using an extra return channel.

The LMDS, having cell radii of less than 12 km, can deliver two-way high-speed data, broadcast video, video-on-demand services, and telephony to residential areas. These systems have a total capacity of 34 to 38 Mbit/s per transport stream, giving high flexibility for inclusion of any type of data. The interactive channel capacity may range from a few kilobits per second to at least 25.6 Mbit/s [14].

The LMDS transmitter should be sited at a high point overlooking the service area. The transmitter covers a sector typically 60° to 90° wide. Full coverage thus requires four to six transmitters. The streams transmitted contain 34–38 Mbit/s of data addressed to everybody in the coverage zone (television), subgroups, or individuals (typical communication is the Internet). The return channel is determined by the needs of the individual user and can be typically up to 8 kbit/s, with possibilities up to 25.8 kbit/s. A capacity comparison of LMDS with other competing broadband access technologies is provided in [14].

The primary disadvantages of both MMDS and LMDS are CCI from other cells and limitations on coverage. Coverage issues are not as great a challenge with MMDS as they are with LMDS. Millimeter-wave radio signals do not penetrate trees and are susceptible to precipitation effects. Thus, line-of-sight propagation paths are required, making antenna placement on subscribers challenging. Even if the transmitter and receiver are placed at fixed points with line of sight, the influence of motion of traffic and foliage creates a hostile fading environment. Reference [2] and the references therein discuss propagation issues that post major impediments for LMDS services. Some possible solutions to these problems, such as overlapping cells, use of repeaters, and reflectors, are discussed in [14].

LMDS technology, first implemented in 2000, is expected to enhance development of broadband services such as e-commerce and distance education.

### 5.4.2   Multipoint-to Multipoint-Systems

As a solution to the propagation and coverage difficulties in point-to-point and point-to-multipoint BWA systems, multipoint-to-multipoint systems such as the one shown in Figure 5.8 have been developed.

Suppose it is desirable to extend the reach of a wireless system by letting every transceiver communicate with any other transceiver in the system. In such a system, multiple logical links exist between one receive/transmit point and its neighbors. Information would be forwarded through the network to the correct destination. Each link between two points may have different characteristics, such as transmit power, data rate, and reliability. All of these factors call for a new approach toward

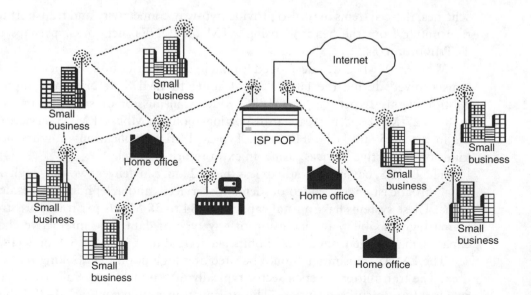

**Figure 5.8**   Point-to-multipoint Internet radio operating system (IROS) scheme from Rooftop Communications. (*Source*: [13]. ©1999 IEEE.)

the physical medium access, network protocols, and even an overall operating system. A few such systems in operation and further research carried out in this area of FWA systems are outlined in [13].

### 5.4.3 Wireless LANs as Broadband FWA Systems

A key development in FWA systems in recent times is the extension of wireless LAN technologies for more general wireless access. Wireless LANs and broadband access systems being developed are the multimedia mobile access communication (MMAC) system in Japan, the broadband radio access networks (BRAN) family of systems in Europe, and the IEEE 802.11 in the United States. The evolutionary trends also show the development of hierarchical wireless networking environments, as illustrated in this section.

#### 5.4.3.1 IEEE 802.11 Wireless LANs

In June 1997, the IEEE approved international interoperability standard IEEE 802.11, specifying both physical and medium access control procedures for wireless extensions to LANs. Three physical layers, two in the 2.4-GHz ISM band using frequency-hopped spread spectrum (FH-SS) and direct-sequence spread spectrum (DS-SS) and one using infrared light (IR) were defined. All physical layers supported a data rate of 1 Mbit/s and optionally 2 Mbit/s. For multiple access, the carrier sense multiple access/collision avoidance (CSMA/CA), a distributed medium access control protocol, was adopted.

User demand for higher bit rates and international availability of the 2.4-GHz band has spurred the development of a higher speed extension to the IEEE 802.11 standard called *IEEE 802.11b*, providing a basic rate of 11 Mbit/s and a fallback rate of 5.5 Mbit/s to be used with the already standardized medium access control. References [15–17] are recommended for more general information on these systems. Yet another physical layer option, which offers higher bit rates in the 5.2-GHz band intended for use in UNII devices, was standardized as *IEEE 802.11a,* offering data rates up to 54 Mbit/s using orthogonal frequency division multiplexing (OFDM). OFDM and reasons for its adoption are described later. Table 5.6 summarizes the IEEE802.11 family of standards.

#### 5.4.3.2 BRAN Systems

The ETSI project BRAN defines a family of high-performance radio access standards expected to be deployed in a hierarchical manner. Table 5.7 summarizes the different BRAN systems. Figure 5.9 shows how the three network categories might be deployed in business and domestic environments.

**Table 5.6** The IEEE 802.11 Family of Standards

| Standard | Frequency Band | Data Rate |
|---|---|---|
| IEEE 802.11 | 2.4 GHz and IR | 1–2 Mbit/s |
| IEEE 802.11b | 2.4 GHz and IR | 11 and 5.5 Mbit/s |
| IEEE 802.11a | 5.7 GHz | 6–54 Mbit/s |

**Table 5.7**  BRAN Systems

| System | Description | Frequency Band (GHz) | Application |
|---|---|---|---|
| HIPERACCESS | Fixed, long range, 25 Mbit/s | 40.5–43.5 | Outdoor fixed network providing access to IP, ATM, and third generation mobile infrastructure |
| HIPERLINK | Fixed, very short range (150m), very high speed (155 Mbit/s), point to multipoint | 17 | Wireless indoor backbone for interconnection of HIPERLAN and HIPERACCESS |
| HIPERMAN | Fixed, point to multipoint or mesh | 2–11 | Fixed wireless access provisioning to small and medium enterprises (SMEs) and residences |
| HIPERLAN/2 | Mobile, short-range, up to 54 Mbit/s | 5 | Wireless indoor LAN with QoS |

(*Sources*: [18, 19].)

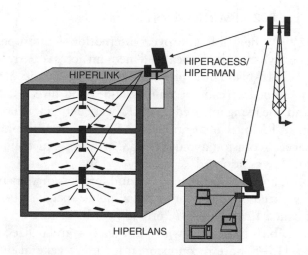

**Figure 5.9**  Scope of the different categories of wireless access systems defined by BRAN (*After:* [18].)

At 40 GHz, in addition to costly front-end technology, attenuation by precipitation is severe. The higher capacity offered at 40 GHz may compensate for these effects in the long run. These bands are expected to be shared among two or three licensees with 500 MHz to 2 GHz per licensee. The attractiveness of this possibility is driving these technology developments for fixed systems, despite propagation and other technical hurdles.

The flexible architecture applied in all BRAN standards defines physical and data link control (DLC) layers, which are independent of the core network. A set of core network–specific convergence layers (CL) are placed at the top of the DLC layer. This allows BRAN systems to be used with a variety of core networks.

### 5.4.3.3  Multimedia Mobile Access Communication System

ARIB's Multimedia Mobile Access Communication System (MMAC) is a high-performance wireless system applicable in stationary, quasistationary, and mobile environments, which can transmit ultra-high-speed, high-quality multimedia information with seamless connections to optical-fiber networks. This provides four categories of broadband services, as shown in Table 5.8. The application concept of MMAC is illustrated in Figure 5.10.

### 5.4.3.4    Harmonization of Standards

ETSI, IEEE 802.11, and MMAC standardization groups have been closely coordinating with each other to harmonize the systems developed by the three fora. There are many similarities between IEEE802.11b and HIPERLAN systems, the main ones being the adoption of orthogonal frequency division multiplexing (OFDM) and link adaptation schemes in the physical layer. The different modes of operation in link adaptation are found in [21]. MMAC also uses OFDM. IEEE802.11b and HIPERLAN systems can operate in infrastructure (centralized) or ad-hoc (direct) modes. Their main difference lies in the medium access control scheme. While IEEE802.11b has a distributed media access control protocol (CSMA/CA), HIPERLAN has a centralized, scheduled media access control protocol based on ATM/TDMA/TDD. MMAC supports both types of media access controls. References [17, 21] provide good comparisons of these systems.

### 5.4.4    IEEE 802.16 Wireless MANs

The IEEE has standardized broadband WLLs for use in, but not restricted to, the LMDS bands as IEEE 802.16 for initial target markets requiring 2 to 155 Mbit/s.

**Table 5.8    MMAC Service Categories**

| Service Category | High-Speed Wireless Access | Ultra-High-Speed Wireless LAN | 5-GHz Band Mobile Access (Wireless Access and Wireless LAN) | Wireless Home Link |
|---|---|---|---|---|
| Service area | Public space (outdoor, indoor) Private space (indoor, premises) | Private space (indoor) | Public space (outdoor, indoor) Private space (indoor, premises) | Private space (indoor) |
| Connected networks Interface | Public network (ATM) Private network (ATM) | Private network (ATM) | Public network (ATM) Private network (ATM , Ethernet) | EEE1394 |
| Information rate | 30 Mbit/s | 156 Mbit/s | 20 to 25 Mbit/s | 30 to 100 Mbit/s |
| Terminal equipment | Notebook-type PCs | Desktop PCs and work stations | Notebook-type PCs, handy terminals | PCs and audio visual equipment |
| Mobility | Still or pedestrian (with handoff) | Still (with handoff) | Still or pedestrian (with handoff) | Still or pedestrian (with handoff) |
| Radio frequency bands | 25-/40-/60-GHz bands | 60-GHz band | 5-GHz band | 5-/25-/40-/60-GHz bands |
| Bandwidth | 500 to 1,000 MHz | 1 to 2 GHz | Greater than 100 MHz | Greater than 100 MHz |

(*Source:* [20].)

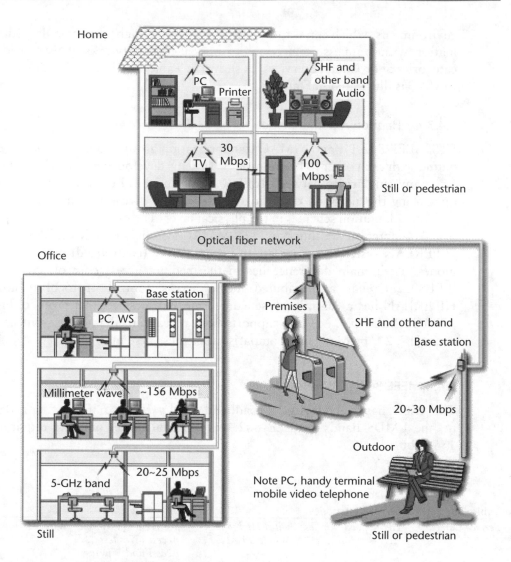

**Figure 5.10**  The application concept of MMAC. (*Source:* [20].)

### 5.4.4.1   IEEE 802.16/16a Overview

The IEEE standard 802.16 WirelessMAN™, "Air Interface for Fixed Broadband Wireless Access Systems," was published in April 2002. It addresses the last mile connection in wireless metropolitan area networks, focusing on the efficient use of bandwidth in the region between 10 and 66 GHz, and defines a common medium access control layer that supports multiple physical layer specifications customized for the frequency of use. Between the physical and media access control layers, a transmission convergence (TC) sublayer is defined, which forms the interface to different physical layers. This standard supports continuously varying traffic levels at many licensed frequencies (e.g., 10.5, 25, 26, 31, 38, and 39 GHz).

The more recent *IEEE802.16a* standard does the same for the frequency band 2 to 11 GHz (licensed and unlicensed). This lower frequency band offers the opportunity to reach many more customers less expensively, although at generally lower data rates. This suggests that such services will be oriented toward individual homes or SMEs, where the higher frequency systems will be geared toward large corporate customers.

Whether the promise of BWA will materialize depends on its appeal to telecom operators from the perspective of deployment economics, where the critical factor is the ease of installation of subscriber units [22]. The ultimate objective is for non-professional installation of integrated all-indoor subscriber units. Consequently, the physical layer has to mitigate the very tough impairments that characterize these nonline-of-sight environments. The IEEE802.16a standard emphasizes this requirement.

Many of the features in 802.16 have already been implemented in the BWA systems discussed in Section 5.3.5 and hence show the gradual evolution of a closely related set of standards. This standard sets the stage for widespread and effective deployment of BWA systems worldwide. The problems in BWA that IEEE802.16 addresses are outlined in [23].

### 5.4.4.2   The Physical Layer

*10 to 66 GHz*

In the design of the physical layer for this region, line of sight was deemed a practical necessity. With this assumption, single-carrier modulation was selected. The air interface is designated *WirelessMAN-SC*. Because it is point to multipoint, the base station transmits a TDM signal. Access in the uplink direction is by TDMA. The TDMA burst design that allows both TDD and FDD was selected. Support for half-duplex FDD subscribers, which may be less expensive because they do not simultaneously transmit, was also added. Both TDD and FDD alternatives support adaptive burst profiles in which modulation and coding options may be dynamically assigned on a burst-by-burst basis. Figure 5.11 shows the TDD and FDD burst structures.

The system uses a frame of 0.5, 1, or 2 ms. This frame is divided into physical time slots for the purpose of bandwidth allocation and synchronization. In the TDD variant, the uplink subframe flows the downlink subframe on the same carrier frequency. In the FDD variant, the uplink and downlink subframes are coincident in time but are carried on separate frequencies.

*2 to 11 GHz*

The current specifications include three alternatives for the physical layer. Design of the 2- to 11-GHz band is driven by the need for nonline-of-sight operation. Because residential applications are expected, rooftops may be too low for a clear line of sight to a base station antenna, possibly due to obstruction by trees. Therefore, significant multipath propagation must be expected. The three alternatives are WirelessMAN-SC2, which uses a single-carrier modulation format; WirelessMan-OFDM, which uses OFDM with a 256-point transform and TDMA access; and WirelessMAN-OFDMA using OFDMA with a 2,048-point transform. In the last one, multiple access is provided by addressing a subset of the multiple carriers to individual receivers. The second alternative is mandatory for license-free bands. An overview of these techniques is provided in Section 5.5. Due to the propagation requirements, the use of advanced antenna systems is supported.

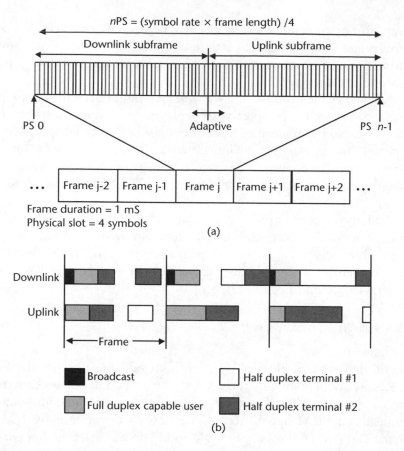

**Figure 5.11** Burst structures in the IEEE802.16 standard (10–16 GHz): (a) TDD, and (b) FDD. (*Source*: [23]. Reproduced with permission of the IEEE.)

### 5.4.4.3   The Medium Access Control Layer

The medium access control layer addresses the need for very high bit rates. Access and bandwidth allocation algorithms must accommodate a large number of terminals per channel with terminals that may be shared by multiple applications. The services required by the end users are varied in their nature and include legacy TDM voice and data, IP connectivity, and packetized VoIP. To support this variety of services, the 802.16 medium access control must accommodate both continuous and bursty traffic. Additionally, these services expect to be assigned QoS in keeping with the traffic types.

The 802.16 medium access control provides a wide range of service types analogous to the classic ATM service categories, as well as newer categories such as guaranteed frame rate (GFR). It also must support a variety of backhaul requirements, including both ATM and packet-based protocols. Convergence sublayers are used to map the transport layer–specific traffic to a medium access control protocol that is flexible enough to efficiently carry any traffic type [24]. The protocol is based on a request-grant mechanism that is designed to be scalable, efficient, and self correcting. The 802.16 access system does not lose efficiency when presented with multiple connections per terminal, multiple QoS levels per terminal, and a large number of statistically multiplexed users. It takes advantage of a wide variety of request mechanisms, balancing the stability of contentionless access with the efficiency of contention-based access [24].

To accommodate the more demanding physical environment and different service requirements of the 2- to 11-GHz frequency range, the 802.16a standard upgrades the medium access control to provide automatic repeat request (ARQ) and support for mesh, rather than only point-to-multipoint architectures. Further details of the IEEE802.16 standard are found in [24].

## 5.5 Enabling Technologies

### 5.5.1 Devices

Realization of millimeter-wave communications systems greatly depends on whether suitable devices are available for operation in the relevant frequency bands. Millimeter-wave components traditionally have been based on two-terminal devices and waveguides as active elements and transmission media, respectively [25]. However, waveguides have disadvantages of bulkiness and rigidity. Technical advancement of high-speed three-terminal devices in millimeter-wave bands in the last two decades has changed components to IC-based planar structures. Compared with waveguides, they have advantages of compactness, light weight, and suitability for commercial production.

Front-end technology is still expensive at millimeter frequencies, but existing high electron mobility transistor (HEMT) modules offer the required performance. HEMT devices have achieved a record high maximum oscillating frequency of 600 GHz [24] and superior low-noise performance. The output power level needed for a 36 Mbit/s per transport beam of an LMDS system, for example, is about 25 dBm [14]. A technology allowing for final-stage amplification of several transport beams would reduce equipment complexity and cost. Multistage millimeter wave low-noise amplifiers have shown high performance (1.7-dB noise figure at 62 GHz) [24]. The front-end technology at 40 GHz is still more expensive.

### 5.5.2 OFDM and OFDMA

#### 5.5.2.1 OFDM

The basic principle of OFDM is to split a high-rate data stream into a number of lower rate streams, which are transmitted simultaneously over a number of subcarriers. Because the symbol duration increases for lower rate parallel subcarriers, the relative amount of time dispersion caused by multipath delay spread is decreased. Intersymbol interference (ISI) is eliminated almost completely by introducing a guard interval (GI) in every OFDM symbol [17]. The OFDM symbol is cyclically extended during the GI.

Figure 5.12(a) shows an example of four subcarriers in one OFDM symbol. All subcarriers differ by an integer number of cycles within the symbol duration, which ensures orthogonality between them. This orthogonality is maintained in the presence of multipath delay spread, as indicated in Figure 5.12. In practice, the most efficient way to generate the sum of a large number of subcarriers is by using the inverse fast Fourier transform (IFFT). At the receiver side, the fast Fourier transform

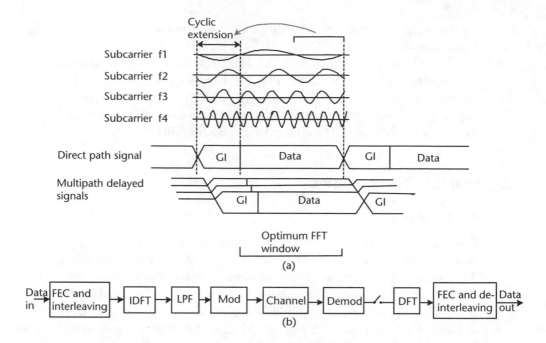

**Figure 5.12** (a) An OFDM symbol (*Source*: [17], ©1999 IEEE), and (b) a simplified block diagram of an OFDM system (*After*: [22]).

(FFT) can be used to demodulate all subcarriers. Figure 5.12(b) shows an OFDM system.

Because of delay spread, the receiver sees a summation of time-shifted replicas of each OFDM symbol. As long as the delay spread is shorter than the guard time, there is no ISI or intercarrier interference within the FFT interval of an OFDM symbol. The only remaining effect of multipath, the random phase and amplitude of each subcarrier, is estimated through the use of pilot symbols, and the carriers are coherently detected. In order to deal with weak subcarriers in deep fades, forward error correction is applied [17].

A key parameter in OFDM is the guard interval ($T_g$). This provides robustness to RMS delay. The symbol duration ($T_s$) is selected by choosing an appropriate balance between the time and power spent on the GI. The subcarrier spacing is the inverse of the symbol duration. The total time spent in transmitting one OFDM symbol is $T_s + T_g$. An OFDM symbol consists of data as well as pilot signals for channel estimation. A large number of subcarriers are present, each carrying data using variable modulation types, from BPSK to 16-QAM, and variable convolutional coding schemes for error correction. Pilots are carried over a number of pilot subcarriers. OFDM is used in IEEE 802.11a, HIPERLAN, and IEEE802.16. References [21, 22] describe the parameters selected for these OFDM schemes. In these implementations, all carriers are transmitted at once. The downstream data is time division multiplexed, and subscribers access the base station in the upstream through TDMA.

### 5.5.2.2 Orthogonal Frequency Division Multiple Access

One physical layer option for the IEEE802.16a standard for 2- to 11-GHz BWA systems is to use orthogonal frequency division multiple access (OFDMA). In this extension of OFDM, the subcarriers are grouped into subchannels, which are used in the downstream for separating data into logical streams. The subchannels use different amplitudes, modulation, and coding schemes to address subscribers with different channel characteristics. In the upstream, the subchannels are used for multiple access [22].

A subchannel is a subset of carriers out of the total set available, as illustrated in Figure 5.13. In order to mitigate the effects of frequency selective fading, the carriers of one subchannel are spread along the channel spectrum. The usable carrier space is divided into a number of $N_G$ successive groups, with each group containing $N_E$ successive carriers after excluding the pilot carriers. A subchannel has one element from each group allocated through a pseudorandom process based on permutations. Each subchannel therefore has $N_G$ subcarrier elements.

OFDMA allows for fine granulation of bandwidth allocation, consistent with the needs of most subscribers, while high consumers of upstream bandwidth are allocated more than one subchannel. A low upstream data rate is consistent with the traffic asymmetry, where the streams from each subscriber add up in a multipoint-to-point regime. In the downstream, all of the subchannels are transmitted together [22]. In essence, OFDMA consists of different users sharing the spectrum, with each transmitting one or more subchannels. This can also be seen as a form of FDMA. With regards to interference, OFDMA subchannels constitute a form of FH-SS.

### 5.5.3 Smart Antennas

Smart antennas (SAs) have intelligent functions, such as suppression of interference signals, auto tracking of desired signals, and digital beam forming with adaptive space-time processing. Because of these characteristics, smart antennas have been considered a key technology for future wireless communications, enabling the reduction of interference and transmission power.

SA technology exploits multiple antennas in transmit and receive with associated coding, modulation, and signal processing to enhance the performance of

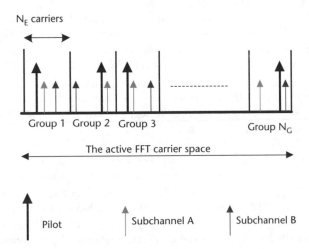

**Figure 5.13**  Subchannels in OFDMA. (*Source*: [22]. Reproduced with permission of the IEEE.)

wireless systems in terms of capacity, coverage, and throughput [26]. In BWA networks, both the base stations and the subscriber can use multiple antennas and therefore SA techniques can be a significant advantage. Reference [26] is suggested for more details on SA technology.

For the implementation of adaptive antenna arrays, complex signal processing is required to estimate the fading channel characteristics and perform adaptive equalization accordingly. Hence, relevant algorithms and processor configurations are important issues being addressed.

## 5.6 The Future

Since the late 1990s, concerted efforts by ETSI, ARIB, and the IEEE have resulted in the development of many standards, particularly for BWA systems in the 2- to 5-GHz bands, as well as at higher frequencies. These were spurred by the need for provisioning Internet and multimedia services to an increasing number of subscribers, complementing or in competition with broadband wireline access methods such as DSL. These standards allow the development of BWA systems in a hierarchical manner suitable for a variety of environments and supporting a limited amount of mobility as well. Technologies such as OFDM and smart antennas have become an integral part of these newest standards.

The success of BWA will depend critically on the ease of installation of subscriber units. The ultimate objective is for nonprofessional installation of integrated all-indoor subscriber units. Consequently, this is an aspect that is strongly emphasized in present standardization efforts.

## References

[1]   Muller, N. J., *Mobile Telecommunications Factbook*, New York: McGraw Hill Professional, 1998.

[2]   Honcharenko, W., et al., "Broadband Wireless Access," *IEEE Communications Magazine*, January 1997, pp. 20–26.

[3]   Webb, W., "A Comparison of Wireless Local Loop with Competing Technologies," *Electronics & Communication Engineering Journal*, October 1998, pp. 205–211.

[4]   Webb, W., *Introduction to Wireless Local Loop, Second Edition: Broadband and Narrowband Systems,* Norwood, MA: Artech House, 2000.

[5]   Drewes, C., W. Aicher, and J. Hausner, "The Wireless Art and the Wired Force of Subscriber Access," *IEEE Communications Magazine,* May 2001, pp. 118–124.

[6]   Lu, W., "Compact Multidimensional Broadband Wireless: The Convergence of Wireless Mobile and Access," *IEEE Communications Magazine*, November 2000, pp. 119–123.

[7]   Yu, C. C., et al.,"Low-Tier Wireless Local Loop Radio Systems—Part 1: Introduction," *IEEE Communications Magazine*, March 1997, pp. 84–92.

[8]   Yu, C. C., et al., "Low-Tier Wireless Local Loop Radio Systems—Part 2: Comparison of Systems," *IEEE Communications Magazine,* March 1997, pp. 94–98.

[9]   Momtahan, O., and H. Hashemi, "A Comparative Evaluation of DECT, PACS and PHS Standards for Wireless Local Loop Applications," *IEEE Communications Magazine*, May 2001, pp. 156–163.

[10] Noerpel, A. R., "PACS: Personal Access Communications System: An Alternative Technology for PCS," *IEEE Communications Magazine*, October 1996, pp. 138–150.

[11] Datapro Information Services Group, *Digital European Cordless Telecommunications (DECT)*, New York: McGraw-Hill Companies Inc., 1997.

[12] Hart, C., "Fixed Wireless Access: A Market and System Overview," *Electronics & Communications Engineering Journal*, October 1998, pp. 213–220.

[13] Dutta-Roy, A., "Fixed Wireless Routes for Internet Access," *IEEE Spectrum*, September 1999, pp. 61–69.

[14] Nordbotten, A., "LMDS Systems and Their Application," *IEEE Communications Magazine*, June 2000, pp. 150–154.

[15] LaMaire, R. O., et al., "Wireless LANs and Mobile Networking: Standards and Future Directions," *IEEE Communications Magazine*, August 1996, pp. 86–94.

[16] Crow, B. P., et al., "IEEE 802.11 Wireless Local Area Networks," *IEEE Communications Magazine*, September 1997, pp. 116–126.

[17] Van Nee, R., et al., "New High-Rate Wireless LAN Standards," *IEEE Communications Magazine*, December 1999, pp. 82–88.

[18] Haine, J., "HPERACCESS: An Access System for the Information Age," *Electronics & Communication Engineering Journal*, October 1998, pp. 229–235.

[19] ETSI, "BRAN Summary," http://portal.etsi.org/bran/Summary.asp.

[20] ARIB, "What is MMAC?" http://www.arib.or.jp/mmac/what.htm.

[21] Doufexi, A., et al., "A Comparison of the HIPERLAN/2 and IEEE802.11a Wireless LAN Standards," *IEEE Communications Magazine,* May 2002, pp. 172–180.

[22] Koffman, I., and V. Roman, "Broadband Wireless Access Solutions Based on OFDM Access in IEEE 802.16," *IEEE Communications Magazine*, April 2000, pp. 96–103.

[23] Marks, R.B., "The IEEE 802.16 WirelessMan Standard for Broadband Wireless Metropolitan Area Networks," http://www.wirelessman.org/docs/03/c80216-03_06.zip.

[24] Eklund, C., R. B. Marks, and K. L. Stanwood, "IEEE Standard 802.16: A Technical Overview of the WirelessMAN™ Air Interface for Broadband Wireless Access," *IEEE Communications Magazine*, June 2002, pp. 98–107.

[25] Ohmori, S., Y. Yamao, and N. Nakajima, "The Future Generations of Mobile Communications Based on Broadband Access Technologies," *IEEE Communications Magazine*, December 2000, pp. 134–142.

[26] Sheikh, K., et al., "Smart Antennas for Broadband Wireless Access Networks," *IEEE Communications Magazine*, November 1999, pp. 100–105.

## CHAPTER 6

# Digital Subscriber Loop

## 6.1 Introduction

With more than a century of infrastructure development, conventional telephone connections are available to a larger fraction of the world population than any other electronic system. During the 1990s, Internet and the World Wide Web–based services proliferated rapidly, allowing people to connect their computers to information resources. Voiceband modems help connect computers to the Internet. Up to about 56 kbit/s, these modems depend on the guaranteed voice bandwidth of the POTS service irrespective of the length of the channel. However, this analog modem and telephone network technology is simply inadequate to cater for the present day demand for new services. Assuming little or no network delays, a 10-Mb data transmission—the equivalent of a four-minute audio/video clip—takes about 95 minutes to download when using a 14.4 kbit/s analog modem and 25 minutes when using a 56-kbit/s modem.

ISDN trials in the early 1980s proved that the "last mile" of the copper pair can carry much higher frequencies than the baseband frequencies of the telephony voice signal. With these trials, engineers were convinced that the last mile of the twisted pair can be a major opportunity to provide high-bit-rate services. Learning from the ISDN experiences, DSL techniques having advanced digital modulation techniques, error handling, and bandwidth optimization combined with digital signal processing techniques evolved into a family of a variety of techniques known as xDSL. This chapter provides an overview of the DSL technology, various versions of DSL systems, and their implementation aspects.

## 6.2 Signal Impairments and the Channel Length

Today's telecom outside plant is dominated with copper pairs of unshielded twisted wire of different gauges to nearly a billion of wireline subscribers on the global telecommunication network. The original aim was to make use of these last mile copper pairs to exchange information within a 4-kHz bandwidth. Trunking and channel multiplexing schemes always considered this limited bandwidth as adequate for basic voice with subscriber identification. Even if the channel could be nearly 45,000 miles long, such as in a satellite link, with many electronic subsystems in between, this limited baseband provides an adequate SNR and other related parameters.

### 6.2.1 The Subscriber Loop

With the proliferation of Internet and multimedia applications, subscribers found that the last mile was a serious bottleneck, despite advances in switching and transmission technologies. Impairments in the copper loop present barriers to high-speed data within the last mile. As depicted in Figure 6.1, the variation of signal strength and noise along a channel causes the SNR to deteriorate with distance.

Alexander Graham Bell's first telephone service was based on a single iron wire and the Earth return path. This solution proved to be unsatisfactory, and customers were sometimes to pour water on their ground rods to improve the reliability. This was later solved by the use of a pair of bare wires strung in parallel a few inches apart. Then, detecting the problem of crosstalk between nearby pairs, Bell invented the twisted wire pair in 1881, four years after the first telephone service.

Figure 6.2 shows an overview of a typical copper access network. The main network consists of large multipair cables radiating out from the main distribution

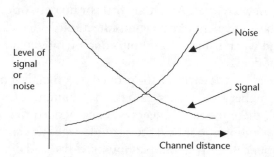

**Figure 6.1**   Signal and noise along a communication channel.

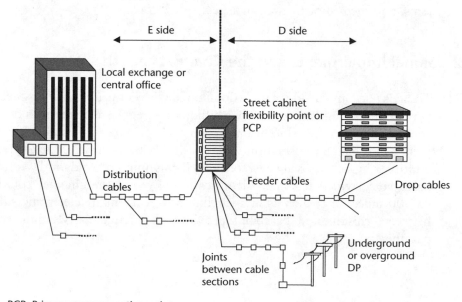

PCP: Primary crossconnection point
DP: Distribution point

**Figure 6.2**   A typical access network topology.

frame (MDF) of the serving local exchange to flexible points known as primary cross-connection points (PCP). Individual cable segments are joined together to form the link from the MDF to the PCP. From the PCP outward, the network is known as a distribution (D-side) network, whereas the MDF to PCP link is known as the exchange (E-side) network. From the PCP, distribution cables radiate out to distribution points (DPs) and from there to the customer premises, via drop wires. The distribution network may be overhead or underground. The cables that are predominantly used in the access network are UTPs of differing gauges. Thinner pairs are normally used nearer to exchanges, and thicker pairs near the customers, due to practical reasons of bundling and handling. Larger gauges closer to the customer premises help achieve a maximum range for a given transmission and signaling resistance requirement.

### 6.2.2   Impairments in the Subscriber Loop

Copper access networks face a variety of barriers to their operation. They can be classified as intrinsic or extrinsic to the cable environment. Noise and crosstalk are the two major impairments. In a DSL environment that exploits the higher frequency capabilities (than the typical 4-kHz bandwidth) of shorter subscriber loops, the environment is further complicated by a large number of contributing factors. Figure 6.3 shows the key environmental factors with which DSL systems should coexist, indicating the intrinsic and the extrinsic environment [1]. DSL environment for one area of a network will be different from another.

Examples of intrinsic impairments are thermal noise, echoes and reflections, crosstalk, noise sources such as surge protectors, RFI filters, bridge taps and loading coils, and conditions generated due to the condition of the cable infrastructure

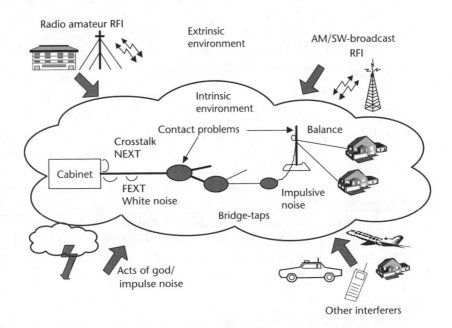

**Figure 6.3**   Intrinsic and extrinsic environment of DSL systems. (*Source:* [1]. Reproduced with permission of the IEE.)

(e.g., faulty pairs, leakages to ground, battery or Earth contacts, and high resistance joints). Examples of extrinsic impairments are impulsive noise (from lightning, electric fences, power lines, machinery, and fluorescent lighting) and interference from radio transmitters.

### 6.2.2.1 Noise

Noise sources can be classified as capacity limiting or performance limiting. Capacity-limiting noise, such as thermal noise and crosstalk, can be characterized statistically and are easy to account for in the design. Performance-limiting noise, such as impulses and radio frequency interference (RFI), is intermittent in nature, geographically variable, and unpredictable. These are used in design safety margins. DSL systems use additional signal-processing techniques, such as error correction, to overcome these. A description of noise and its effect on signals are given in Chapter 1.

Impulsive noise is random noise events from intrinsic sources such as tip/ring, loop disconnect, or extrinsic sources, such as lightning or power surges. These difficult-to-predict, random incidents are variable from a few hundred nanoseconds to over a millisecond or so.

### 6.2.2.2 Crosstalk

Along the channel, crosstalk, with its two elements of near-end crosstalk (NEXT) and far-end crosstalk (FEXT), creates the largest contribution to capacity limitations in DSL systems. NEXT is created by unwanted coupling between signals propagating in opposite directions, and FEXT is created by signals in the same direction. NEXT increases according to $(\text{frequency})^{\frac{3}{2}}$ and FEXT increases according to $(\text{frequency})^2$. The NEXT and FEXT environment depends on the characteristics of the services in the different pairs, cable construction, service mix, and fill among the pairs in the bundle. ANSI standards provide details for NEXT and FEXT for different services. More details can be found in [2].

### 6.2.2.3 RFI

The metallic copper network, without major shielding against RFI, can work as transmit or receive elements for radio frequencies. Because DSL technology makes use of the high-frequency capabilities of the loop, electromagnetic compatibility becomes an important issue. Some DSL versions (discussed later) have their spectra overlapping with different RF systems, such as medium-and short-wave AM stations, public safety and distress bands, and amateur radio. Due to the property called *cable imbalance* (discussed later), these radio frequencies can be received or transmitted by copper pairs. This creates impairments when certain types of DSL modems attempt to use their higher frequency capabilities.

#### Cable Balance

The cable balance $B$ is the degree of signal coupling between the common mode (cable to ground) and differential mode (between the two wires of the pair). Numerically B is given as,

$$B = 20\log\left(\frac{V_{cm}}{V_d}\right) \qquad\qquad (6.1)$$

where $V_{cm}$ and $V_d$ are the common-mode and differential-mode voltages, respectively.

The balance of UTP cables decreases with increasing frequency, with values of over 50 dB below 1 MHz and less than 30 dB at higher frequencies. Balance is variable from pair to pair, is dependant on environmental factors, and has sharp peaks at some frequencies [1]. The effect of balance on RFI depends on the cumulative effect of many factors, such as characteristics of the interference source, cable orientation, polarization effects of the source and the pair, and cable nonuniformity.

*RFI Ingress and Egress*
Copper cables picking up RF noise in high-speed DSL systems, such as very-high-bit-rate DSL (VDSL), described later, is termed RFI *ingress*. This is dependant on the shielding, twist, and physical orientation of the cable. The reciprocal effect of the electromagnetic emission of VDSL signals traveling on a cable pair is known as RFI *egress* and depends on the shielding and the length of exposed parts (such as in overhead lines). Untwisted wiring and inferior shielding within customer premises can influence the levels of RFI ingress and egress.

Some parametric rules for computing ingress voltage could be based on the case of an amateur RF transmitter modeled as an omnidirectional RF source. The induced $V_d$ in the drop wire can be expressed as [3],

$$V_d = 5.48\sqrt{P_t}\ /\ d\sqrt{B} \qquad\qquad (6.2)$$

where $P_t$ is the power of the omnidirectional antenna source, $d$ is the distance from the RF power source, and $B$ is the cable balance.

Induced voltages is inversely proportional to cable balance, as shown in (6.2). The expression is used to estimate the ingress from both AM and radio amateur stations. Similarly, the $V_{cm}$ also can be expressed. For details Chapter 3 of [2], as well as [3, 4] are suggested.

### 6.2.2.4  Cable Loss

Most common cable pairs are UTPs of diameter 0.32 mm, 0.4 mm [American wire gauge (AWG) 26], and 0.5 mm (AWG24). Cable loss, the attenuation along the cable length, results in the degradation of the amplitude of the transmitted signal. Figure 6.4 indicates the typical loss characteristics of 0.4-mm and 0.5-mm UTP cables of 1 km sections terminated with 100-$\Omega$ resistance. As the cable loss increases with the frequency, it adversely affects the transmission of higher frequencies.

Cable loss estimation sometimes becomes very difficult with the in-premises wiring, as the quality of the in-premises wiring is not usually governed by the service provider. The indications in the Figure 6.4 are an overall effect of the line resistance/km ($R$), inductance/km ($L$), and the capacitance/km ($C$) variations with the frequency. $R$ increases with frequency, while $L$ decreases with frequency, and $C$ is almost constant with frequency with typical values given in Table 6.1. For details, [2] is recommended.

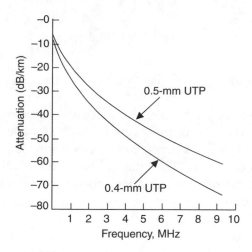

**Figure 6.4**   Cable loss versus length for 1-km length of cable terminated with 100Ω. (*Source:* [1]. Reproduced with permission of the IEE.)

**Table 6.1**   Typical Approximate Line Parameters for 0.5-mm UTP Cables

| Parameter | Value at Different Frequencies | |
|---|---|---|
| | dc | 1 MHz |
| Resistance/km *(R)* | 280Ω | 570Ω |
| Inductance /km *(L)* | 600 $\mu$H | 500 $\mu$H |
| Capacitance /km *(C)* | 50 nF | 50 nF |

### 6.2.2.5   Bridge Taps

Bridge taps are unterminated copper pair tails or sections of open circuited lengths created on copper pairs in service. These are simply depicted in Figure 6.5(a). In some countries (like the United States), it is common to have a sheathed bundle deployed past the customer premises and left unterminated at the end of the cable run. Sometimes to provide the service to a customer, the sheath is breached, a pair is selected and tapped, and a separate drop wire is coupled to the tap providing the service to the customer. This leaves the loop's original end unterminated. As shown in Figure 6.5(b), this situation causes variable-length bridge taps dependant on the customer distribution in the area. Another possibility is at the customer premises with more than one socket outlet for the phones, with some left unconnected to telephones. Another example of a bridge tap created by a repair is shown in Figure 6.5(c).

At baseband frequencies (300–3,400 Hz), bridge taps do not create difficulties. However DSL systems (which make use of frequencies up to a few megahertz) could be badly affected by these, as they create peaks of insertion losses at different frequencies, as depicted in Figure 6.5(d). Note that the cables without bridge taps have uniform drop of insertion loss with frequency, while bridge taps cause peaks at specific frequencies.

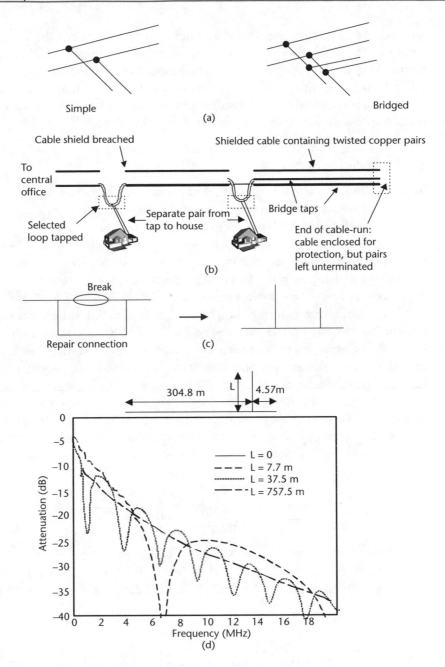

**Figure 6.5**  Unterminated bridge taps and their effect on insertion loss at higher frequencies: (a) bridge taps of different kinds, (b) schematic illustration of tapping copper pairs creating bridge taps, (c) example of creating two bridge taps by a repair, and (d) insertion loss due to different bridge lengths. (*Source:* [1]. Reproduced with permission of the IEE.)

## 6.3  DSL Techniques

A DSL modem (transceiver) is an all-digital device, in contrast to voiceband modems. The voiceband modem translates the computer's digital signals to voiceband signals carried across the switched telephone network, irrespective of the length of the channel or the number of switched connections in between. By contrast, a DSL modem establishes a digital circuit between the customer and the nearest CO only. Over this circuit, any combination of services may be employed, as long as they do not exceed the bandwidth limit of the modem. Figure 6.6 compares the reference models for voiceband modems and DSL modems.

### 6.3.1 Evolution of DSL

Around 1975, it was believed that 20 kbit/s was the highest data rate achievable on telephone lines. Advances in very-large-scale integrated circuits (VLSI) and DSP during 1975 to 1990 helped designers to develop higher speed modems with better error handling. By the mid 1980s, basic rate ISDN systems were allowing 144 kbit/s along the last mile of a loop, and during early 1990s, advanced voiceband modems were introduced with higher rates. In 1995, 33.6-kbit/s modems were introduced, making the Internet access faster for subscribers. By 1996, 56 kbit/s modems based on the V.90 standard were entering the market gradually. All voiceband modems were designed to operate over an end-to-end PSTN connection having only 4-kHz bandwidth. Bell Communications Research, Inc., developed the first DSL system in 1987 to deliver video on demand and interactive television over copper pairs. With the first efforts stalled, around the mid 1990s, interest in DSL regained momentum after it became clear that installing fiber-based broadband loops was too costly and time consuming. Another boost for DSL came with the U.S. Telecommunications Reform Act of 1996, which ended the local service monopolies and allowed competition

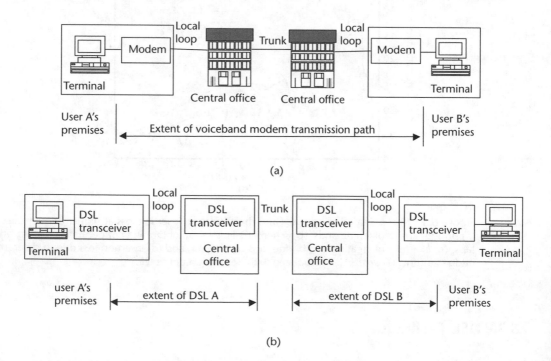

**Figure 6.6** Reference models for voiceband modems and DSL modems: (a) voiceband modems providing end-to-end service, and (b) DSL modem pairs at each local loop involved in a session.

among local phone companies, long-distance carriers, cable companies, radio and television broadcasters, and ISPs. As a consequence, local exchange carriers found themselves in need of a broadband service for the local loop to counter the competition from cable service providers. From the mid 1990s, DSL technology started maturing rapidly. Table 6.2 illustrates the progress in data transmission rates with approximate service introductions.

### 6.3.2   Basic System Components of DSL

Figure 6.7 indicates the basic arrangement of a DSL modem at the subscriber premises. Compared to a voiceband modem, DSL provides simultaneous voice and data using a larger spectrum of frequencies, by the use of suitable filters. Figure 6.7(a) indicates the arrangement of a highpass filter (HPF) and a lowpass filter (LPF) separating the voiceband and the higher frequencies. Figure 6.7 (b) indicates the case of multiple filters.

#### 6.3.2.1   The DSL Transceiver

DSL systems come in many variations generally referred to as xDSL. All xDSL systems use advanced signal processing techniques that force Claude Shannon's 1948 theory of channel capacity to its limit through the use of innovative modulation concepts, creating data rates at several megabits per second. Designing xDSL systems combines advanced DSP subsystems together with analog-front-end (AFE) circuitry, as depicted in Figure 6.8(a). With reference to Figure 6.6(b), DSP blocks work with the host computer or the network at the subscriber premises or with the broadband network interfaces at the network end. Figure 6.8(b) indicates a typical driver/receiver pair working with a line transformer converting signals into a single duplex channel within the last mile of the loop.

The AFE, or the blocks between the local loop and the DSP subsystem, consists of a line-transformer/hybrid interface, analog filters, line drivers and receivers, and the ADC and DAC. AFE, often implemented as an ASIC, usually provides most of

**Table 6.2**   Timeline of Voiceband Modems, ISDN, and DSL Introductions

| Year | Version | Data Rate |
|------|---------|-----------|
| 1955 | Bell 103 modem | 300 bit/s |
| 1970 | Bell 202 modem | 1,200 bit/s |
| 1981 | V22 bis modem | 2,400 bit/s |
| 1986 | Basic rate ISDN | 144 kbit/s |
| 1992 | High speed digital subscriber line (HDSL) | 1.5/2.0 Mbit/s |
| 1993 | V34 modem | 28.8 kbit/s |
| 1996 | PCM modem | 56 kbit/s |
| 1997 | Asymmetric digital subscriber line (ADSL) | Up to 7 Mbit/s |
| 1999 | Very-high-bit-rate DSL (VDSL) | Up to 52 Mbit/s |

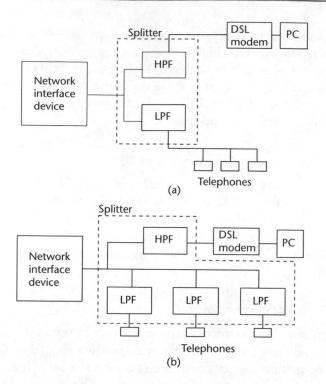

**Figure 6.7**   Filter (splitter) arrangements of a DSL service at the subscriber premises: (a) single LPF coupling several phone handsets, and (b) multiple LPFs.

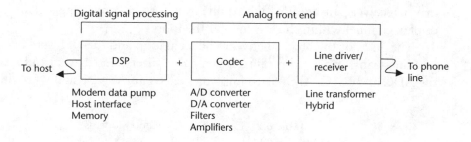

**Figure 6.8**   Concept of a DSP subsystem and an AFE in a xDSL system: (a) DSP working with AFE, and (b) hybrid circuit example. (*Source:* [5]. EDN: Reproduced with permission of Reed Business Information.)

the electronic circuit blocks, except for the line transformer. The DSP subsystem acts as a data pump, host interface, and the memory for the signal processing with a large set of advanced processing algorithms. Some of these techniques will be discussed later.

*AFE*

AFE circuitry linking the loop and the DSP subsystem plays a key role in successful and practical implementation of any xDSL system. The AFE must digitize or recreate analog signals with the following attributes:

- Excellent dynamic range;
- Drive capability for high-speed (fast-slew-rate) signals into the loop;
- Capture of weak incoming analog signals within the presence of strong transmission signals;
- Handling of complex loop impairments with predictable intrinsic and unpredictable extrinsic situations;
- Minimum power consumption.

Until the early 1990s, semiconductor technology did not provide cost effective mixed-signal electronic components suitable for the AFE. While the cost-effective and customizable DSP technology was maturing in the early 1990s, ASICs for AFEs were introduced by many semiconductor companies. The AFE in all xDSL versions needs to pump significant signal power on to the loop to ensure that the receive end power is sufficient for data recovery. Typical average levels of 100 to 400 mW are driven, with possible peaks of 5 to 10W.

The AFE design is a challenging task due to following reasons:

- Unpredictable the loop lengths (100 to 20,000 ft) and different standards and practices by telecommunications companies, making difficult specifications for the filters and programmable gain amplifier design;
- Intrinsic and extrinsic noise situations with unpredictable specifications, particularly for extrinsic ingress;
- With simultaneous voice and data transfer (in contrast to voice modems), phone-generated situations like off hook, dial tone, and ringing must be tackled while the modem is in data-transfer mode;
- Frequencies handled are few megahertz with the *crest factor* in the range of 12 to 15 dB (the crest factor of a waveform is the peak to rms value of a waveform);
- Tight total harmonic distortion (THD) figures on the order of –70 dBc;
- Limitations on power consumption.

Delivering low-distortion power to the load side is critical to the transmitting side of the xDSL link. Similarly, there are also important issues regarding the receiving end of the data path. Depending on the loop length, the receiver may have to attenuate signals up to 13 dB (for very short loops) or provide a gain up to 40 dB (for very long loops), creating a difficult design specification. Noise and resolution specifications of the data converters (ADC and DAC) lower the achievable data rates for longer loop lengths. Figure 6.9 depicts the limits [5].

### 6.3.2.2   Splitters

Many xDSL versions are expected to work in parallel with the POTS for the subscriber to use same loop for voice and data simultaneously. Baseband voice is at the lower end of the spectrum, while the xDSL uses frequencies well beyond 1 MHz. This demands a splitter circuit (see Figure 6.7), which has one or several filters to separate the POTS and xDSL signals. In practice, the filters are used at the network interface device (which separates the telephone company wiring from the

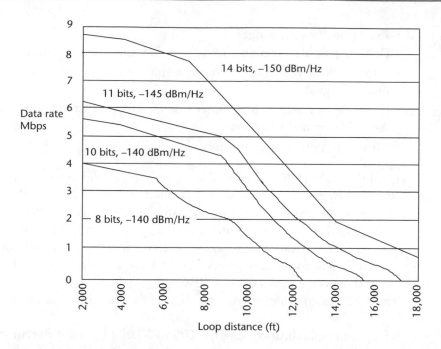

**Figure 6.9** Bidirectional distance as function of ADC/DAC resolution and receiver noise floor. (*Source:* [5]. EDN: Reproduced with permission of Reed Business Information.)

homeowner's wiring). However, some versions, such as G.Lite (discussed later), can work without splitters, minimizing the cost of installation and the expertise needed at the subscriber's premises, with the penalty of lower data rates.

### 6.3.3 xDSL Modulation Techniques and Modem Implementation

All transmission channels are analog and thus impose a variety of transmission impairments. All modems use some form of modulation/demodulation to convert a stream of bits into analog signals, and vice versa. In the transmission path, the bit stream from the host computer end is considered a message source and then goes through an encoding process. It is finally modulated before transmission over the analog pair. Let's discuss this complex processes in more detail.

Early DSL systems such as ISDN and HDSL used baseband transmission with multilevel signal encoding commonly known as pulse amplitude modulation (PAM). An example of a PAM scheme is the two-binary, one quaternary (2B1Q) coding used in ISDN.

Passband techniques include both single-carrier modulation techniques and multicarrier modulation techniques. Carrierless amplitude and phase (CAP) and quadrature amplitude modulated (QAM) schemes are single carrier systems. Discrete multitone (DMT) and discrete wavelet multitone (DWMT) are multicarrier systems. Different DSL systems use different line codes or modulation schemes, based on applicable standards. Emergence of DSP technology with FFT algorithm capability has allowed the implementation of these schemes, particularly the multicarrier modulation (MCM) schemes. Let's first discuss the essential fundamentals of CAP, QAM, and DMT.

### 6.3.3.1  QAM

QAM is a multilevel digital modulation scheme made up of a combination of amplitude and phase modulation. To illustrate the principles of QAM, let us take 16-level QAM (16-QAM) as an example.

The 16-QAM scheme divides the data stream into two parallel streams at half the original rate and converts each of the two streams to a four-level digital signal (PAM). These two streams are then modulated on two carriers of the same frequency, but 90° out of phase with respect to each other (i.e., *quadrature carriers*). This is illustrated in Figure 6.10(a).

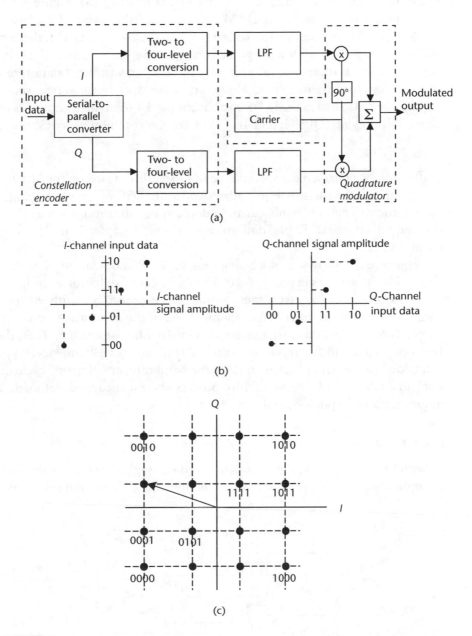

**Figure 6.10**  The QAM concept: (a) 16-QAM modulator, (b) constellation encoding, and (c) the constellation.

The constellation encoder generates four-level signals on the *I*- (in-phase) and the *Q*-channel (quadrature), as in Figure 6.10(b). Each of these signals amplitude modulates each of the quadrature carriers, resulting in four-carrier amplitudes. The summation of the two-quadrature carriers result in 16 different amplitude-phase combinations of the carrier at the modulated output. Each point in this *constellation* corresponds to a group of four consecutive input data bits. The constellation is shown in Figure 6.10(c). An arrow drawn to each point in the constellation from the origin in Figure 6.10(c) represents the corresponding carrier amplitude and phase.

The QAM scheme can be implemented in *M* levels in general, where $M = 2^{2L}$. These are in general referred to as M-ary QAM or M-QAM. Figure 6.10(a) can be taken as a modulator for M-QAM by replacing the two-to-four-level converter block with a $2L$ level converter, where $L = 1, 2, 3,...$ For M-QAM, the constellation will have *M* points, with each point corresponding to a group of $2L$ bits. As the number of levels increases, the spectral efficiency (transmitted data rate in a given channel bandwidth) increases. However, for a specific bit error rate, the SNR requirement also increases. In DSL applications, 4-QAM [also known as quadrature phase shift keying (QPSK)] and 16-QAM are the common versions.

### 6.3.3.2 CAP

CAP could be considered a special case of QAM. It is based on the phase-amplitude constellation, but the carrier is embedded in the DSP process that implements the modulator. It employs nonlinear time domain equalization, such as decision feed-back equalization (DFE) or Tomlinson precoding. Chapter 7 of [2] is suggested for more details.

Figure 6.11 shows a block diagram of CAP modulation. The constellation encoder is the same as in Figure 6.10. The *I* and *Q* constellation encoder outputs are sent through two filters with same amplitude response, but with 90° phase difference (i.e., quadrature filters). After summation of the quadrature signals, the data is locked into an implied carrier embedded in the filter algorithm. Then the result is D/A converted and filtered to smooth the DSP sampling frequency. Typically, the embedded carrier frequency is equal to the baud rate, creating one cycle of the "carrier" in each symbol interval. Entire process shown is carried out using DSP techniques with no explicit modulation.

### 6.3.3.3 MCM

In both QAM and CAP, the modulation is on a single carrier. It is real in QAM and is implicit in CAP. By contrast, MCM divides the data stream into multiple lower

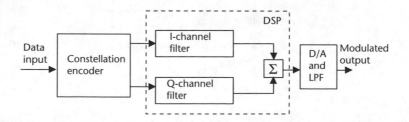

**Figure 6.11**  The CAP concept.

rate streams modulated on different carriers and transmitted in parallel. Figure 6.12(a) compares a single carrier and a multicarrier system in spectral terms. In effect, MCM is a form of FDM. MCM divides the loop bandwidth into several narrow channels, each carrying a portion of the data.

MCM techniques are attractive for DSL, as attenuation and crosstalk are frequency dependent. Let's recall that the Shannon's channel capacity theorem (1.2) relates the information capacity of a channel to its SNR and the bandwidth. If single carrier modulation is used in a DSL, the channel capacity is governed by the average SNR across its entire bandwidth. However, the SNR varies across the loop bandwidth due to frequency dependency of impairments. If the loop bandwidth is divided into a number of narrow channels, Shannon's theorem is applied to each channel separately. This allows the selective use of different QAM techniques. At a frequency where impairments are high, a simpler modulation technique such as QPSK can be used, while at frequencies where the impairments are low, a spectrally efficient scheme such as 16-QAM can be used. In effect, the modulation technique can be adapted to suit local conditions of the loop. In extreme cases, where the channel impairments are severe, such frequencies can be eliminated from use.

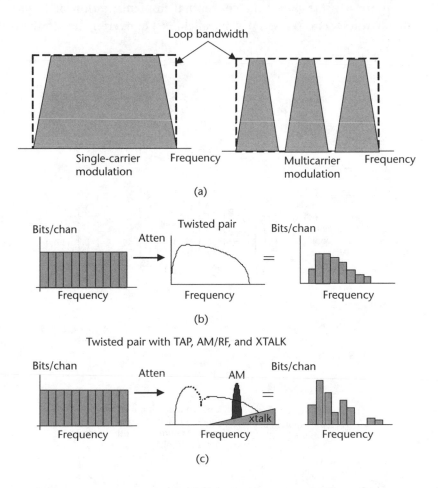

**Figure 6.12**  MCM: (a) frequency use comparison in single carrier and multicarrier schemes, (b) data rate allocation on a basic local loop with MCM, and (c) data rate allocation on a local loop with impairments.

By distributing the transmit data among channels according to local conditions, with adaptive modification of the data allocations according to changing loop conditions, MCM techniques allow higher data rates compared to single-carrier modulation schemes in the presence of loop impairments. DMT modulation is an MCM technique using a large number of separate carriers on the last mile of a loop.

Figures 6.12(b) and (c) explain the concept of bit allocations in DMT. As in the leftmost sketch of Figure 6.12(c), a relatively wide bandwidth channel (i.e., a short loop) can be used as several FDM channels, where each carry equal bits per channel. However, if the pair has attenuation characteristics, such as in the middle of the Figure 6.12(c), the bits per channel could vary according to individual subchannel capacities. This is shown in the right-hand sketch, where each channel's bit rate follows the attenuation (or SNR) versus frequency characteristics of the channel. Channels with attenuation is low, SNR is high, can carry a higher data rates. However, the case becomes more complex when there are bridge taps, RFI, and crosstalk, as in the case of Figure 6.12(c). The overall effect is to have different number of bits per subchannel, dependant on the channel characteristics, and to reach a better overall bit rate compared to QAM or CAP schemes.

Figure 6.13(a) shows the conceptual implementation of MCM. The transmit data stream is sent to a serial to parallel (S/P) converter that converts the bit stream

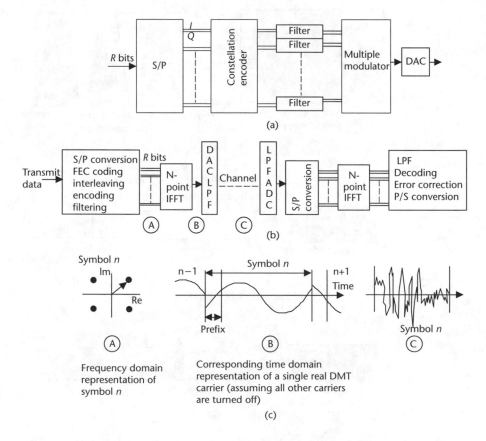

**Figure 6.13**  Multitone concept implementation: (a) block diagram of a MCM transmitter, (b) basic multicarrier modem, and (c) transmit signal from the DMT modem (*Source:* [1], reproduced with permission of the IEE).

to multiple parallel streams. As in Figure 6.13(a), input to the S/P converter is a sequence of symbols of $B$ bits each. The output of S/P converter for each symbol is $N_c$ groups of $b(n)$ bits each. That is,

$$B = \sum_{n \le N_c} b(n) \tag{6.3}$$

The groups of $b(n)$ are then constellation encoded, filtered, and modulated onto $N_c$ subcarriers. The modulated signals are then summed and transmitted on the loop. More details are available in [6, 7].

A block diagram of a multicarrier modem is shown in Figure 6.13(b). The main processing in the transmitter and receiver involve digital FFT and IFFT algorithms, which are implemented with the present DSPs and ASICs. The group of modulated signals are generated by the FFT and then converted to the time domain using IFFT.

Figure 6.13(c) is an example where the lowest frequency carrier has four-state (two bits per carrier, or QPSK) encoding, and the state in the top right quadrant has been selected. The incoming raw data from the digital interface is scrambled, interleaved, and mapped to relevant frequency domain signals and then transmit gain scaled. Then, this encoded data on each carrier is processed through an IFFT. A cyclic prefix is added, which protects against ISI and obviates the need for time domain equalization. The right end of figure shows the corresponding time domain of a single real DMT carrier and assumes that all other carriers are turned off (for explanation purposes only). The signal then passes through a DAC and is filtered before transmission. The right-hand end of the figure shows schematically a symbol with all carriers turned on.

Table 6.3 indicates strengths and weaknesses of the DMT scheme. This technique can cope well with unpredictable RFI environments, but performance depends on several factors, including the exact position of an interferer relative to a DMT carrier. Details are available in [1].

## 6.4   Different Versions of DSL

All DSL technologies run on existing copper loops and use advanced modulation techniques, compared to voice grade modems. Different variations of DSL as per Table 6.4 are available for different applications. These are:

**Table 6.3**   Advantages and Disadvantages of DMT

| Advantages | Disadvantages |
| --- | --- |
| Dynamic bit allocation ability to maximize the available spectrum despite interferers | A large crest factor |
| Flexible power spectral density control | Long symbol length giving an inherent latency and processing delay |
| Longer symbol length with resilience to short noise impulses | A high perceived conceptual complexity |
| An inherent programmable data rate capability | A long start-up (training) procedure of several seconds |

(*Source:* [1].)

**Table 6.4**  Characteristics of Different DSL Systems

| Acronym | Standard or Document[1] | Number of Wire Pairs | Modulation | Payload Data Rate (Mbit/s) | Mode | Distance[2] | Applications | Splitter |
|---|---|---|---|---|---|---|---|---|
| HDSL | G.991.1(first generation) G.991.2 (second generation) | 1–3 | 2B1Q/CAP | 1.544–2.048 | Symmetric | ≤ 5 km, ≤ 12 km with repeaters | T1 or E1 service access | None |
| - do - | T1E1.4 Tech report 28 | 2 | 2B1Q/CAP | 1.544–2.048 | Symmetric | ≤ 5 km, ≤ 12 km with repeaters | T1 or E1 service access | None |
| ADSL (G.dmt) | G.992.1 or T1.413 Issue 2 | 1 | DMT | Downstream: ≤ 6.144, Upstream: ≤ 0.640 | Asymmetric | 3.6 km at maximum data rate | Internet access, video on demand, simplex video, LAN access, interactive multimedia | At entrance |
| ADSL Lite (G.Lite) | G.992.2 | 1 | DMT | Downstream: ≤ 1.5, Upstream: ≤ 0.512 | Asymmetric | Best- effort service | Internet access | No entrance splitter, but micro filter is used |
| VDSL[3] | G.vdsl (working title) | 1 | DMT,QAM | 2–26 upstream, ≤ 26 or 52 downstream | Symmetric or asymmetric | ≤ 300m at maximum data rate | To be used in *hybrid fiber/ copper* systems to connect optical network units | Not decided |
| SDSL | G.shdsl (working title) | 1[4] | Trellis-coded pulse-amplitude modulation (TC-PAM) | 0.192–2.32 | Symmetric | 2 km at maximum data rate | Feeder plant, LAN, WAN, and server access | None |
| HDSL2[3] | T1E1.4 | 1[4] | TC-PAM | 1.544 – 2.048 | Symmetric | 5 km | Feeder plant, LAN, WAN, and server access | None |

(*Source*: [8].)

1. G numbers are for ITU documents; T1 numbers are from ANSI T1.413 Committee.
2. Loop reach for 0.5-mm (AWG 26) wire with no bridged taps.
3. The standard is still under development; parameters may change.
4. Two-wire pairs may be used for longer loops.

- High-bit-rate-DSL (HDSL);
- ADSL;
- Rate-adaptive DSL (RADSL);
- VDSL;
- Single-line DSL (SDSL).

### 6.4.1  HDSL

High-bit-rate, high-data-rate, or high-speed DSL systems are based on a two-pair access technology for achieving symmetrical data transmission rates confirming to 1.544 Mbit/s (T1) or 2.048 (E1) standards using either baseband 2B1Q or pass band CAP modulation schemes. The HDSL spectrum in the copper loop is between 0 and 300 kHz for CAP (CAP64) and 0 to 425 kHz for 2B1Q data transmission. HDSL provides two-way 1.5- or 2-Mbit/s rates over 3.7 km (12,000 ft) of 0.5-mm

(24 AWG) twisted pairs. While over 95% of the HDSL systems use no repeater, a repeater in the mid span can increase the loop length. Practical HDSL systems provide typical BERs of $10^{-9}$ to $10^{-10}$ with the assurance of $10^{-7}$ BER similar to ADSL systems.

T1 systems use two pairs of wires, with each pair conveying 768-kbit/s net data rate in both directions. This technique is called dual duplex transmission, as depicted in Figure 6.14. E1 systems have the option of using two or three pairs. Dual-duplex 2B1Q transmission is used for almost all HDSL systems worldwide, while some DMT and CAP systems are used in some parts of Europe. For T1 systems, a 768-kbit/s data rate and 16-kbit/s overhead make the total figure of 784-kbit/s net rate on each pair. For more details, see [2, 9, 10, 31].

## 6.4.2 ADSL

ADSL technology provides data rates between 640 kbit/s and 1 Mbit/s over copper loops of length between 12,000 ft and 18,000 ft from the customer premises to the local exchange (upstream). The bit rate towards the customer (downstream) is much greater than the upstream direction (up to 9 Mbit/s); hence the term *asymmetric*. Analog voice is simultaneously transmitted at baseband frequencies and combined with the data via a LPF (splitter). In addition to splitters, an ADSL system consists of an ADSL transmission unit (ATU) at the CO side (ATU-C), and an ATU at the remote side (called ATU-R). Figure 6.15 shows the reference model for ADSL. Different interfaces defined in the ADSL reference model, shown as V, U-C, U-R, T, and POTS-C and POTS-R make the systems compatible with multivendor equipment.

The ADSL concept definitions began in 1989, primarily at Bellcore. Early developments began at Stanford University and AT&T Bell labs in 1990 with prototype designs available by 1992 and field trials by 1995. At first, ADSL was conceived at a fixed rate of 1.5 Mbit/s downstream and 16 kbit/s upstream for video applications

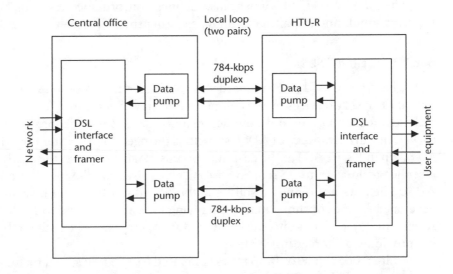

**Figure 6.14** HDSL dual duplex transmission system. (*After:* [2, 31].)

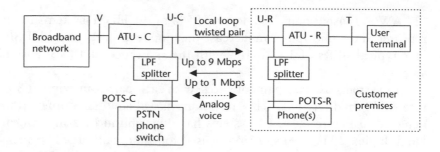

**Figure 6.15**  ADSL reference model.

and were called ADSL1. Consequently, ADSL 2 (with 3 Mbit/s down stream and 16 kbit/s upstream) and ADSL 3 (with 6 Mbit/s downstream and 64 kbit/s upstream) versions were developed. ADSL 3 became the basis for issue 1 of the ANSI T1.413 ADSL standard. Terms ADSL1, ADSL2, and ADSL 3 were replaced by ADSL after this standard.

### 6.4.2.1   ADSL Standard—ANSI T1.413

The ANSI T1.413 standard defines ADSL modems that use the DMT technique, a subset of MCM concepts [6]. The technique uses 256 tones (or subcarriers), with each conveying several bits. The exact number depends on impairments. The ANSI standard specifies categories I and II of performance. To comply with the standard, a design must meet category I of the basic feature set. Category II enables higher bit rates, longer distances, and better performance under poor loop conditions and accomplishes these using techniques such as trellis coding, echo cancellation, and transmission-power boost. More details are available in [11]. Category I allows bit rates up to 6.144 Mbit/s (upstream) and 640 kbit/s (downstream), with category II allowing 8 Mbit/s and 1 Mbit/s, respectively.

The ANSI T1.413 document also defines conformance tests for DMT ADSL modems simulating impairments and measuring the BER (see Chapter 8).

### 6.4.2.2   G.Lite ADSL

ADSL based on the T1.413 standard was considered more complex and too depend-ant on the PSTN services by the computer industry. Therefore a less complex, "lite" technique called G.Lite and now designated G.992.2 was proposed. As discussed earlier, one disadvantage of ADSL systems is the need for a splitter (or a diplex filter, as shown in Figure 6.7) at the customer premises. This requires the costly element of a service technician to be sent to customer premises for the installation. G.Lite or ADSL Lite does not require a splitter and is a medium-bandwidth version of ADSL developed for the consumer market, allowing Internet access up to 1.5 Mbit/s down-stream and up to 500 kbit/s upstream. G.Lite operates over existing telephone wir-ing providing a simpler installation.

G.Lite modems use only 96 tones compared to 256 tones used by full-rate ADSL modems (see Figure 6.16). The fewer tones limit the upper frequency limit to 420 kHz, a data rate of about 1 Mbit/s (upstream) and 512 kbit/s (downstream).

G.Lite modems are designed to be compatible with the full-rate ATU-C systems at the CO side by not allowing the ATU-C to assign data to channels outside those shown in Figure 6.16. Due to less bandwidth used, G.Lite modems use less power—almost 3 dB lower than full-rate ADSL.

### 6.4.3    G.Lite and Microfilters

Theoretically, the G.Lite service could coexist with POTS without customer-premise splitters, thus eliminating the visits of technicians (*truck roll*). However, approximately 80% of the G.Lite field trials around 1999 had required one or more of so-called *microfilters* [12, 13]. These small LPFs that replace the large splitter filters help eliminate the objectionable phone noise or severe modem performance degradation. Figure 6.17 shows the comparison of ADSL, splitterless G.Lite, and the use of microfilters.

To explain the filtering needs, let's look at the simplified spectrum use in POTS and ADSL depicted in Figure 6.17(d). ADSL frequency use commences around 25 kHz and reaches over 1.1 MHz. Although simple spectral compatibility is apparent within the band, in practical systems each band exhibits impedance effects,

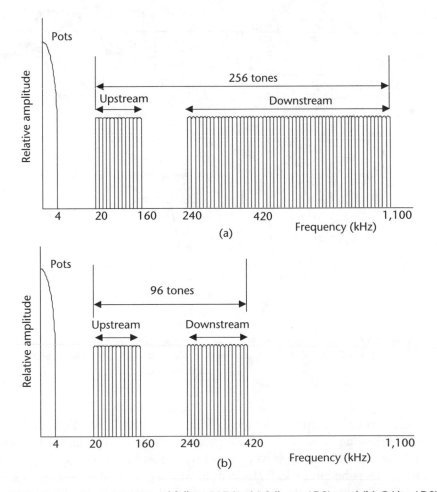

**Figure 6.16**    Comparison of G.Lite and full-rate ADSL: (a) full-rate ADSL, and (b) G.Lite ADSL.

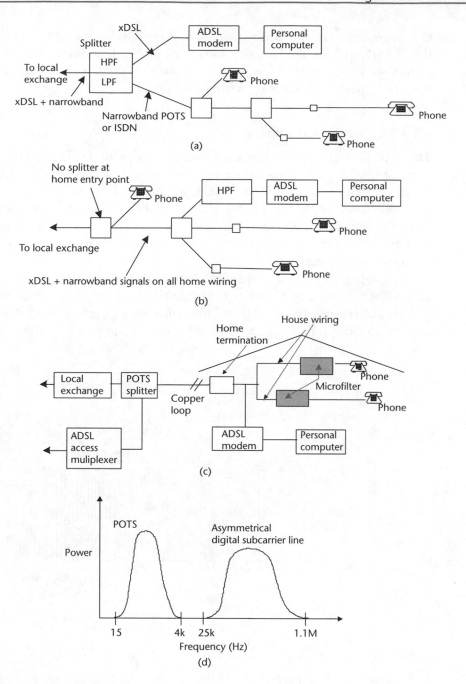

**Figure 6.17** Use of filters in ADSL systems and spectrum usage: (a) traditional ADSL with splitters, (b) splitterless G.Lite, (c) use of microfilters in G.Lite, and (d) simplified spectrum use in POTS and ADSL.

which interfere with broadband data. Standards (like Federal Communications Commission part 68 regulations) clearly state that a telephone shall not transmit signals beyond 4 kHz. However, these do not mention anything about possible impedance effects beyond 4 kHz. Indeed, most telephones demonstrate a substantial load on the tip/ring interface at frequencies as high as several hundred kilohertz. The impedance in the POTS band to the phone line changes from a virtual open circuit (when on hook) to a few hundred ohms when off hook. Within the ADSL frequency range, the off-hook impedance of some telephony devices can present $600\Omega$ starting from 25 kHz to less than $10\Omega$ towards the upper range of ADSL signals around 1 MHz.

The filters in Figure 6.17(a) and (b) basically assist in keeping each side (POTS device and the ADSL) isolated from the unintended frequencies from the other band. Many telephones could dissipate a small percentage of the ADSL modem energy without a noticeable impact on voice. However, some telephones dissipate this energy as background noise in the ear piece. With splitterless G.Lite, each telephone (irrespective of its immunity to noise pickup) could change the circuit impedance substantially during the transitions from off hook to on hook, forcing the ADSL modem to retrain. This could typically take over 10 seconds, with the possible interruption of ADSL data. To prevent objectionable disconnections, the developers of G.Lite conceived a fast retraining within about two seconds, by relying on previous experiences of impedance change [12]. Unfortunately, for some applications, such as voice over DSL (VoDSL), even a two-second training and associated disconnection may be unacceptable.

Because noise and modem problems are potential customer support problems, many carriers have decided to use microfilters on G.Lite systems to prevent retraining and noise pick up problems. For details, [12] is recommended.

### 6.4.4 RADSL

RADSL is a term that applies to ADSL systems capable of automatically determining the transport capacity of the individual loop and operating at the highest possible rate. The T1.413 standard provides for rate-adaptive operation. RADSL exploits the adaptive nature of the DMT technique, and the operation is similar to the voiceband modems. RADSL provides faster bit rates on loops with less cable loss and better SNR. RADSL implementations support maximum downstream rates from 7 to 10 Mbit/s with corresponding upstream rates from 512 to 900 kbit/s. Long loops with over 18,000 feet may support only 512 kbit/s downstream with 128 kbit/s upstream.

### 6.4.5 VDSL

VDSL is an extension of ADSL technology to higher rates up to 52 Mbps. At high bit rates, the copper loop must be so short that a combination of optical fiber and copper loop should be used. Compared to the case of ADSL/HDSL using copper loops from the CO, VDSL will primarily be used for loops from an optical networking unit (ONU), located within a kilometer from the customer. Optical fiber is used to couple the ONU to the CO. Figure 6.18 indicates the connection between CO, ONU, and the customer premises using a VDSL modem.

VDSL supports data rates of 6.5 to 52 Mbit/s downstream and 1.6 to 6.5 Mbit/s upstream for asymmetrical services. Over short copper loops, 6.5- to 26-Mbit/s rates are possible for symmetrical services. Spectral usage for VDSL extends from 300 kHz up to 30 MHz. Table 6.5 indicates the VDSL requirements developed by the T1E1.4 standards working group.

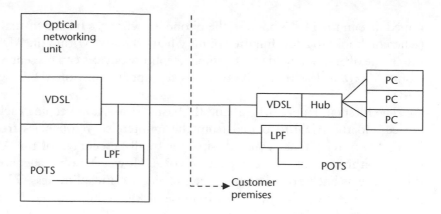

**Figure 6.18**  The VDSL model.

**Table 6.5**  VDSL Requirements

| Loop Length (m/kilofeet) | Downstream Rate (Mbit/s) | Upstream Rate (Mbit/s) |
|---|---|---|
| 300/1,000 | 52 | 6.4 |
| 800/2,500 | 26 | 3.2 |
| 300/1,000 | 26 | 26 |
| 600/1,800 | 13 | 13 |
| 1200/3,750 | 13 | 1.6 |

(*Source:* [2].)

Candidate modulation schemes for VDSL include M-ary CAP, M-ary QAM, and DMT. To allow simultaneous transmission of voice (0–4kHz) and ISDN signals (0–80 kHz), T1E1.4 suggests placing the modulated spectrum between 1 MHz and 2 MHz for the upstream and 2 MHz and 18 MHz onwards for the downstream band. For details, see [2, 7, 9, 14]. Symmetric VDSL modems are used for business services such as data networks or LAN extensions, while asymmetric systems are used for Internet or video on demand services.

### 6.4.6  SDSL

SDSL (or symmetric DSL) is an umbrella term for a number of supplier-specific implementations over a single copper pair, providing variable rate symmetric service from 160 kbit/s to over 2 mbit/s. The service may be with or without POTS. SDSL is optimized to support applications such as Web hosting, collaborative computing, LAN access, and video conferencing, which require higher upstream capacity than ADSL provides. SDSL is generally aimed at business markets rather than the residential use.

In this single-pair implementation, 2B1Q or CAP techniques are used with state-of-the-art echo cancellation and adaptive equalization techniques. This is to achieve symmetric data transmission rates such as 384 kbit/s, 768 kbit/s, 1 Mbit/s, 1.5 Mbit/s, or 2 Mbit/s, with the sub-T1/E1 rates. With the low-cost DSPs with gigaflop capabilities, SDSL modem implementation is becoming practical and may compete with HDSL modems in the future. SDSL modulation schemes and data transfer rates are not standardized yet.

### 6.4.7  Second Generation HDSL

In all DSL implementations, effective data rates reduce with reach. This trade-off is critical in defining the coverage area for a DSL service. The spectral compatibility issue is particularly complex as the DSL deployment reaches critical mass. Twisted-pair loops are in groups of 25 or more, and these are called *binder groups*. Inside binder groups, when there is a mix of ADSL, HDSL, and telephony transmissions, the systems must not interfere with each other. As for the number of pairs required for a given transmission, specially for home services, the ideal is to use only one, as it is the cheapest in cost.

To accommodate these growing demands, during the mid 1990s, service providers suggested that ANSI come up with a robust, symmetrical DSL service that can reach an overwhelming customer base, with the key criteria of guaranteed performance. Second generation HDSL (HDSL2) was the outcome around 2000. A more generalized version, G.shdsl, was also developed by ITU. G.shdsl is a rate-adaptive service (with rates from 192 kbit/s to 2.3 Mbit/s) primarily aimed at the consumer and SOHO markets.

HDSL2/G.shdsl standards provide reliable services over 20,000 ft on a single pair. The symmetric bandwidth and low latency of these standards are able to provide multiple digital voice channels within the data stream. Complexity differences compared to HDSL systems are:

- A higher transmitter power of 3 dB;
- Complex echo canceler schemes;
- A higher computational power required of DSPs for error correction.

## 6.5  Spectral Considerations in DSL

DSL evolved from the concept that the copper pair can carry signals up to a few megahertz, provided the impairments due to crosstalk and RF interference are tackled by error-handling techniques. This concept has been proven by the implementation of access techniques described in Section 6.4. This section discusses the spectral issues arising out of the loop impairments seen by DSL signals, as well as those arising out of the need to share the loop with the POTS service.

### 6.5.1  Duplexing Variants of ADSL Systems

There are two duplexing variants for ADSL systems. The first one is the case where echo cancellation (EC) is required to separate the upstream and downstream channels, which occupy overlapping frequency bands, as shown in Figure 6.19(a). The alternative is to use FDD with separate bands for the upstream and downstream, as shown in Figure 6.19(b). The former is called echo-canceled ADSL (EC-ADSL), and the latter is called FDD-ADSL.

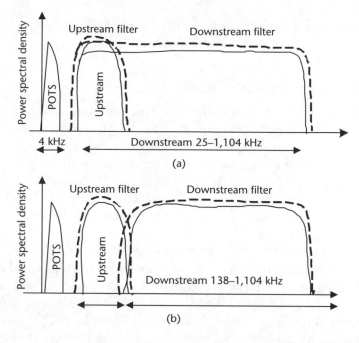

**Figure 6.19**  Comparison of EC-ADSL and FDD-ADSL: (a) echo-canceled version, and (b) FDD version.

The FDD implementation requires simpler filters and lesser demand from the analog components as well as DSP components. In theory, EC-ADSL can achieve a higher downstream bandwidth but on shorter loops. In practice, on longer lines there is a critical point where FDD could support a higher data rate than EC. Furthermore, the FDD systems suffer from FEXT talk rather than the more damaging NEXT [15]. Installing additional FDD-ADSL modems do not harm the existing ones.

With the technology developments related to semiconductors and software, newer services are utilizing lower power levels, which minimize crosstalk but increase bandwidth. Upstream ADSL uses more power than downstream ADSL. This assumes that the nominal background noise on a twisted pair is no larger than −140 dBm.

### 6.5.2   Spectral Compatibility

Copper pairs grouped into binders may carry different types of services, from POTS and ISDN to xDSL systems, causing crosstalk, which could limit the bit rate and the loop reach. Management of crosstalk requires care in the bandwidth and signal power of transmitters and out-of-band signal rejection by receivers. This is often referred to as *spectral compatibility* and is reminiscent of the management of RF broadcast systems with regard to RF power and spectrum use. In DSL systems, spectral compatibility is a commonly used term referring to the degree of mutual crosstalk between various services. It also refers to RF emissions.

Figure 6.20 indicates the spectrum usage of DSL systems compared with POTS and ISDN. The process, referred to as *unbundling*, allows alternative

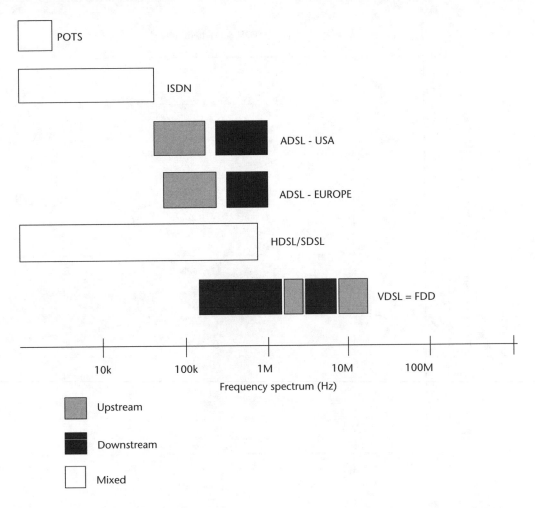

**Figure 6.20**  Existing and tentative spectrum usage in DSL systems compared with POTS/ISDN (*Source:* [8]. ©2000 IEEE.)

operators to use the copper pairs installed and owned by the local, national, or monopoly telephone operators. This makes the binders carry many different types of services within a given local exchange area.

With focus turning to broadband access based on xDSL services, spectral compatibility has become in important concern [8]. The strongest influence of unbundling on the engineering of equipment is on the issue of compatibility or the coordination of the power spectral emissions.

As broadband services and DSL connections proliferate, the important concern is that one type of service will lead to errors in another type of modem sharing the same cable. With ISDN, ADSL, HDSL, and VDSL deployed on the same binder, the possibility increases that a given situation of unacceptable crosstalk might be overlooked. For example, Figure 6.21 indicates the typical levels of NEXT and FEXT compared to the received and transmitted signal power for a 500-m cable with nine VDSL disturbers.

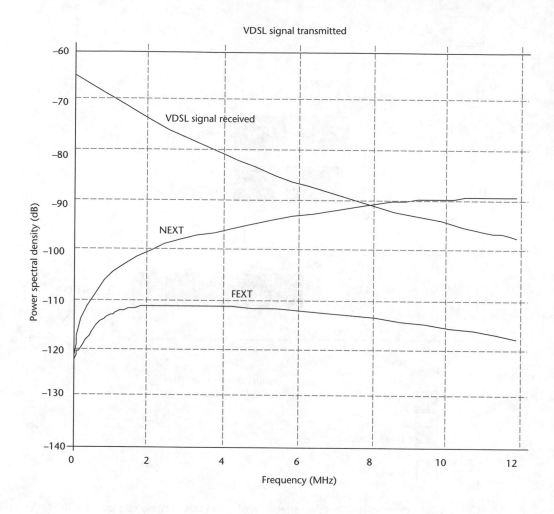

**Figure 6.21**   Crosstalk levels compared with VDSL transmit/receive signals (*Source*: [8]. ©2000 IEEE.)

As Figure 6.21 shows, at higher frequencies, crosstalk can be relatively large enough to disturb or disrupt another service. For these reasons, DSL standards define the power spectral density (PSD) masks for different cases of xDSL services, for upstream and downstream ADSL. A comprehensive discussion on this subject is available in [2].

## 6.6   Operation of DSL

Practical implementation of DSL systems were possible with the advances of semi-conductor families such as DSPs, ADCs, and DACs for the digital blocks and the basic analog parts such as current feedback op amps [16] for the AFEs . As details of xDSL modem designs are beyond the scope of this chapter, the following paragraphs highlight a few important operational aspects of DSL modems.

Establishing and maintaining a DSL session requires heavy processing in a DSP or an ASIC, tackling many complex activities, such as initialization, adaptation, timing recovery, and performance measurements.

### 6.6.1   Initialization

Initialization is usually accomplished through a handshaking startup sequence that executes the following steps in order:

- Activation;
- Gain setting and control;
- Synchronization;
- Channel identification;
- Equalization;
- Secondary channel handling.

Echo cancellation, if used, occurs within any segment following the gain setting. Early systems with lower performance executed only a subset of these, while ADSL and VDSL execute all steps. A generalized sequence diagram is shown in Figure 6.22.

#### 6.6.1.1   Activation

Activation is to notify each end of the modem pair to establish the service. This occurs by one of the modems transmitting an initialization request and the other modem acknowledging the request. Once the handshake commences, the initialization process continues based on variations applicable to each type of the xDSLs. For example, in T1.413-based ADSL systems, 32-ms duration tones selected from four possible values (207, 189.75, 224.25, or 258.75 kHz) are transmitted downstream and responded by a different frequency in the upstream. The signals are first sent at a higher amplitude of –4 dBm and then lowered down to –28 dBm for periods of 16 ms each from the ATU-C, and acknowledgments are awaited from the ATU-R with tones selected from frequencies of 43.125, 34.5, or 60.375 kHz for 16-ms periods, each at –2 dBm and –22 dBm, respectively. Each tone is followed by 32-ms silent

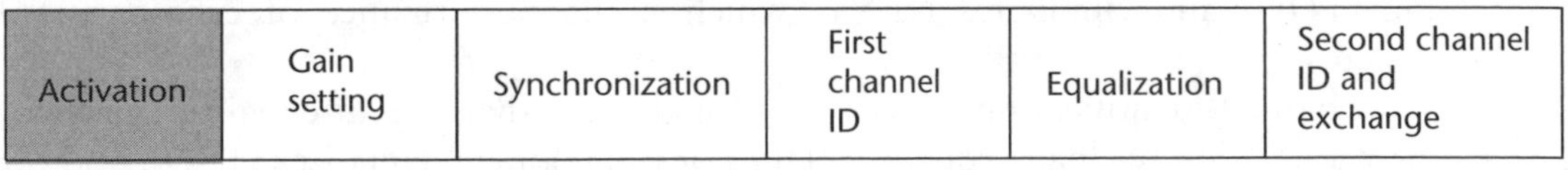

| Activation | Gain setting | Synchronization | First channel ID | Equalization | Second channel ID and exchange |
|---|---|---|---|---|---|

Figure 6.22   General timeline for initialization of a DSL system.

periods to avoid overlap and echo problems. In the case of HDSL activation, a different process is used [2].

### 6.6.1.2   Gain Setting and Control

Setting the gain of AFE blocks is important, as it affects cost and the radiated emissions. DSL systems can span a wide dynamic range of attenuation, with nine orders of magnitude. Hence the transmitted signal amplitude varies, and the receiver can accommodate signals within 30-dB range. On shorter loops, transmitters may need to reduce the power up to 20 dB to prevent saturation and to minimize crosstalk.

A DSL transmitter commences with a low level, increasing gradually until the receive end responds with a sufficient level. On achievement of correct level, the transmitter power is maintained. This operation is referred to as *power control*. For ADSL systems, a standardized transmission system uses measured upstream power level versus a downstream PSD [2].

Receive end uses automatic gain control (AGC) to scale the received signals for optimal ADC use. The AGC output level is typically set to a level equal to the peak to average ratio (PAR) below the maximum possible voltage at the ADC input using DSP-assisted gain control algorithms.

### 6.6.1.3   Synchronization

For accurate performance, synchronization of the clocks at either end is very crucial. During the initial synchronization a midband sinusoid is transmitted in the downstream, and the zero crossings are detected using a phase locked loop (PLL) technique for clock recovery.

### 6.6.1.4   Channels and Channel Identification

DSLs use several channels of information simultaneously for different applications or services. For example, ISDN has two B channels for information, a D channel for signaling, and an embedded operations channel (EOC) for control and maintenance. ADSL has data channels, an EOC, and a separate band for analog voice. HDSL has one wideband channel and an EOC. For multiple channel usage TDM, FDM, or separate physical channels (space division multiplexing) can be used. ADSL places voice in the lowest frequency band and data in a higher frequency band using FDM. HDSL applies space division multiplexing, with two pairs for 1.5 Mbps and three pairs for 2 Mbps. TDM is the most commonly used method. In addition to sending multiple channels in the same direction, it can also be used for duplexing. Alternate transfer of information in upstream and downstream is called time division duplexing (TDD) and eliminates NEXT, which limits performance of echo-canceled systems.

Channel identification is necessary for designs that use an equalizer, particularly a decision feedback equalizer (DFE) or time domain equalizer (TEQ), and for DMT. Channel identification, in a single step for HDSL or twice for some ADSL, measures the channel impulse response and noise PSD or SNR. Multitone channel identification directly estimates signal and noise parameters for each of the subchannels.

### 6.6.1.5    First Channel Identification

As an example, let's discuss the first channel identification in T1.413 ADSL modems with DMT. First, the impulse-response DFT values and the noise-sample variances at the subchannel center frequencies are identified. This process allows the calculation of SNR. The pulse-response DFT value measurement is called the *gain estimation,* and the noise variance measurement is termed the *noise estimation.* Details are discussed in [2].

### 6.6.1.6    Second Channel Identification

A second channel identification is performed, after the equalizer training and setting (discussed later). This second step allows the SNR measurement with the equalizer in place, compared to theoretical projections of gain and noise estimations. In this process, several parameters and imperfections are measured. Secondary channel is used for feedback of channel characteristics for setting of any channel-dependant transmit parameters using a low-speed and highly reliable transmission format to convey information without error. For example, in T1.413, two sets of four tones independently carry the same information at 2 bits/tone (one byte) with an additional CRC check.

### 6.6.1.7    Equalization

As discussed in Section 6.1, due to attenuation and noise at higher frequencies, successively transmitted symbols interfere with one another. This phenomenon, called intersymbol interference (ISI), is a major impairment in DSL systems. Equalization is a term used to denote techniques to reduce the mean square ISI.

   Channel equalization for DSL is usually separated into two parts: analog and digital. The overall attempt is to minimize the mean-square error characteristics for the channel. As depicted in Figure 6.23, an analog filter with low-pass characteristics in general is placed within the channel to create a flat frequency response.

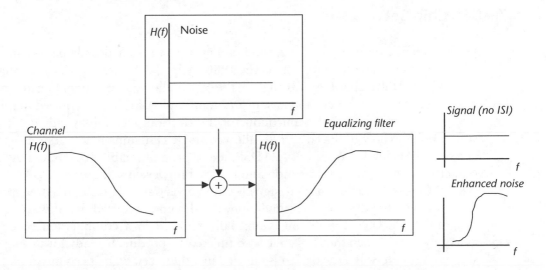

**Figure 6.23**    Illustration of analog filter (linear equalizer) and the associated noise enhancement.

However, this creates increased noise levels. This effect, called *noise enhancement,* makes analog filtering inadequate for error-free operation. To minimize the noise-dependant errors, digital filtering techniques are used. For details, Chapters 7 and 8 of [2] are recommended. Information on DMT-based systems is available in [17].

### 6.6.2   Adaptation of the Receiver and Transmitter

By the use of adaptive algorithms and other advanced techniques [2, 4], the receiver and transmitter are continually adapted to perform better, despite such characteristics as:

* Slow variations of characteristics of the loop due to temperature and other environmental factors;
* Abrupt changes of crosstalk and RFI;
* Channel impedance changes due to situations such as off hook (specially in G.lite).

### 6.6.3   Performance Measurements

For QoS in DSL systems, the service provider should guarantee the loop performance by a series of field tests based on the possible worst-case scenarios. Such performance measurements are discussed in Chapter 8.

### 6.6.4   Timing Recovery

Timing recovery is the extraction of the symbol rate and the phase by the receiver. A timing recovery process, known as *loop timing,* often occurs at the customer premises modem, and the recovered clock is used for both data detection as well as modulation in the reverse direction. For loop-timed DSLs, only one modem needs to recover symbol timing; the other modem uses the same clock for transmission and reception. Discrete-time PLL techniques are used for this process [2].

## 6.7   DSL Chip Sets

Development experience on voiceband modems over a four-decade period and the rapid advances in semiconductors in the 1980s helped designers tackle data rates up to 56 kbit/s. DSP technology advances in the 1990s allowed error handling and compression in 56-kbit/s voice modems. In 1993, multiple DSPs were required to implement ADSL transceivers. Semiconductor fabrication technology advances and software knowhow in the late 1990s allowed single DSP implementations as well as custom VLSI chips or ASICs. While DSPs allow programmability, ASICs allow cost reduction in mass-scale fabrication and lowered power consumption.

ASIC-based or DSP-based designs begin in a very similar fashion. After specification, algorithm design, and simulation, the design is realized in software. For a programmable DSP, the design is normally realized in a combination of C and assembly code. When the DSP code is complete, the product is ready for testing and shipment. In an ASIC, the design is realized in either very-high-speed integrated circuit (VHSIC) hardware description language (VHDL) or the Verilog language.

When the Verilog or VHDL code is completed, the design must go through synthesis, timing analysis, and layout prior to chip fabrication. Then the fabricated chip is integrated into the system and tested for overall performance.

### 6.7.1 An Example of a DSP Implementation—A DMT System

In an xDSL chipset, total power consumption is based on the digital circuit blocks and the AFE. During 1980s to 2000, power consumption of DSP chips has decreased by three orders of magnitude [18]. When the consumption of the digital circuit blocks drops below the AFE's power consumption, both DSP and ASIC solutions for xDSL hardware becomes very practical.

Several xDSL systems use DMT modulation technology. Figure 6.24 depicts the general block diagram of a DMT system for ADSL and VDSL. In the transmit direction, data from the digital interface is scrambled, Reed-Solomon encoded, and interleaved. Interleaved data is then mapped onto a sequence of constellation points with one constellation point for tone in a DMT symbol. In ADSL, the mapping may include a 16-state trellis encoder. Each tone is then multiplied by a gain-scaling coefficient to control the subchannel's power. The frequency domain symbol is then transformed into the time domain by an IFFT. A cyclic extension is added to the time domain symbol, and, in VDSL, a nonrectangular window is applied to reduce out-of-band energy [18, 19]. The resulting digital signal is converted to analog by a DAC, and the signal is transmitted on to the loop.

The received analog signal is converted to digital by means of the ADC. In echo-canceled ADSL, as upstream and downstream bands overlap, an echo-cancellation block is needed to remove the echo of the transmit signal from the received signal. In ADSL, a TEQ is used to shorten the channel impulse response [18, 20], and the received time domain signal is windowed (using a simple rectangular window) and converted to individual DMT symbols. In VDSL, it is

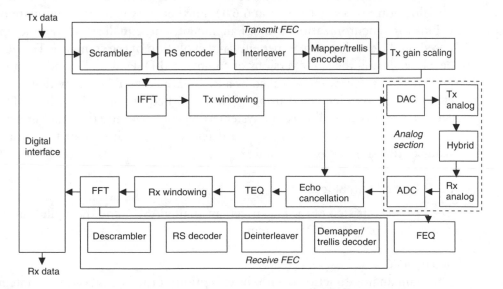

**Figure 6.24** A DMT system block diagram applicable for ADSL and VDSL. (*Source:* [18]. ©2000 IEEE.)

important to use a nonrectangular window to reduce spectral leakage due to side lobes [19]. Time-domain symbols are converted to the frequency domain using an FFT. The demapper function makes corrective decisions on the noisy constellation points. If trellis coding is used in ADSL , the demapper function includes trellis decoding. Following the demapper, the output byte stream is deinterleaved, the Reed-Solomon code words are decoded, and the data is unscrambled and sent to the digital interface. For DSP implementation details, [18] is suggested.

### 6.7.2 ASIC Solutions

Early developments (1970–1990) of digital telephony were mainly by large telecom companies. In the 1990s, software value addition became an important developmental element, and third parties were able to make serious contributions. However, hardware implementation allowed large telecom companies to own intellectual property rights of essential basic semiconductors as well as ASICs. The following section indicates the essentials of developing ASIC-based ADSL modems.

#### 6.7.2.1 Design Process

As an example, let's discuss the ASIC design process of the FDD-type ADSL transceiver discussed in Section 6.5.1 and based on [15], which summarizes the design experience of a telecom product design team in the early 1990s.

An emulator system was built using off-the-shelf DSPs, dedicated ICs (for FFT function), a Reed-Solomon encoder/decoder, ADCs and DACs, line drivers, and an analog ASIC, as indicated in the block diagram in Figure 6.25(a). At the top level, the transceiver consists of three main functional blocks: the driver, the analog functions, and the digital functions. The choice of FDD simplifies each of these functions for an ASIC-based design.

The design approach was to build a programmable hardware system while accommodating the details of the evolving ADSL standard with the then-undefined initialization process (see Section 6.6.1). Further, it was to accommodate the algorithms for synchronization, symbol alignment, and equalization, and be optimized based on the development experience and evolution of the standard. The design was developed using the early versions of commercial DSPs with limited computational power. The system was controlled by a PC and the digital board, with memories and custom logic (implemented using FPGAs) worked as the link with the DSP susbsytems. This process allowed the system to synchronize, equalize, measure noise performance, track signal changes, and detect error conditions of the entire system while optimizing the AFE as well.

#### 6.7.2.2 Building Blocks

The system consisted of the line driver, an analog ASIC, and the digital subsystem to works as a complete ADSL system.

*Analog ASIC*

The main analog functions, with the exception of the line driver, were integrated in an analog ASIC shown in the middle part of the Figure 6.25(a). The same analog part was used at either end of the subscriber loop. In the upstream direction, a higher

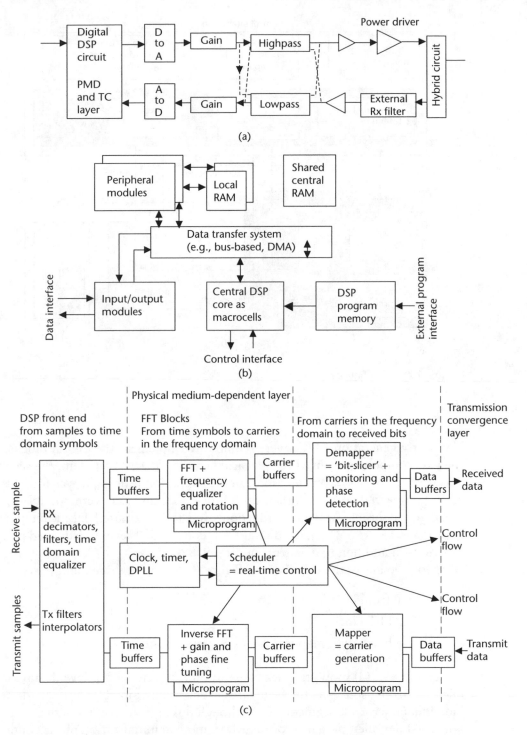

**Figure 6.25** ASIC design process of an FDD-type ADSL: (a) overall block diagram, (b) signal processing core and peripheral functions in the ASIC, (c) block diagram of digital ADSL signal processor (DASP), and (d) interaction of software modules in microprocessor and DASP. (*Source:* [15]. ©2001 IEEE.)

sampling speed and two filters were included for FDD separation. The filters were switched from upstream receive to downstream transmit when necessary.

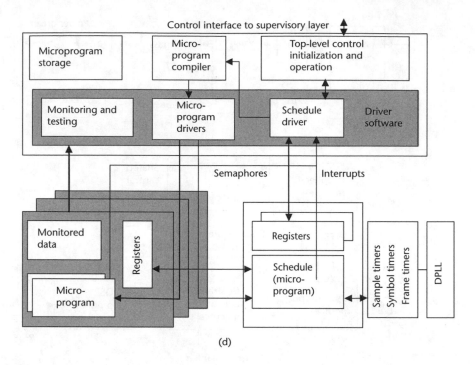

**Figure 6.25**   (continued).

*Digital ASIC*

Two early generation ASICs were required to implement the digital blocks. While the first component realized the physical layer in the DMT modem, the second one was used for the TC layer functions. In the emulation process, it was found that the early commercial DSPs with their prohibitive price and size were not able to implement all of the physical layer functions and the associated DMT bit pump. As depicted in Figure 6.25(b), the digital ASIC consisted of several dedicated peripheral functions in addition to a general-purpose DSP.

More details of the DASP are in Figure 6.25(c) with three main parts:

1. DSP front end;
2. The FFT block;
3. The block with mapper, demapper, and monitor.

Figure 6.25(d) shows the clear separation between the top-level behavior of the microcontroller software, the firmware needed to download the microcode, and the individual microcode segments. Even though this appears complex, the entire transmit scheduler microprogarm for the ADSL modem initialization phase contains less than 256 reduced instruction set computer (RISC) commands. For details of the design process, [15] is suggested.

This is a classic example of an ASIC design for an ADSL modem, where a number of real-time DSP functions are realized in silicon. Evolution related to the standard was accommodated by the modifications of microprograms. The DASP was to run on a slow master clock speed of 35 MHz to keep the power consumption to a minimum, with overall processing power achieved by parallel processing techniques, and the equivalent computational power (MIPS rating) of the system per silicon area was much higher than a programmable DSP solution.

### 6.7.2.3 Evolution of ASICs and Their Impact on ADSL Designs

The early implementation of this ASIC solution for the digital block was by two large digital ASICs in 0.7-$\mu$m complementary metal-oxide semiconductor (CMOS) technology. These were merged into a single component in the second generation, and in the third generation, the DMT-DSP (called SACHEM) was achieved with a transistor count of 4.8 million in 0.35-$\mu$m CMOS with a power dissipation of 800 mW using a slow master clock speed of 35 MHz.

A simplified block diagram of the third generation ASIC solution is shown in Figure 6.26. For details, [21] is suggested. In this situation, the SACHEM device performs the digital functions, and the block with data converters and filters (indicated as ADSLC in Figure 6.26) implements the AFE while discrete components are used for line drivers.

In a later version (2001), the components on each side of the subscriber loop are differentiated, and multiple units (SACHEMs) have been integrated in a single chip with power consumption of 100 mW or less per transceiver. This indicates a classic example of a design effort typical of the telecom industry. The specific example is by Alcatel.

### 6.7.3 Practical Building Blocks—Standard Components for DSL Implementation

The 1990s were the maturity period for the xDSL technology, with several component manufacturers introducing many catalogue semiconductors for DSL systems. This included both digital parts and the mixed-signal components. Some examples are depicted in a representative sense in the following sections, and the reader is expected to visit Web sites of the chip manufacturers and designers for up-to-date information.

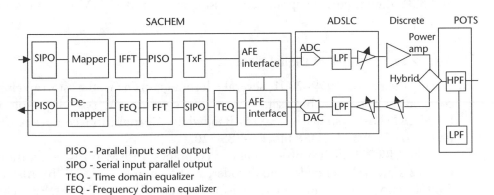

PISO - Parallel input serial output
SIPO - Serial input parallel output
TEQ - Time domain equalizer
FEQ - Frequency domain equalizer
TxF - Transmit framing

**Figure 6.26** Simplified block diagram of ADSL modem in third generation. (*Source:* [21]. ©1999 IEEE.)

### 6.7.3.1  Line Drivers and AFEs

DSL signals are in the range of megahertz compared to voiceband modems, and the SNR issues become complex. Varying characteristics of the loop pose challenges in xDSL systems, due to loading coils (some U.S. companies install them every 6,000 ft to improve voiceband performance), bridge taps, crosstalk, and RFI. For example, the crosstalk effect is worse for upstream than for downstream signals in ADSL. Also, the impact of a bridge tap is a function of the tap length and the signal wavelength [22]. For these reasons, amplifiers and drivers for xDSL systems need be configured using special component families such as current feedback op amps [16]. Some solutions use discrete line driver chips, while some AFEs integrate the line drivers within the ASIC.

While no single component set simultaneously optimizes all factors, the range of integration in chip sets varies from discrete driver amplifiers (such as AD8016 from Analog Devices, Inc.) to chip sets that comprise drivers, receivers, filters, converters and data pumps, and bus interfaces. Power dissipation is a major factor for line drivers. Vendors often package the drivers separately to allow semiconductor manufacturing process optimizations.

Common line interface designs include dual drivers and receivers forming balanced input/output channels behind a hybrid transformer. Examples are National Semiconductor's components such as LMH 6672 (a dual high-speed line driver) and LMH6654/6655 (single/dual low noise xDSL receivers for CO or CPE).

Figure 6.27(a) depicts a fully integrated AFE from Silicon Laboratories [22], where the following blocks are fully integrated:

- Transmit/receive digital and analog filters;
- A 14-bit DAC and ADC;
- A line driver and receiver;
- A voltage controlled crystal oscillator (VCXO).

With a handful of passive components, a crystal, and a transformer, the device supports 8 Mbit/s downstream and 800 kbit/s upstream, supporting ITU G992.1 (G.dmt) and G992.2 (G.Lite) specifications. The device operates from 5V analog and 3.3V digital power supplies and dissipates around 1W. More details are found in [22].

### 6.7.3.2  An Example of Digital Blocks

During the period 1999–2001, several telecom giants as well as startup chip design companies released chip sets for xDSL systems. A two-chip solution for HDSL2/G.shdsl systems by Excess Bandwidth Corporation comprises a tunable AFE (EBS 710) and a DSP block—EBS 720.

The EBS 720 digital processing end combines a PAM transceiver, framer, and the 512-state trellis encoder and decoder, as shown in Figure 6.27(b). Also included are the echo canceler, precoder, feed-forward equalizer, and the DFE . The device uses a R4000, 32-bit embedded 75-MIPS RISC processor. This CMOS chip consumes few hundred milliwatts and supports speeds between 192 kbit/s and 304 Mbit/s. For details, [23] is suggested.

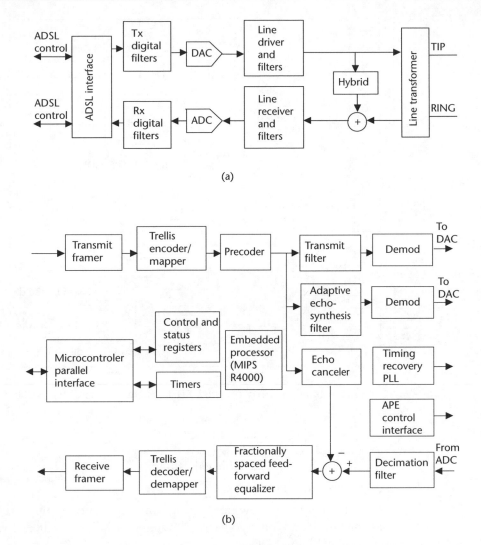

**Figure 6.27** ICs for DSL systems: (a) an AFE from Silicon Laboratories—Si3101 (*Source:* [22]; EDN reproduced with permission of Reed Business Information), and (b) EBS 720 digital block for HDSL-2 systems (*Source:* [23]; courtesy of *Electronic Design Magazine*).

Host-based architectures, where part of the xDSL processing happens in the CPU of portable equipment, such as laptop computers or personal digital assistants, are also being introduced into the market. One such example is the Integrated Telecom Express, Inc.'s scalable ADSL modem chip set [24].

### 6.7.4  Higher Levels of Integration in ASICs for ADSL

#### 6.7.4.1  Highly Integrated AFEs

One such example is an AFE by Analog Devices, Inc., which allows the use of the same IC, both at the CO and the RT sides, by the use of programmable components such as filters under the command of a separate DSP. Figure 6.28 indicates the chip architecture. It has four major components:

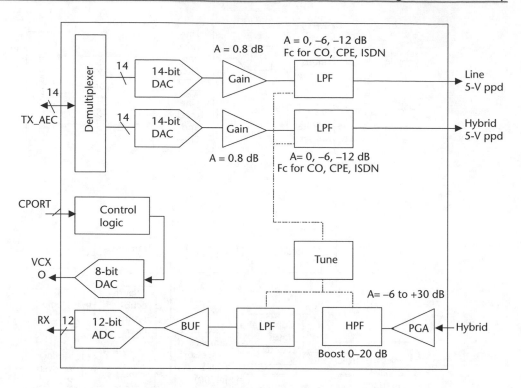

**Figure 6.28**  An AFE suitable for Category I and II ADSL systems. (*Source:* [25].  ©2000 IEEE.)

1. Transmit channel with variable gain filters;
2. Analog echo cancellation;
3. Receive channel with a low noise, programmable gain amplifier (PGA) of a gain range of 36 dB, an equalizer, a lowpass antialiasing filter, a wideband buffer, and a 12-bit ADC;
4. Auxiliary and support circuitry—a DAC to drive a VCXO and a serial interface to the DSP.

This 0.6-$\mu$m CMOS IC consumes only 525 mW on a 5-V power supply [25]. Another IC that integrates a four-channel AFE for CO ADSL modems is discussed in [26].

### 6.7.4.2   System–on-Chip for ADSL

With the semiconductor processing reaching systems-on-chip (SOC) capability [27], several DSP blocks could be within a single IC. These allow designers to realize multiport ADSL data DSPs for CO and DSL access multiplexer (DSLAM) applications. One such example is the ADSL-Lite Data DSP (ALiDD), a four-channel system for G.Lite.

Figure 6.29 shows the details of G.lite system realized with the AliDD, based on a DMT technique, with a maximum of 128 tones, where 2 to 15 bits per tone is implemented.

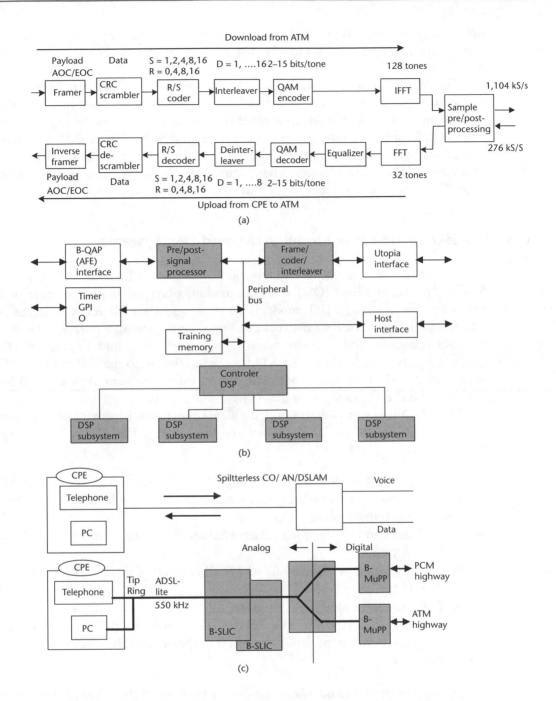

**Figure 6.29** ADSL data DSPs for G.Lite system: (a) G.Lite system architecture, (b) simplified AliDD architecture, and (c) integrated voice and data line card. (*Source:* [28]. ©2001 IEEE.)

The ALiDD has five programmable 16-bit DSPs with a licensable core from DSP group [29]. Four of these are associated with each G.Lite data pump, and the fifth is a controller DSP. The controller DSP is used for initialization sequences, management layer functionality, administration, and control, and it also performs the arbitration of common memory interface. This architecture provides the flexibilityto adapt to any future standards. Figure 6.29(b) shows the simplified AliDD architecture.

The AliDD is one of the devices of the B-MuSLIC chipset targeted for CO and DSLAM applications in the broadband market segment. It combines voice and broadband services in a highly integrated chip set and eliminates the need for a discrete POTS splitter by integration of the broadband SLIC. Figure 6.29(c) indicates the target application. For more details, [28] is recommended. At the time of writing, multiport full-rate ADSL system architecture was also under development by Infineon Technologies, developer of the AliDD chip set.

## 6.8   The ISO Seven-Layer Model, ATM, and DSL Systems

The International Standards Organization (ISO) seven-layer model, the Open Systems Interconnection (OSI) reference model, plays an important role in the xDSL modem designs. ADSL modems work frequently with ATM switches (see Chapter 3) in the public data networks or broadband systems. Figure 6.30(a) shows the ISO reference model, where networking equipment could include an xDSL modem. Figure 6.30(b) shows the ATM stack in relation to the OSI model. Digital blocks of an xDSL system work with the physical medium dependent (PMD) sublayer and the TC sublayer of the ATM.

The PMD sublayer contains only the PMD functions. It provides bit transmission capability, including bit alignment and line coding. The TC sublayer handles five functions:

1. Generation and recovery of the transmission frame, which contains the cells;
2. Transmission frame adaptation, which adapts the cell flow according to the physical transmission;
3. Cell delineation function, which enables the receiver to recover the cell boundaries;
4. The header error correction, which allows errors in the cell header to be detected;
5. Cell rate decoupling, which inserts idle cells in the transmission direction in order to adapt the rate of the ATM cells to the physical payload capacity of the transmission system. It surpasses all idle cells in the receiving direction.

Framing of the data and associated details are beyond the scope of this chapter. For more details, Chapters 10 through 14 of [2] is suggested.

## 6.9   End-to-End ADSL Systems

ADSL systems allow broadband services to conveniently connect a computing device to a broadband network. Figure 6.31 indicates the systems and responsibilities involved. The layers indicated in the previous section can be very broadly visualized here.

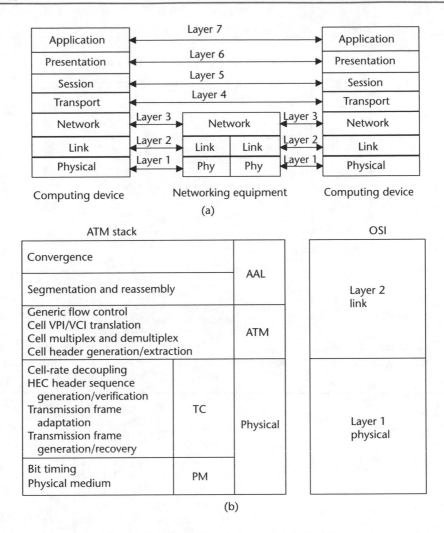

**Figure 6.30** ISO seven-layer model and ATM reference model: (a) ISO seven-layer model, and (b) OSI link and physical layers and ATM reference model.

In this situation, there are three realms or domains of responsibilities: service provider network realm, access network realm, and customer premises realm. This is shown in Figure 6.31(b). In these systems, the broadband service providers such as ISPs and video on demand providers need to work closely with PSTN service providers.

Figure 6.32 indicates how an ATM access network can work in the DSL environment. In this case the DSLAM can reside in the nearest PSTN local exchange.

## 6.10 Standards

With the unprecedented growth of data communications and broadband services, development of standards have become a complex, time-consuming, and evolving process.

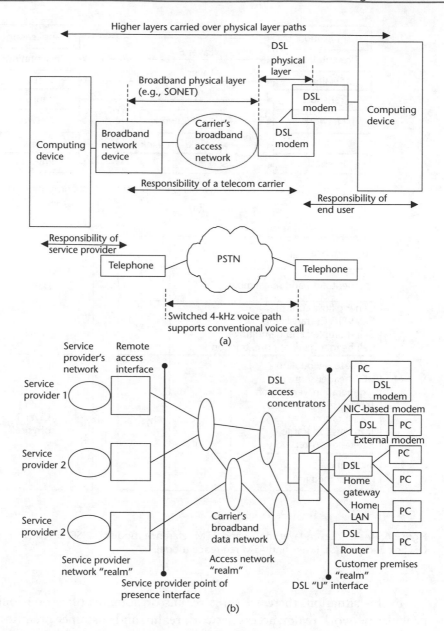

(a)

(b)

**Figure 6.31** DSL systems in an end-to-end environment and generic access architecture: (a) DSL modems in an end-to-end environment, and (b) generic DSL access architecture.

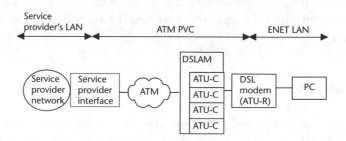

**Figure 6.32** DSL access for ATM networks.

The DSL community has generated an array of standards, based on the rate of development of semiconductors. Table 6.6 summarizes some significant ones, which are close to reaching ITU formal approval [30]. ADSL2 increases data rates by devoting twice the frequency spectrum from 1.1 to 2.3 MHz, allowing 25 and 20 mbit/s on local loops up to 3,000 and 5,000 ft, respectively.

Figure 6.33 shows the groups involved in the standardization processes.

Four base groups are international organizations, regional organizations, national organizations, and implementation forums. A good overview of this process and the applicable and developing standards are indicated in [2]. As it is

**Table 6.6**   ADSL Flavors

| ADSL Flavor | Top Downstream Speed (Mbit/s) | Description |
|---|---|---|
| ADSL (G.922.1) | 8 to 10 | Standard ADSL |
| ADSL2 (G.992.3, G.dmt.bis) | ~12 | Enhanced ADSL; offers slight rate and reach increases over ADSL, plus rate adaptation and diagnostic features |
| ADSL2+(G.992.5) | ~25 | Double-spectrum version of ADSL2 |
| ADSL2++(G.992.?) | ~50 | Quad-spectrum version of ADSL2; not yet standardized |

(*Source:* [30].)

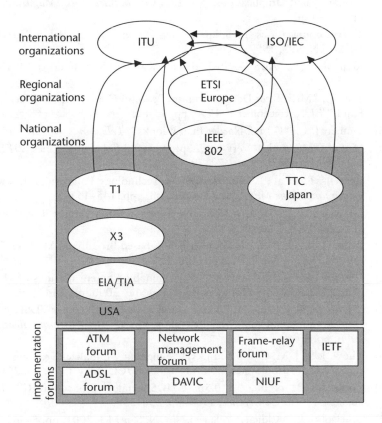

**Figure 6.33**   DSL-related standards organizations. (*Source:* [2].)

beyond the scope of this chapter to discuss these details, an interested reader should keep in touch with the current literature on the subject.

# References

[1] Czajkowski, I. K., "High Speed Copper Access: A Tutorial Overview," *Electronics and Communications Journal,* IEE, June 1999, pp. 125–148

[2] Starr, T., J. M. Cioffi, and P. J. Silverman, "Understanding Digital Subscriber Line Technology," Englewood Cliffs, NJ: Prentice Hall, 1999.

[3] Foster, K. T., and J. W. Cook, "The Radio Frequency Interference (RFI) Environment for Very High-Rate Transmission over Metallic Access Wire Pairs," ETSI TM3 TD29, Bristol, U.K.

[4] Clerck, L., et al., "Mitigation of Radio Interference in xDSL Transmission," *IEEE Communications Magazine,* March 2000, pp. 168–173.

[5] Schweber, B., "Analog Front Ends Bridge the xDSL-to-Real World Chasm," *EDN,* April 1,1999,pp. 48–64.

[6] Bingham, J. A. C., "Multicarrier Modulation for Data Transmission: An Idea Whose Time Has Come," *IEEE Communications Magazine,* May 1990, pp. 5–14.

[7] Bingham, J. A. C., "ADSL, VDSL and Multicarrier Modulation," New York: Wiley Interscience, 2000.

[8] Ödling, P., B. Mayr, and S. Palm, "The Technical Impact of the Unbundling Process and Regulatory Action," *IEEE Communications Magazine,* May 2000, pp. 74–80.

[9] Bhagavath, V. K., "Emerging High-Speed xDSL Access Services: Architectures, Issues, Insights, and Implications," *IEEE Communications Magazine,* November 1999, pp. 106–114.

[10] Telecommunications Techniques Corporation, "HDSL Basics," Technical note TBOSP HDSL TN-5/97, 1997.

[11] Kempainen, S., "ADSL—The End of the Wait for Home Internet," *EDN,* October 10, 1996, pp. 52–70.

[12] Sun, T., "Microfilter Design Promises Peaceful Coexistence Between ADSL and the Voice Band," *EDN,* December 9, 1999, pp. 55–62.

[13] Tanner, J. C., "G.Lite: Harder than It looks," *Telecom Asia,* July 1999, pp. 42–44.

[14] Cioffi, J. M., et al., "Very-High-Speed Digital Subscriber Lines," *IEEE Communications Magazine,* April 1999, pp. 72–79.

[15] Reusens, P., et al., "A Practical ADSL Technology Following a Decade of Effort," *IEEE Communications Magazine,* October 2001, pp. 145–151.

[16] Kularatna, N., "Modern Component Families and Circuit Block Design," Butterworth-Newnes, 2000, Chapter 2.

[17] Pollet, T., et al., "Equalization for DMT-Based Broadband Modems," *IEEE Communications Magazine*, May 2000, pp. 106–113.

[18] Wiese, B. R., and J. S. Chow, "Programmable Implementations of xDSL Transceiver Systems," *IEEE Communications Magazine,* May 2000, pp. 114–119.

[19] Mestdagh, D. J. G., M. R. Isaksson, and P. Ordling, "Zipper VDSL: A Solution for Robust Duplex Communication over Telephone Lines," *IEEE Communications Magazine,* May 2000, pp. 90–96.

[20] Chow, J. S., J. M. Cioffi, and J. A. C. Bingham, "Equalizer Training Algorithms for Multicarrier Modulation Systems," *Proc. of IEEE ICC,* May 1993, pp. 761–765.

[21] Cornil, J. P., et al., "A 0.5m CMOS Analog Front-End IC," *Digest of Technical Papers—ISSCC Conference,* 1999, pp. 238–239.

[22] Israelsohn, J., "Addicted to Speed," *EDN,* Sept. 13, 2001, pp. 54–62.

[23]   Mannion, P., "Chip Set for Symmetrical DSL Doubles Twisted Pair's Reach," *Electronic Design*, June 26, 2000, pp. 70–76.

[24]   Mannion, P., "Host-Based ADSL Modem Halves Power and Cost," *Electronic Design*, May 1, 2000, pp. 72–74.

[25]   Guido, J., et al., "Analog Front End IC for Category I and II ADSL," *IEEE Symp. on VLSI Circuit, Digest of Technical Papers*, 2000, pp. 178–181.

[26]   Kenney, J., et al., "A 4 Channel Analog Front End for Central Office ADSL Modems," *IEEE Custom Integrated Circuits Conf.*, 2000, pp. 307–310.

[27]   Kularatna, N., *Digital and Analogue Instrumentation-Testing and Measurement*, IEE Publishing, 2003, Chapter 2

[28]   Jain, R. K., et al., "System-on-Chip Design of a Four Port ADSL-Lite Data DSP," *The 2001 IEEE Symposium on Circuits and Systems*, Vol. 4, May 6–9, 2001, pp. 242–245.

[29]   Cravotta, R., "2002 DSP Directory," *EDN*, April 4, 2002, pp. 33–57.

[30]   Miller, M., "Bandwidth Bloom," *EDN*, May 29, 2003, www.e-insite.net/edn.

[31]   ITU-T: "G999.1–High Bit Rate Digital Subscriber Line (HDSL) Transceivers," 10/98, 1998.

## CHAPTER 7

# Transmission Techniques

## 7.1 Introduction

Transmission media utilized in modern telecommunications systems are a mix of unshielded copper cables, coaxial cables, microwave links, fiber-optic cables, and wave guides operating over a wide range of frequencies from a few kilohertz to gigahertz and terahertz. While the basic copper loop carries baseband voice frequencies, with the exception of DSL, where frequencies up to several megahertz are transmitted, most metallic cable systems carry frequencies of the order of megahertz aided by multiplexing techniques.

In present digital telecommunications networks, technologies such as DSL, plesiochronous digital hierarchy (PDH), synchronous digital hierarchy (SDH), and synchronous optical network (SONET) are used for transmission. In these systems, digital bit streams are formed by stacking digitized information channels in an organized hierarchical way, coupled with digital modulation techniques. While lower bandwidth systems are based on metallic cables or radio systems, higher bandwidth systems are mostly optical based.

This chapter is a review of basic of transmission system fundamentals and an introduction to digital microwave and optical systems.

## 7.2 Transmission Line Behavior

### 7.2.1 Lumped and Distributed Circuits

When the physical dimensions of electrical components are insignificant for the operation of an electrical circuit, the elements are *lumped*. When the physical dimensions of an element significantly affect the propagation of signal information, the element is *distributed*. In electrical circuits, lumped elements permit the use of Kirchhoff's circuit laws, and the analysis of circuits is quite straightforward. However, with distributed elements, Kirchhoff's laws fail, and the mathematics of the system become more complex. For example, consider applying Kirchhoff's current law to a transmitting antenna where the current enters at the feeder but does not exit in the classical sense.

Distributed parameter models are used in real transmission lines, such as feeder cables or waveguides, in the true sense of transporting RF energy from one location to the other. However, with the usage of high-speed digital circuits as well as microwave frequency signals on printed circuit boards (PCBs), designers are forced to consider even the metal interconnections on PCBs as transmission lines. In these situations, even though it is hard to distinguish transmission lines from simple interconnections, some rules can be effectively utilized for determining whether a signal path is lumped or distributed [1–3]. A practical criterion by which a system can be judged as lumped or distributed for the purposes of digital electronics is discussed in [4].

### 7.2.2 Propagation on a Transmission Line

The classical way of modeling transmission lines that have distributed parameters is to assume a structure of resistances ($R$), inductances ($L$), and capacitances ($C$), as in Figure 7.1. By studying this equivalent circuit, several characteristics of the transmission line can be determined. If the line is infinitely long, $R$, $L$, $G$, and $C$ are defined per unit length, and several useful parameters as per Table 7.1 can be defined and used to characterize the line.

The coefficients $\rho$ and $\sigma$ indicate the level of matching achieved in a transmission line and can be measured with transmission test equipment. Again, if a system

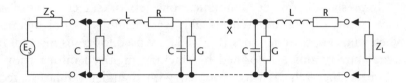

**Figure 7.1**   Classical model of a transmission line.

**Table 7.1**   A Summary of Transmission Line Parameters and Their Significance

| Parameter | Symbol | Expression | Significance |
|---|---|---|---|
| Characteristic impedance | $Zo$ | $Zo = \sqrt{\dfrac{R + jwL}{G + jwC}}$ | Used to match input and output ends of a transmission line |
| Attenuation coefficient | $\alpha$ | | Gives the attenuation per unit length |
| Phase shift coefficient | $\beta$ | $V_p = \dfrac{w}{\beta}$ | Gives the phase shift per unit length |
| Velocity of propagation | $V_p$ | $V_p = \dfrac{Vc}{\sqrt{E_r}}$ | Propagation velocity along a transmission medium is a fraction of the velocity of light $V_c$ and related to the relative permittivity of medium $E_r$ |
| Propagation constant | $\gamma$ | $\gamma = \dfrac{\alpha + j\beta}{\sqrt{(R + jwL)(G + jwC)}}$ | Indicates the level of attenuation and the phase change along the line |
| Voltage at a distance ($x$) | $E_x$ | $E_x = E_{in}e^{-\gamma x}$ | Provides the actual voltage at a given distance from the feeding point |
| Current at a distance ($x$) | $I_x$ | $I_x = I_{in}e^{-\gamma x}$ | Provides the actual current at a given distance from the feeding point |
| Voltage reflection coefficient | $\rho$ | $\rho = \dfrac{Z_L - Z_O}{Z_L + Z_O} = \dfrac{e_r}{e_i}$ | Indicates the level of reflection along the transmission line where $e_i$ and $e_r$ are forward and reflected voltages, respectively |
| Voltage standing wave ratio (VSWR) | $\sigma$ | $\sigma = \dfrac{1 + |\rho|}{1 - |\rho|}$ | The magnitude of the steady state sinusoidal voltage along a line terminated in a load other than $Z_0$ varies periodically as a function of distance between a maximum and minimum value; this variation is caused by the phase relationship between incident and reflected waves and provides a picture of the degree of signal reflection present on the line |

consists of a connector, a short transmission line, and a load, the measured VSWR indicates only the overall quality of the system. It does not indicate which of the system components is causing the reflection. It also does not indicate if the reflection from one component is of such a phase as to cancel the reflection from another. The engineer must make detailed measurements at many frequencies before it is found what must be done to improve the broadband transmission quality of the system.

## 7.3  Decibel Measurements

The language of *decibels* originated in the telephone industry. As the response of human senses to stimuli such as sound and light are closely proportional to the logarithm of the power level of the stimuli, the use of decibels in telecommunications is justified. Moreover, logarithmic measures simplify many calculations that are commonly needed in communication systems. Such systems consist of distinct units such as transmission lines, amplifiers, attenuators, and filters. Each of these units changes the signal power level by some factor. The ratio of power delivered to the terminal load and the power supplied by the signal source is the product of all of these factors. If the effect of each of these units is expressed by the logarithm of the factor by which it modifies the signal power, then the logarithm of the effect of all of the units combined is the sum of the logarithms of the effects for each of the separate units.

### 7.3.1  Bel and Decibel

The bel ($B$), named in honor of Alexander Graham Bell, is defined as the common logarithm of the ratio of two powers, $P_1$ and $P_2$. Thus, the number of bels $N_B$ is given by

$$N_B = \log(P_2 \, / \, P_1) \tag{7.1}$$

If $P_2$ is greater than $P_1$, $N_B$ is positive, representing a gain in power. If $P_2 = P_1$, $N_B$ is zero, and the power level of the system remains unchanged. If $P_2$ is less than $P_1$, $N_B$ is negative, and the power level is diminished (i.e., attenuated).

A smaller and more convenient unit for engineering purposes is the decibel, whose magnitude is 0.1 $B$. Thus,

$$N_{dB} = 10 N_B = 10 \log(P_2 \, / \, P_1) \tag{7.2}$$

An absolute meaning can also be given to the decibel value of a power level by making $P_1$ a standard reference value. If $P_1$ is one watt, the decibel value of a power level $P_2$ is designated by $\pm N_{dB}$ dBW (e.g., 0 dB $\equiv$ 1W, 10 dB $\equiv$ 10W). When one milliwatt is used as a reference value (i.e., $P_1 = 1$mW), then the power level is designated by $\pm N_{dB}$ dBm.

The power level change, expressed in decibels, is correctly given in terms of voltage and current ratios alone only for the special case for which impedances across the circuit are equal, as per Figure 7.2, with $Z_1 = Z_2$ and hence $R_1 = R_2$,

$$N = 20 \log \left| \frac{V_2}{V_1} \right| \text{dB} \tag{7.3}$$

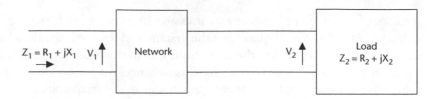

**Figure 7.2**  Two-port passive network terminated with arbitrary impedance.

When this condition of equal impedances does not exist, caution should be taken to interpret values accurately. For example, it is not correct to state that a voltage amplifier with voltage amplification of 1,000 has a 60-dB gain in a case where input impedance and the load impedance have different values. For details, [5] is suggested.

## 7.4  Basic TDM Techniques and Digital Transmission Systems

In the previous sections, we discussed the analog and high-frequency aspects of transmission systems. These are applicable from the analog subscriber loop to radio transmission systems. During the past 2 decades, many terrestrial microwave and fiber-optic systems were introduced to carry digitally multiplexed high-speed bit steams using TDM techniques. The following is a summary to review the essentials of such systems.

With voice telephony embracing the digital technology in the late 1970s, worldwide switching and transmission systems gradually commenced the conversion process from analog systems to digital transmission systems. The basic voice digitization process used is known as PCM, and systems for the bundling of many voice channels to one single data stream is known as TDM.

### 7.4.1  PCM

PCM is a sampling, quantization, and coding process that converts a voice conversation into a 64-Kbps standard rate. This basic transmission level is known as digital signal—level zero (DS0).

In PCM, the voice signal is time sampled at a rate of 8,000 Hz, (where the incoming signal is band limited to 4,000 Hz). These samples are then converted to pulses using a process known as PAM. In the next stage, each pulse is assigned an equivalent eight-bit binary value based on its amplitude, through the process known as quantization and coding. The resulting output is a digital representation of the pulse and, by extension, the sampled analog waveform at the data rate of 64 kbit/s.

The actual processes in telecommunications applications make use of a nonuniform coding process (companding) based on the $A$-law and the $\mu$-law prescribed by telecom standardization bodies such as CCITT. Chapter 2 discusses these processes in detail.

### 7.4.2  E1 and T1 Systems

Once digitized, voice and data signals from many sources can be combined (i.e., multiplexed) and transmitted over a single high-speed link. This process is made possible by TDM. TDM divides the link into 24 or 30 discrete 64-kbit/s timeslots. An identical number of DS0 signals (representing multiple separate voice or data calls) are assigned to each timeslot for transmission within the link, as shown in Figure 7.3.

In the United States and Japan, a 24-channel systems known as T1 is used, while a 30-channel system called E1 standardized by the CCITT is used in countries in Europe and most other parts of the world.

#### 7.4.2.1  E1 Systems

A typical TDM/PCM transmission concept is shown in Figure 7.4.

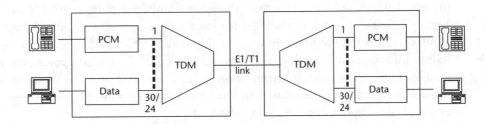

**Figure 7.3**  TDM.

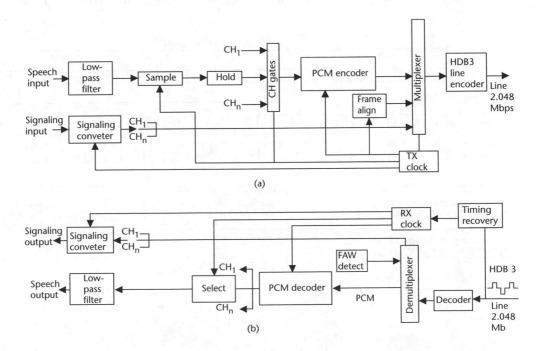

**Figure 7.4**  The transmission concept in E1/T1 systems: (a) transmit path, and (b) receive path.

*The Transmit Path*

The speech signal is first band limited by the LPF so that only the frequency band 300 to 3,400 Hz remains. It is then sampled at the rate of 8 kHz to produce the PAM signal. The PAM signal is temporarily stored by a hold circuit so that it can be digitized in the PCM encoder. Samples from a number of telephone channels (24 or 30) can be processed by the encoder within one sampling period of 125 $\mu$s. These samples are applied to the PCM encoder via their respective gates selected by the transmit timing pulses. At the output of the encoder, the speech samples exit as eight-bit PCM codewords. These codewords from the speech path are combined with the frame alignment word, service bits, and the signaling bits in the multiplexer to form frames and multiframes. They are then passed on to the high-density bipolar three (HDB3) line encoder, which converts the binary signals into bipolar (pseudoternary) signals for transmission over wireline, digital microwave radio (DMR), or optical fiber links.

In the European CCITT system, each frame contains 32 time slots of approximately 3.9-$\mu$s duration. The time slots are numbered from 0 to 31, as shown in Figure 7.5. Time slot 0 is reserved for the frame-alignment signal and service bits. Time slot 16 is reserved for multiframe alignment signals and service bits and for signaling information of each of the 30 voice channels. Each multiframe consists of 16 frames. Hence, the time duration of one multiframe is 2 ms. The purpose of formation of multiframes is to allow the transmission of signaling information for all 30 channels during one complete multiframe.

The signaling information for each telephone channel is processed in the signaling converter. This converts the signaling information into a maximum of four-bit codes per channel. These bits are inserted into time slot 16 of each PCM frame, except frame 0. The 16 frames in each multiframe are numbered 0 to 15. Because in

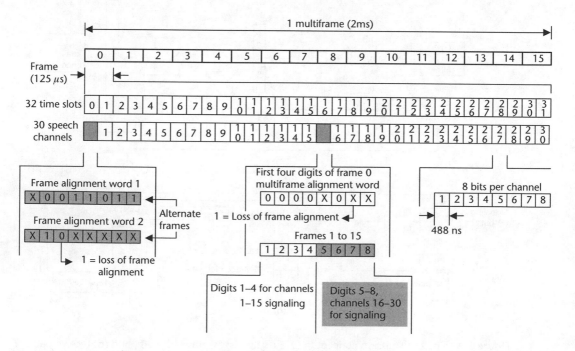

**Figure 7.5**  Thirty-channel PCM frame and multiframe details.

each frame, signaling information from two telephone channels is inserted into time slot 16, signaling information from the 30 telephone channels can be transmitted within one multiframe.

The transmission rate of the PCM signals is 2,048 kbit/s (2.048 Mbit/s). This is determined by the timing clocks in the transmission end, which control the processing of the speech, signaling, synchronizing, and service information. This TDM/PCM signal is commonly referred to as 2 mbit/s or E1 data stream.

*HDB3 Line Encoding*

The TDM signal is line encoded at the final stage in the transmit path. The purpose of the HDB3 code is to limit the number of zeros in a long sequence of zeros to three. This assures clock extraction in the regenerator of the receiver. This code is recommended by the CCITT (Recommendation G.703). An example is shown in Figure 7.6. Longer sequences of more than three zeros are avoided by the replacement of one or two zeros by pulses according to specified rules. These rules ensure that the receiver recognizes that these pulses are replacements for zeros and does not confuse them with code pulses. This is achieved by selecting the polarity of the alternative mark inversion (AMI) code [6]. Also, the replacement pulses themselves must not introduce an appreciable dc component. The rules for HDB3 coding can be summarized as follows:

1. Invert every second *one* for as long as a maximum of three consecutive *zeros* appear.
2. If the number of consecutive *zeros* exceeds three, set the *violation pulse* in the fourth position. The violation pulse purposely violates the AMI rule.
3. Consecutive violation pulses shall be of opposite polarity. If this rule cannot be applied, set a *one* according to the AMI rule in the position of the first *zero* in the sequence.

*The Receive Path*

Figure 7.4(b) shows the receive path. The 2,048-kbit/s pseudoternary HDB3 signal, which comes from the line, is first decoded by the HDB3 decoder into a binary signal. This signal is then separated by the input demultiplexer or separator into the respective speech channels, together with supervisory information (e.g., signaling). The speech segments are sent to the PCM decoder, the signaling bits are sent to the signaling converter, and the frame-alignment bits and service bits for alarms are sent to the frame-alignment detector and alarm unit. The timing signals for the receiver are recovered from the line codes and processed in the receiver timing unit to

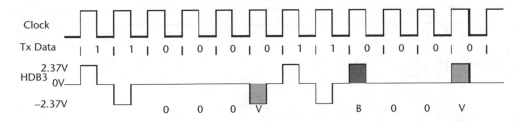

**Figure 7.6**   The HDB 3 code.

generate the clock signals for processing the received signals. In this manner, the receiver is kept synchronized to the transmitter. Synchronization between the transmitter and receiver is vital for TDM systems.

The PCM decoder reconstructs the original analog speech patterns. The bits belonging to signaling are converted into signaling information by the receive signaling converter and sent to the respective telephone channels. The frame-alignment word and service bits are processed in the frame-alignment and alarm units. Frame-alignment word (FAW) detection is done here. If a FAW error is detected in four consecutive frames, the frame-alignment loss (FAL) alarm is initiated. Some of the service bits are used to transmit and receive alarm conditions.

*Frame Alignment*

As shown in Figure 7.5, a time slot of eight bits per frame is available for frame alignment. This means that 64 kbit/s is reserved for this purpose. The basic principle of frame alignment is that the receiver identifies a fixed word and then checks its location at regular intervals. This makes it possible for the receiver to orient itself to the incoming bit flow and to distribute the correct bits to the respective channels. In addition to frame alignment, the assigned time slot is also used for transmission of information concerning the alarm status in the near end terminal to the remote end terminal. Spare capacity is also available for both national and international use. The words in time slot 0 in frames with even numbers are often called frame-alignment word 1, while those in odd frames are called frame-alignment word 2. For details, [6] is suggested.

In time slot 16 of frame number 0 only, the multiframe-alignment word (0000) is sent. In time slot 16 of frame numbers 1 to 15, the signaling information of channel pairs 1/16, 2/17, and so on, are sent in the respective frame order.

### 7.4.2.2  T1 Systems

Instead of the 30-channel primary systems used in Europe and most other parts of the world, North America and Japan uses a system that has 24 channels for the primary PCM system. The basic PCM system is the same, though with a different companding scheme, as discussed in Chapter 2. The frame structure is the major difference, as indicated in Figure 7.7. The 24-channel frame is 125 $\mu$s long, as in the case of the 30-channel frame. However, the frame contains only 24 times slots, each having eight bits. The first seven bits are always used for speech, while the eighth bit is for speech in all frames except the sixth frame, where it is used for signaling. At the

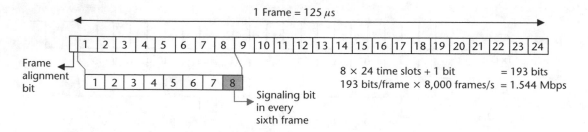

**Figure 7.7**  The 24-channel T1 system.

start of every frame, one bit is included for frame- and multiframe-alignment purposes. Each frame therefore contains $24 \times 8 + 1 = 193$ bits. As the sampling rate is 8 kHz, there are 8,000 frames per second, giving $193 \times 8,000 = 1.544$ Mbit/s. The signaling bit rate is $(8,000/6) \times 24 = 32,000$ bps.

The 24-channel system also has a 1.5-ms multiframe consisting of 12 frames. The frame- and multiframe-alignment words are transmitted sequentially, by transmitting one bit at the beginning of each frame. They are sent bit by bit on the odd and even frame cycles, and their transmission is completed only after each multiframe has been transmitted. The frame and multiframe alignment words are both six-bit words. A comparison of the construction of the frame and multiframe for the CCITT 30-channel (E1) and U.S./Japan 24-channel (T1) systems is summarized in Table 7.2. For details, [7] is proposed.

## 7.5   Plesiochronous Higher-Order Digital Multiplexing or PDH

The 30- or 24-channel PCM systems are only the first, or primary, order of digital multiplexing as designated by standardization bodies such as CCITT and ANSI. Up to the mid 1990s, most countries used digital hierarchical transmission systems known as PDH that build on the T1/E1 systems. If it is necessary to transmit more than 30 or 24 channels, the system is built up hierarchically, as in Figure 7.8, in different parts of the world as indicated. In the following sections, some details of these systems and the reasoning behind the conversion to newer and more flexible systems are discussed.

In CCITT-based systems, four primary (E1) systems are multiplexed to form an output having 120 channels. This is called the second order of multiplexing, or E2. Similarly, four 120 E2 systems are multiplexed to give an output of 480 channels (E3) in third-order multiplexing. Table 7.3 indicates these levels and corresponding bit rates for each asynchronous multiplexer levels for CCITT systems. Similarly, for 24-channel T1 systems, Table 7.4 depicts the levels and corresponding bit rates. For details of these systems, [6] is suggested.

**Table 7.2**   Comparison of 24- and 30-Channel (TDM) Systems

|  | 24-Channel Systems | 30-Channel Systems |
|---|---|---|
| Sampling frequency (kHz) | 8 | 8 |
| Duration of time slot ($\mu$s) | 5.2 | 3.9 |
| Bit width ($\mu$s) | 0.65 | 0.49 |
| Bit transfer rate (Mbit/s) | 1.544 | 2.048 |
| Frame period ($\mu$s) | 125 | 125 |
| Number of bits per word | 8 | 8 |
| Number of frames per multiframe | 12 | 16 |
| Multiframe period (ms) | 1.5 | 2 |
| Frame-alignment signal in | Odd frames | Even frames |
| Frame-alignment word | 101010 | 0011011 |
| Multiframe-alignment word | 001110 | 0000 |

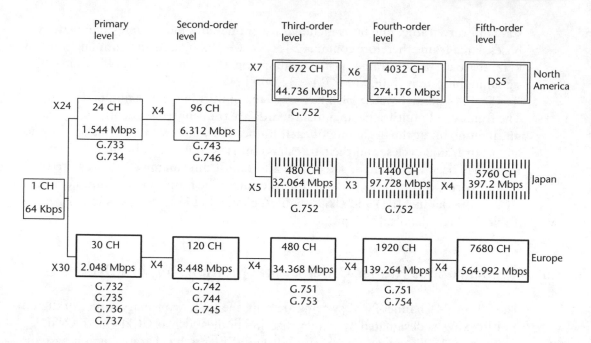

**Figure 7.8**  The plesiochronous digital hierarchy.

**Table 7.3**  Plesiochronous Multiplexer Levels and Bit Rates for CCITT System

| Level | Number of Channels | Bit Rate (Mbit/s) |
|-------|--------------------|-------------------|
| First (E1) | 30 | 2.048 |
| Second (E2) | 120 | 8.448 |
| Third (E3) | 480 | 34.368 |
| Fourth (E4) | 1,920 | 139.264 |
| Fifth (E5) | 7,680 | 565.992 |

**Table 7.4**  Plesiochronous Multiplexer Details for North American System

| Level | Number of Channels | Bit Rate (Mbit/s) |
|-------|--------------------|-------------------|
| DS-1 | 24 | 1.544 |
| DS-1C | 48 | 3.152 |
| DS-2 | 96 | 6.312 |
| DS-3 | 672 | 44.736 |
| DS-4 | 4,032 | 274.176 |

Plesiochronous systems do not synchronize the switches, they merely use highly accurate clocks at all switching nodes so that slip rate between the nodes is acceptably low. This mode of operation is easier to implement, as it avoids distributing timing through the network. Plesiochronous networks, however, carry the cost of burden of highly accurate and independent timing sources.

The PDH systems described have the advantage of operating independently without a master clock to control them. Each lower rate multiplexer (such as a 2-Mbit/s E1 or a 1.5-Mbit/s T1 link) has its own clock. These plesiochronous transmission systems could have minute differences in frequency from one multiplexer to another, so when each provides a bit stream for the next hierarchical level, bit stuffing (justification) is necessary to adjust for these frequency differences.

## 7.6  Synchronous Digital Multiplexing

Despite the attractive aspects of asynchronous multiplexing, there is one major drawback. If, for example, a 140-Mbit/s system is operating between two major cities, it is not possible to identify and gain access to individual channels at intermediate points en route. In other words, drop and insert capability requires a complete demultiplexing procedure.

Recognizing these disadvantages in PDH systems, in the late 1980s industry and standardization bodies commenced working on synchronous systems leading to the SDH and SONET systems. In 1988, the CCITT established a worldwide standard for the SDH in the form of Recommendations G.707, 708, and 709. In addition to being a technical milestone, this agreement also unifies the bit rates so that this new synchronous system does not have the existing interface problems between T1 and E1 systems. The resulting recommendations were intended for application to optical fiber transmission systems originally referred to as the SONET standard. Although SDH now supersedes the SONET description, they both refer to the same subject matter and are sometimes used interchangeably in the literature.

### 7.6.1  SDH

SDH standards approved in 1988 define transmission rates, signal format, multiplexing structures, and tributary mapping for the network node interface (NNI)—the international standard for the SDH. In addition to defining standards covering the NNI, CCITT also embarked on a series of standards governing the operation of synchronous multiplexers (Recommendations G.781 and G783) and SDH network management (G.784). It is the standardization of these aspects of SDH equipment that will deliver the flexibility required by network operators to cost effectively manage the growth in bandwidth and provisioning of new customer services.

The SDH specifications define optical interfaces that allow transmission of lower rate signals (e.g., PDH) at a common synchronous rate. A benefit of SDH is that it allows multiple vendors' optical transmission equipment to be compatible in the same span. SDH also enables dynamic drop-and-insert capabilities on the payload without having to demultiplex and remultiplex the higher rate signal, causing delays and requiring additional hardware. As the overhead is relatively independent of the payload, SDH easily integrates new services, such as ATM and FDDI, along with existing T- and E- PDH systems.

## 7.6.2   SONET

In 1985, Bellcore proposed the idea of an optical carrier-to-carrier interface that would allow the interconnection of different manufacturers' optical equipment. This was based on a hierarchy of digital rates, all formed by the interleaving of a basic rate signal. SONET attracted the interest of carriers, regional Bell operating companies (RBOCs), and manufacturers alike and quickly gained momentum. Interest in SONET by CCITT (now ITU-T) expanded its scope from a domestic to an international standard, and by 1988 the ANSI committee had successfully integrated changes requested by the ITU-T and was well on its way towards the issuance of the new standard. Today, the SONET standard is contained in the ANSI specification T1.105 "Digital Hierarchy—Optical Interface Rates and Formats Specifications (SONET)," and technical recommendations are found in Bellcore TR-NWT-000253 "Synchronous Optical Network (SONET) Transport System: Common Generic Criteria." The SONET specifications define optical carrier (OC) interfaces and their electrical equivalents to allow transmission of lower rate signals at a common synchronous rate.

As in the case of SDH, the overhead is relatively independent of the payload, and SONET is also able to integrate new services, such as ATM and FDDI, in addition to existing DS3 and DS1 services. Another major advantage of SONET is that the operations, administrations, maintenance, and provisioning (OAM&P) capabilities are built directly into the signal overhead to allow maintenance of the network from one central location.

### 7.6.3   SDH/SONET Multiplexing

#### 7.6.3.1   SDH Multiplexing and Frame Formats

SDH multiplexing combines lower speed digital signals such as 2-, 34-, and 140-Mbps signals with required overhead to form a frame called synchronous transport module at level one (STM-1). Figure 7.9(a) shows the STM-1 frame, which contains nine segments of 270 bytes each. The first nine bytes of each segment carry overhead information, the remaining 261 bytes carry payload. When visualized as a block, the STM-1 frame appears as nine rows by 270 columns of bytes, as shown in Figure 7.9(b). The STM-1 frame is transmitted row by row, with the most significant bit (MSB) of each byte transmitted first.

In order for SDH to easily integrate existing digital services into its hierarchy, it operates 125 $\mu$s per frame, so the frame rate is 8,000 frames per second. The frame capacity of a signal is the number of bits contained within a signal frame. Figure 7.9 shows that,

Frame capacity = 270 bytes/row × 9 rows/frame × 8 bits/byte = 19,440 bits/frame

The bit rate of the STM-1 signal is calculated as follows:

Bit rate = frame rate × frame capacity

Bit rate = 8,000 frames/second × 19,440 bits/frame = 155.52 Mbit/s

Three transmission levels (STM-1, STM-4, and STM-16) have been defined for the SDH hierarchy. As Figure 7.10 shows, the ITU has specified that an STM-4

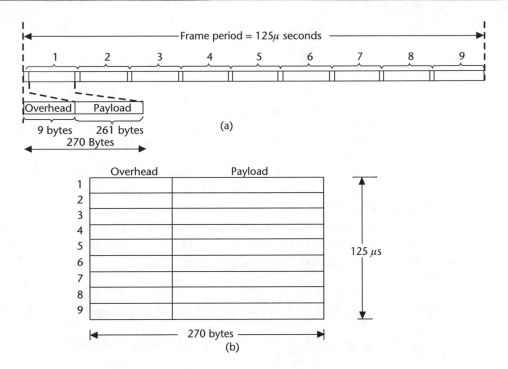

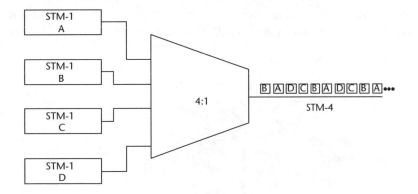

**Figure 7.9**  Two views of the STM-1 frame format: (a) time sequence, and (b) as a block.

**Figure 7.10**  SDH multiplexing.

signal should be created by byte interleaving four STM-1 signals. The basic frame rate remains 8,000 frames per second, but the capacity is quadrupled, resulting in a bit rate of 4 × 155.52 Mbit/s, or 622.08 Mbit/s. The STM-4 signal can then be further multiplexed with three additional STM-4 tributaries to form a STM-16 signal. Table 7.5 lists the defined SDH frame formats, their bit rates, and the maximum number of 64-kbit/s telephony channels that can be carried at each level

### 7.6.3.2   SONET Multiplexing and Frame Formats

SONET multiplexing combines low-speed digital signals such as DS1, DS1C, E1, DS2, and DS3 with required overhead to form a building block called synchronous

**Table 7.5**   SDH Levels

| Frame Format | Bit Rate | Maximum Number of Telephony Channels |
|---|---|---|
| STM-1 | 155.52 Mbit/s | 1,920 |
| STM-4 | 622.08 Mbit/s | 7,680 |
| STM-16 | 2.488 Gbit/s | 30,720 |

transport signal level one (STS-1). Figure 7.11(a) shows the STS-1 frame, which is organized as nine rows by 90 columns of bytes. It is transmitted row by row, with the MSB of each byte transmitted first.

In order for SONET to easily integrate existing digital services into its hierarchy, it was defined to operate at the basic rate of 125 $\mu$s per frame. Hence the frame rate is 8,000 frames per second. This is similar to the case of SDH. Figure 7.11(a) shows that

Frame capacity = 90 bytes/row × 9 rows/frame × 8 bits/byte = 6,480  bits/frame

Now the bit rate of the STS-1 signal is calculated as follows:

Bit rate = 8,000 frame/second × 6,480 bits/frame = 51.840 Mbit/s

Higher rate signals are formed by combining multiples of the STS-1 block by interleaving a byte from each STS-1 to form an STS-3, as shown in Figure 7.11(b). The basic frame rate remains 8,000 frames per second, but the capacity is tripled,

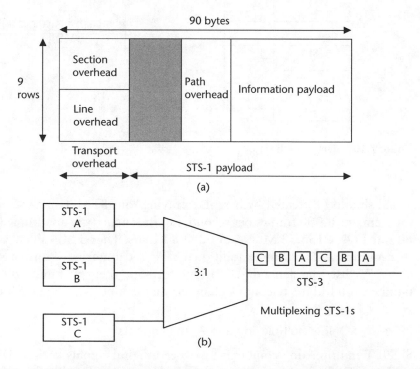

**Figure 7.11**   SONET frame format and multiplexing: (a) frame format, and (b) multiplexing.

resulting in a bit rate of 155.52 Mbit/s. The STS-3 may then be converted to an optical signal (OC-3) for transport, or further multiplexed with three additional STS-3 tributaries to form an STS-12 signal, and so on. Table 7.6 defines common SONET optical rates, their equivalent electrical rates, and the maximum number of DS0 voice channels that can be carried at that rate. Comparing the contents of Table 7.4 and Table 7.6, STS-1 carries the same number of DS0s as a DS3 signal. OC-3, OC-12, and OC-48 are the most popular transport interfaces today.

Table 7.7 indicates relationship between SONET and SDH transmission rates.

### 7.6.3.3  Transport capabilities of SDH/SONET

All of the tributary signals that appear in today's PDH networks can be transported over SDH. The list includes CCITT 2-, 34-, and 140-Mbit/s tributary signals, and North American DS1, DS2, and DS3 signals. This means that SDH is completely backwards compatible with the existing transmission networks. SDH can be deployed, therefore, as an overlay network supporting the existing network with greater flexibility, while the transfer to SDH takes place. In addition SDH/SONET transport capabilities have the flexibility to accommodate more advanced network signals such as:

- ATM (the standard for broadband ISDN);
- FDDI (a high-speed LAN standard);
- DQDB (a metropolitan area network standard).

It is beyond the scope of this chapter to describe more details of SDH/SONET. The reader may obtain more information from [6, 8–11] and relevant ITU or North American standards.

**Table 7.6**  SONET Rates

| Frame Format | Optical | Bit Rate | Maximum DS0s |
|---|---|---|---|
| STS-1 | OC-1 | 51.84 Mbit/s | 672 |
| STS-3 | OC-3 | 155.52 Mbit/s | 2,016 |
| STS-12 | OC-12 | 622.08 Mbit/s | 8,064 |
| STS-24 | OC-24 | 1.244 Gbit/s | 16,128 |
| STS-48 | OC-48 | 2.488 Gbit/s | 32,256 |
| STS-192 | OC-192 | 9.953 Gbit/s | 129,024 |

**Table 7.7**  SONET/SDH Transmission Rate Comparison

| SONET Signal | SDH Signal | Transmission Rate |
|---|---|---|
| STS-1 | | 51.84 Mbit/s |
| STS-3 | STM-1 | 155.52 Mbit/s |
| STS-12 | STM-4 | 622.08 Mbit/s |
| STS-24 | | 1,244.16 Mbit/s |
| STS-48 | STM-16 | 2,488.32 Mbit/s |

### 7.6.4   Timing and Synchronization in Digital Networks

Based on the discussion as per previous sections, modern SDH/PDH networks transport two basic entities—data and timing—as part of a service. Figure 7.12 illustrates a simple SDH/PDH network model showing the role of the timing component. A PDH circuit is transported over an SDH path while being multiplexed with other PDH circuits, cross connected with other SDH payloads, and regenerated. This model network is synchronized from a logically separate synchronization network, although it is likely that the sync signals will be physically carried on parts of the SDH network.

In such synchronous networks, the incoming STM-N channels (of higher data rates) can have slow phase movement (wander) relative to the reference. The causes of such wander in a synchronous system are described in Chapter 12 of [5].

#### 7.6.4.1   Synchronization of Clock Distribution Networks

The timing distribution chain in an SDH network is designed to minimize the phase movement between synchronization references. A typical timing distribution chain is shown in Figure 7.13 as per relevant standards such as ETSI 300 462-2 [12].

The primary reference clock (PRC) is very stable; its frequency is generally accurate to one part in $10^{11}$, and it has very little short-term wander. The timing signal is distributed (by either STM-N sections or PDH paths) to slave clocks called *synchronization supply units* (SSUs). There is usually one SSU at a physical site, called a *node*, that supplies timing to all the *SDH equipment clocks* (SECs) at that node. More details are available in Chapter 12 of [5].

In broadband networks, timing has always been very carefully specified, controlled, and distributed within networks, across network interfaces, and between customers. To deliver the timing part of the service, the network must be properly synchronized. Compared to PDH technology, the new SDH technology in public

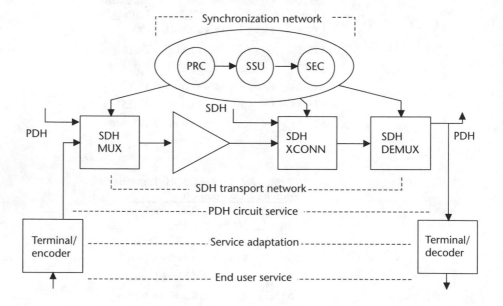

**Figure 7.12**   A model broadband network.

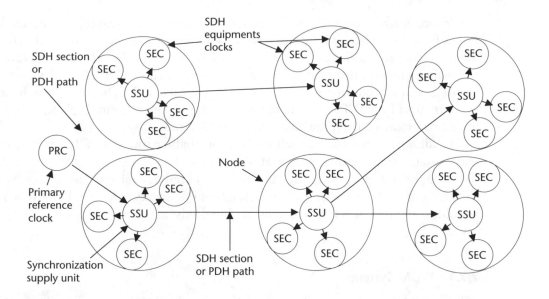

**Figure 7.13**   Synchronization network architecture for SDH systems. (Courtesy of Tektronix, Inc., United States)

networks around the world represents a quantum leap in performance, management, and flexibility. As a consequence, timing and synchronization are of strategic importance to network operators as they work in the new deregulated environment of the 1990s. More details on this will be discussed in Chapter 8.

## 7.7   Optical Networks

The physical transport of data streams has typically been done over microwave links, coaxial cables, and optical fibers. But as the data rates increases, copper- and microwave-based systems rapidly run out of bandwidth, giving way to optical fibers. Optical networks are those in which the dominant physical layer technology for transport is optical fiber. They can be opaque or all optical. Optical fiber may carry signals on a single wavelength or on many wavelengths using wavelength division multiplexing (WDM) or DWDM. Fiber systems provide the streams with much wider bandwidth. While frequency as a parameter is widely used to distinguish the location of wireless signals below 300 GHz, at light frequencies, the wavelength parameter is the preferred measure. Wavelength is expressed in nanometers (nm) or micrometers ($\mu$m), and the frequencies are expressed in terahertz (THz). Based on the relationship of $C = f\lambda$, wavelength is calculated, where the speed of light, $C$, in vacuum is $2.998 \times 10^8$ m/s and in fiber $2.99 \times 10^8$ m/s.

Data to be transmitted in a fiber-optic network modulate a laser-generated infrared beam by on-off keying (OOK). Infrared signals best match the light-carrying characteristics of fiber-optic cable, which has an attenuation response of about 0.2 dB/km in the two narrow bands of frequencies centered on 1,310 nm and 1,550 nm.

In opaque networks, the path between end users is interrupted at intermediate nodes by optical-electronic-optical conversion operations. The traditional SONET/SDH systems are opaque single-wavelength systems. While SONET can operate in a point-to-point link, the most common topology is a ring. Multiple nodes comprise the ring with an add-drop multiplexer (ADM) at each node. One of the key features and benefits of a SONET system, the ADM permits data from multiple sources to be added or extracted from the data stream as required.

In all-optical networks, each connection (light path) is totally optical, or at least totally transparent, except at the nodes. The dominant role assumed by the all-optical approach applies not only in long-haul telecom networks, where the light paths can be hundreds to thousands of kilometers in length, but also in metropolitan environments where tens to hundreds of kilometer length paths are involved.

### 7.7.1 WDM Systems

WDM enables several optical signals to be transmitted by a single fiber. Its principle is essentially similar to FDM systems, where several signals are transmitted using different carriers, occupying nonoverlapping parts of a frequency spectrum. In the case of WDM, the spectrum band used is in the region of 1,310 or 1,550 nm. Today's widely installed WDM systems are in the opaque systems category where the intermediate nodes are either ADMs or digital cross connects. The first-course WDM systems used two channels operating on 1,310 nm and 1,550 nm. Later, four channels of data were multiplexed. More details on these can be found in [13–16]. Figure 7.14 illustrates this concept.

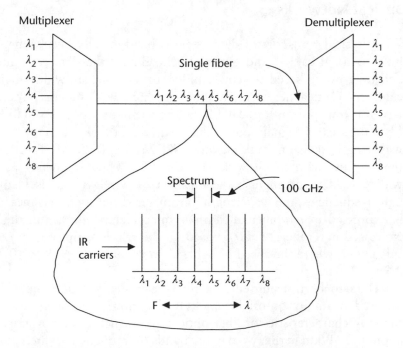

**Figure 7.14**   Concept of WDM.

### 7.7.2   DWDM Systems

When larger numbers of different wavelengths are multiplexed in a fiber system, the systems are termed DWDM systems. In these systems, eight, 16, 32, 64, or even 100 data channels are multiplexed. Standard channel wavelengths have been defined by the ITU as between 1,525 and 1,565 nm with 0.8-nm (100-GHz) channel spacing approximately. The band within 1,525 to 1,565 nm is called the conventional band, or C band. Another block of wavelengths from 1,570 to 1,610 nm is called the long-wavelength band, or L band, while the wavelengths from 1,525 to 1,538 nm is called the S band. Figure 7.15 depicts the spectrum of a narrowband optical filter for DWDM systems with a channel spacing of 100 GHz or 0.8 nm.

### 7.7.3   Photonic Networks

Key components in a photonic network based on WDM systems are shown in Figure 7.16 with optical add-drop multiplexers (OADM) and optical cross-connect (OCX) systems. Figure shows both point-to-point WDM transmission system as well as ring systems coupled by OADM and OCX units.

Early optical networks were point-to-point WDM transmission systems, while the newer systems are more commonly ring types with DWDM techniques. Key components in a DWDM system are the multiplexers and demultiplexers. Numerous methods have been developed over the last two decades to add and separate optical signals. Optical couplers are used for multiplexing while optical filters are used for the demultiplexing. One technique that has become very useful is the arrayed waveguide grating. In the process of multiplexing and transmission on to the fiber media, optical amplifiers play an important role. In an erbium doped fiber amplifier (EDFA), optical isolators, laser pumps, optical couplers, and erbium doped fiber coils are used inside. Another new kind of optical amplifiers are Raman amplifiers (RAs). More details on optical filters, arrayed waveguide gratings, and optical amplifiers can be found in [13, 16].

With higher bandwidth demands for faster network access and VoIP-based systems, optical networks are the only way to cope up. Expanding and retrofitting the

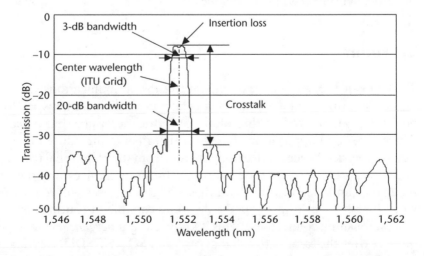

**Figure 7.15**   Typical spectrum of a narrowband optical filter for DWDM systems.

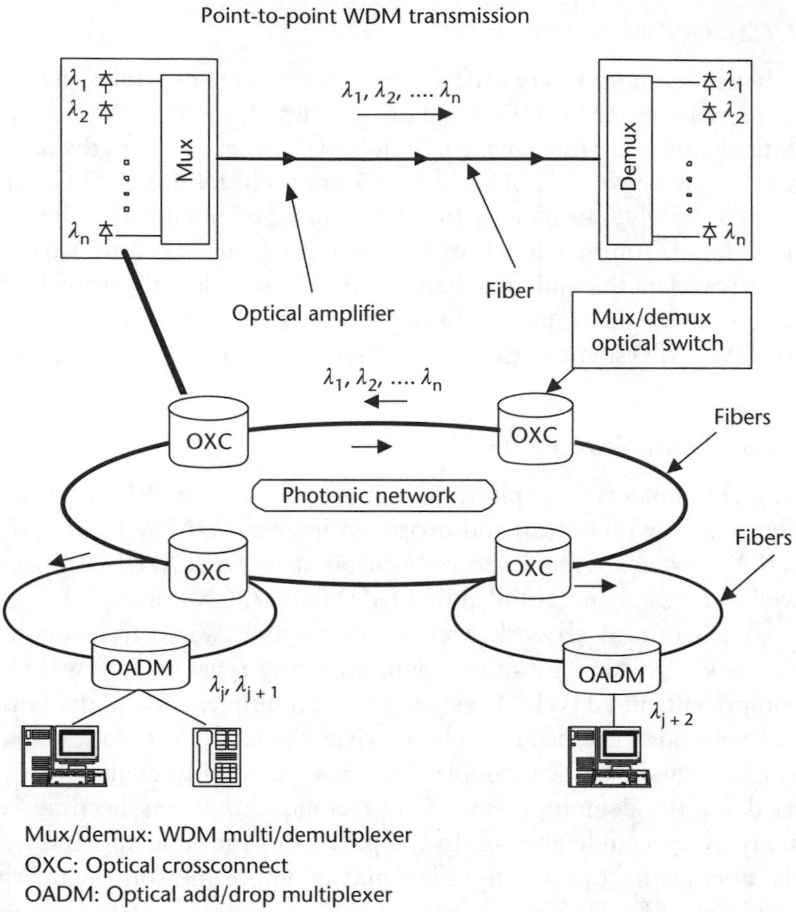
Point-to-point WDM transmission
$\lambda_1$
$\lambda_2$
$\lambda_n$
Mux
$\lambda_1, \lambda_2, .... \lambda_n$
Demux
$\lambda_1$
$\lambda_2$
$\lambda_n$
Fiber
Optical amplifier
Mux/demux optical switch
$\lambda_1, \lambda_2, .... \lambda_n$
OXC
OXC
Fibers
Photonic network
Fibers
OXC
OXC
OADM
$\lambda_j, \lambda_{j+1}$
OADM
$\lambda_{j+2}$
Mux/demux: WDM multi/demultplexer
OXC: Optical crossconnect
OADM: Optical add/drop multiplexer

**Figure 7.16** A typical photonic network configuration. (*Source:* [16]. ©2000 IEEE.)

existing SONET systems can be performed by three fundamental methods: by lighting up the unused (dark) fibers, increasing transmission speeds in existing fibers, or by implementing DWDM techniques.

## 7.8 The Future

Though voice has played the dominant role in communications, the recent explosion in Internet traffic implies a paradigm shift from the voice network to a data-centric network [17]. Not only is the volume of traffic increasing, but we see new multimedia services and applications based on them being introduced on telecommunications networks. Voice services may change from the traditional to nonreal-time store and forward and to real time over asynchronous services. Internet will carry real-time voice and video. As a result, these diverse services will require one global network that consists of programmable nodes capable of routing any type of payload and establishing connectivity. This is known as *convergence*. Convergence of diverse services over the same network that can offer SONET/SDH, ATM, and IP is becoming the direction of the communications industry [18].

The foremost important requirement of future transmission networks is the large transport bandwidth capability. High reliability will be indispensable for the future society based on multimedia networks and services. These create the technology need for processing, switching, and transmission at optical rates over 40 Gbit/s or higher. Photonic networks employing DWDM appear to be the solution for such robust, efficient networks. An infrastructure based on DWDM allows for a mix of technologies, including SONET, IP, and ATM. However, to make DWDM efficient, several other supporting developments are essential.

The bottleneck posed by optical-to-electrical conversion needed for amplification and switching in transport networks is removed by the development of optical amplifiers and optical-switching technologies. Optical switching based on micro-optical electromechanical systems (MOEMSs) has emerged as the leading technology in realizing transparent optical switching subsystems [19]. Another advancement towards the implementation of all-optical networks is optical wavelength routing switches (WRSs) or OXCs [20]. These deal with the switching functions involved in the assignment of routes and wavelengths for the establishment of light paths across a network. References [16–23] provide more details on the state of the art of these technologies and future trends.

# References

[1]   Royle, D., "Rules Tell Whether Interconnections Act Like Transmission Lines," *EDN*, June 23, 1988, pp. 131–136.

[2]   Royle, D., "Correct Signal Faults by Implementing Line-Analysis Theory," *EDN*, June 23, 1988, pp. 143–144.

[3]   Royle, D., "Quiz Answers Show How to Handle Connection Problems," *EDN*, June 23, 1988, pp. 155–159.

[4]   Hart, B. L., *Digital Signal Transmission Line Technology*, London: Chapman and Hall, 1988.

[5]   Kularatna, N., *Digital and Analogue Instrumentation—Testing and Measurement*, London: IEE Press, 2003.

[6]   Winch, R. G., *Telecommunication Transmission Systems*, New York: McGraw Hill, 1993.

[7]   Telecommunications Techniques Corporation, "T1 Basics," Technical Note, TIBTN, November 1994.

[8]   Hewlett Packard, "Introduction to SDH," HP application note 5091-3935E, February 1992.

[9]   Telecommunications Techniques Corporation, "The Fundamentals of SDH"; TTC application note I-SDH TN May 1995.

[10]  Telecommunications Techniques Corporation, "The Fundamentals of SONET," TTC application note SONET TN, July 1995.

[11]  Bellamy, J., "Digital Telephony, Second Edition," New Jersey: John Wiley, 1991.

[12]  Tektronix, Inc., "Performance Assessment of Timing and Synchronization in Broadband Networks," application note-FL 5336/xbs 2GW-11145-0, April 1997.

[13]  Cavendish, D., "Evolution of Optical Transport Technologies: From SONET/SDH to WDM," *IEEE Communications Magazine*, June 2000, pp. 164–172.

[14]  Frenzel, L., "Optical Networks Will Come to the Rescue as Bandwidth Demands Increase," *Electronic Design*, November 20, 2000, pp. 87–102.

[15]  Green, P., "Progress in Optical Networking," *IEEE Communications Magazine*, January 2001, pp. 54–61.

[16]  Hibino, Y., "An Array of Photonic Filtering Advantages," *IEEE Circuits and Devices*, November 2000, pp. 21–27.

[17]  Yoshimura, H., K. Sato, and K. Takachio, "Future Photonic Networks Based on WDM Technologies," *IEEE Communications Magazine*, February 1999, pp. 74–81.

[18]  Kartaloupoulos, S. V., *Understanding SONET/SDH and ATM*, IEEE Press, 1999.

[19]  Mechels, S., et al., "1D MEMS-Based Optical Switching Susbsystem," *IEEE Communications Magazine*, March 2003, pp. 88–93.

[20]  Zang, H., et al., "Dynamic Lightpath Establishment in Wavelength-Routed Networks," *IEEE Communications Magazine*, September 2001, pp. 100–108.

[21]  Bursky, D., "Of Hollow Decks and Transporters: The Promise of Unlimited Bandwidth," *Electronic Design*, January 10, 2000, pp. 56–64.

[22]  Israelson, J., "Switching the Light Fantastic," *EDN*, October 26, 2000, pp. 113–120.

[23]  Benjemin, D., "Optical Services over the Intelligent Optical Network," *IEEE Communications Magazine*, September 2001, pp. 73–77.

## CHAPTER 8

# Telecommunication Systems Testing

## 8.1 Introduction

Quality assurance of a telecommunications system depends on two basic aspects: the selection of the most appropriate technology and the level of qualitative and quantitative measurements on the systems. The first is a longer term planning-based task; the second is a regular process service providers and commissioning engineers need to follow.

Modern telecommunications systems are an interesting mix of copper and wireless subscriber loops, coaxial cable systems, fixed and mobile microwave radio links, and optical fiber systems. These carry a mix of analog signals and precise pulse timing–based digital signals. Different types of measurement needs exist for the measurement of power, field strength, relative signal amplitude, signal reflections, and loss in these environments. As the length of a transmission medium becomes long compared with the wavelengths of the signals transmitted, it becomes very important to measure these parameters accurately.

With systems such as PDH, SDH, and SONET, accuracy of pulse timing and system synchronization have become the key to QoS. For DSL systems, the end hardware product testing and subscriber loop performance at higher frequencies are the key areas.

This chapter is an overview of telecommunications system testing. The reader is expected to have a basic understanding of test and measurement aspects as applied to electronic systems environments. References [1, 2] are proposed for the background information.

## 8.2 Measurement Areas

In telecommunications transmission systems, there are three measurement areas:

1. Basic transmission parameters and their measurements;
2. Spectrum and frequency measurements;
3. Pulse parameter measurements and synchronization.

Basic transmission parameters such as power, signal level, and reflections are well-established analog measurements for the assurance of primary signal levels. Frequency and spectrum measurements assure the noninterference and quality of transmitted signals in congested RF environments. Measurement of pulse and timing parameters are fully digital, using bit error rate testers (BERTs), digital scopes, and other similar instruments. These deal with picosecond or fractional picosecond accuracy levels, sometimes based on statistical analysis of multiple measurements.

## 8.3    Measurement of Power Levels in Telecommunications Circuits

### 8.3.1    System Level and Zero Transmission Level Point

A multichannel telephone system may be long, containing many stages of gain and attenuation. The power level of a signal measured in such a system is stated relative to a reference point. The reference point is designated as the *zero transmission level point* (zero TLP). It is the point in the system at which the standard test tone has an absolute power of 1 mW or 0 dBm. The level of any point in the system expressed in decibels relative to the zero TLP is referred to as the *relative transmission level* of that point. It is commonly designated by dBr or TL. For example, a –33 dBr (or –33 TL) point in a system will be 33 dB below the zero TLP. Because the standard test tone level is 0 dBm (1 mW) at the zero TLP, the actual power of the standard test tone at any other point in the system when specified in dBm is numerically equal to the transmission level at that point. dBm0 indicates the level of a signal, referred to the circuit's relative zero level.

### 8.3.2    Noise Level Measurements

As per CCITT recommendations, evaluation of noise in telephone circuits is done using a *psophometer* [3]. This measures channel noise in relation to its effect as perceived by the human ear, rather than absolute noise level. Noise measured with a psophometer is designated by the suffix *p*. Thus, dBm becomes dBmp. DBm0p expresses the psophometric noise value in reference to the circuit's relative zero level. In the United States, the common system is *C-message weighting*, where the term used is dBrn [3, 4]. Figure 8.1 indicates the filters used in these two systems. Commercial transmission testers use such filters to evaluate the overall subjective effects of the noise.

### 8.3.3    Power Level Measurements and Terminating Resistance

In telecommunications systems, the power level of a given point is determined by an indirect measurement. The circuit is terminated at the point in question by a

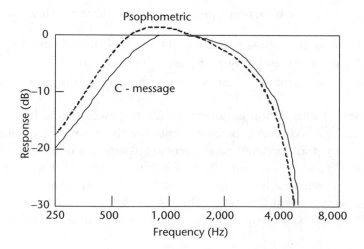

**Figure 8.1**    C-message and psophometric weighting.

resistance. The voltage developed across this resistance is measured by a level-measuring instrument, effectively a voltmeter calibrated in voltage level. The value of the terminating resistance is chosen according to the nominal impedance of the circuit at the measurement point. The standardized value for speech frequencies is 600 Ω. When defining voltage level, we choose the reference voltage to be just that which produces 1 mW in 600 Ω. That is,

$$V_{ref} = \sqrt{1 \times 10^{-3} \times 600} = 775 \text{ mV} \qquad (8.1)$$

When $L_1$ is the measured voltage across 600Ω (measured in millivolts), then

$$\text{Voltage level} = 20 \log_{10} \frac{L_1}{775} \qquad (8.2)$$

The power level at that point is defined as the voltage level described earlier. For a circuit impedance of 600 Ω, the voltage level and power level are identical.

If the terminated impedance differs from 600 Ω a correction $K$, is added. This is given by

$$K = 10 \log \frac{(600 \, \Omega)}{Z} \qquad (8.3)$$

where $K$ is measured in decibels. So, the power level is expressed as

$$\text{Power level} = \text{Voltage level} + K \qquad (8.4)$$

The value of $K$ for some common values of impedance can be found from Table 8.1.

Because the analog loop will survive for a long time, the reader should review the basics and practical measurements on the subscriber loop, as per guided in [5]. Measurements on digital subscriber loops are discussed in [6] and are summarized in Section 8.9.

### 8.3.4    Envelope Delay Measurements

Compared to voice signals, where time delays associated with transmission do not affect the voice quality, modem performance is affected by time delays in the loop. ISI occurs due to this variable time delay for different signal frequencies, which is

**Table 8.1** Approximate Values of $K$ for Some Common Values of Impedance

| $Z(\Omega)$ | $K$ |
|---|---|
| 600 | 0 |
| 300 | 3 |
| 150 | 6 |
| 75 | 9 |
| 50 | 10.8 |

termed *envelope delay* and *group delay* in the United States and Europe, respectively. Modem designers need to measure the amount of delay variation to verify the design of compensating circuits. The traditional method for measuring the envelope delay is the single-tone method where a single carrier frequency $(f_c)$ is amplitude modulated and transmitted for measuring the phase shift at the two sideband frequencies. The sidebands $(f_l$ and $f_u)$ are placed at $f_c \pm 83.33$ Hz and $f_c \pm 41.66$ Hz in U. S. and European systems, respectively. Envelope delay is estimated by measuring the phase difference between the two side bands $(\Delta\Phi)$ where,

$$\Delta\Phi = \Phi_u - \Phi_1 \qquad (8.5)$$

and envelope delay $\tau_{ed}$, is given by,

$$\tau_{ed} = \frac{\Phi_u - \Phi_1}{2\pi(f_u - f_1)} \qquad (8.6)$$

The single tone method can introduce errors if the frequency response is not flat in the neighborhood of $f_c$. Figure 8.2 indicates the normalized envelope delay, which is nonlinear due to the channel's filtering characteristics.

With the availability of inexpensive DSPs for test instruments, two new methods for envelope delay measurements are available: the 23-tone method and the network impulse response (NIR) method. These techniques provide faster results over the entire band. In the former, the envelope delays for 22 frequencies are provided within 5 sec. The NIR technique provides 127 measurements in about 30 sec. Reference [7] is suggested for details.

## 8.4   High-Frequency Power Measurements

Instruments used to measure power at VHF-to-microwave frequencies are of two types: absorption power meters, which use their own load, and through line power meters, which use the (actual) load, which may be remote from the meter.

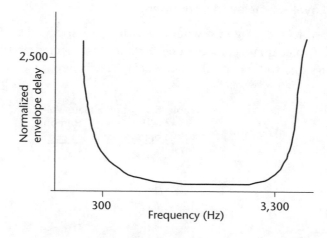

**Figure 8.2**   Frequency versus normalized envelope delay.

### 8.4.1   Absorption Power Meters

Absorption types are more accurate and usually contain a $50\Omega$ or $75\Omega$ load at RF. The main instruments currently used for laboratory work are based on thermistor, thermocouple, or diode sensors. These instruments cover a power range from –70 dBm to a few watts and are similar in use though based on different detecting principles. A power sensor accepts the RF power input, as the meter connected to the sensor measures the power that has been absorbed. However, due to mismatch between impedances of the RF source and sensor, some of the power is reflected back. Due to such nonideal circumstances in RF detection elements, calibration factors are used. For details, [1, 8–12] are suggested.

### 8.4.2   Throughline Wattmeters

Most of the absorption power measurement techniques are generally limited to a few watts. For higher power measurements up to about 1 kW in the frequency range from about 25 to 1,000 MHz, through line wattmeters become very practical. For lower frequency ranges (from 200 kHz to 30 MHz), measurements up to 10 kW are possible.

   The arrangement of such an instrument is shown in Figure 8.3. The main RF circuit of the instrument is a short section of a uniform air-type line whose characteristic impedance is matched to the transmission line (usually $50\Omega$). The coupling circuit that samples the traveling wave is in the plug-in element. The coupling element absorbs energy by mutual inductance and capacitive coupling. Inductive currents within the line section will flow according to the direction of the traveling wave. The capacitive portion of the line is independent of the traveling wave. The element is therefore designed to respond directly to the forward or reverse wave only. Thus, the meter could be used to measure forward and reflected powers separately. The elements could be designed for different ranges of forward or reflected power, usually varying from a few milliwatts to over 1 kW. The plug-in element is designed to be rotated by 180° in its socket in the line section, allowing the measurement of both forward and reverse power.

## 8.5   RF Insertion Units for SWR and Reflection Coefficient Measurements

Transmission line parameters such as VSWR and reflection coefficients are discussed in Chapter 7. Many insertion units are available for their measurement with RF millivoltmeters and RF power meters. Usually these units are available for 50-$\Omega$ and 75-$\Omega$ coaxial cable systems for use in the approximate frequency range 100 kHz to 2 GHz. RF insertion units for measuring voltage range from a few 100 $\mu$V to about 100V. These units can be inserted in the lines at suitable places for measuring the voltage along the line without affecting the VSWR. Dual directional couplers are available for measuring the forward and reflected power range from a few microwatts to about 2 kW, in the frequency range of a few 100 kHz to about 100 MHz. In modern units, the calibration parameters may be stored on the individual units, allowing the elimination of errors due to nonlinearities, level-dependent temperature effects, and frequency response.

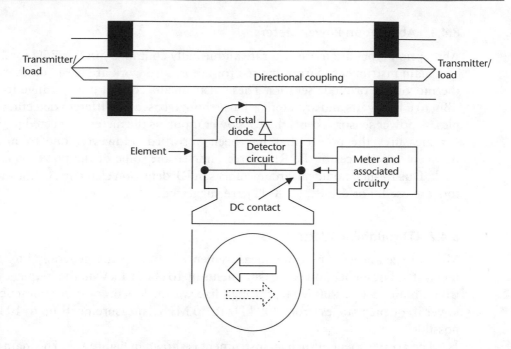

**Figure 8.3**   Construction of a thruline wattmeter[1]. (*Source:* Bird Electronic Corporation.)

Sensor characteristics, temperature response, and frequency response are all stored for each sensor at the factory (measured parameters for each individual head) in an electrically programmable read only memory (EPROM) device, as shown in Figure 8.4. The level-dependent temperature effects are stored as two-dimensional characteristics with a large number of measuring points. Each head comprises a temperature sensor, the signal of which is evaluated in the meter at regular intervals. From the measured temperature and the level values, the respective stored characteristic yields the correction for the measured output voltage. The input parameter is then calculated from this correction. Subsequently, a frequency response correction is carried out by using a stored correction factor for the signal frequency. This comprehensive correction technique has such advantages as exchanging of measuring heads, better accuracy, and traceability of calibration.

With more advanced digital techniques such as TDMA and CDMA, the traditional power-measurement techniques need to be modified for fast tracking of RF peaks, peak-to-average ratio of the signals, statistical analysis such as calculation of the cumulative distribution function, and many other specific needs. These details are available in [12].

## 8.6   Built-In RF Power Measurements

Most of these techniques are useful when applied in analog RF links or in digital systems, which can be tested as simple RF links without live traffic. However, digital

1.   *Thruline* is a registered trade mark of Bird Electronic Corporation.

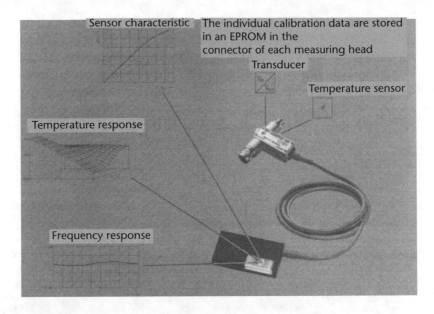

**Figure 8.4**  A modern sensor carrying calibration data on an EPROM in the connector of the measuring head. (*Source:* Rhode and Schwarz.)

RF links, particularly those exploiting spread-spectrum techniques, challenge traditional measurement methods. In these systems, processor-based RF meters are fine tuned to perform these tasks, building on the earlier measurement principles.

Digital techniques such as CDMA demand onboard power measurements. Semiconductor chip manufacturers have delivered chips to answer this need. Unlike FDMA or TDMA, CDMA systems can tune their performances, trading between capacity and voice quality. This elasticity in performance depends on both base and mobile stations carefully assessing their received signal and controlling their transmitted RF power, as described in Chapter 4.

RF-power measurement and control is critical for both cost and performance. The challenge is to implement a measurement function that can update the system every 1.25 ms or faster with fractional decibel resolution. Also, the circuits should take bare minimum space. To achieve these, diode detectors and logarithmic amplifiers are used in practical systems. Figure 8.5 indicates a circuit used in the heart of a transmitter power control circuit. For more details, [13] is proposed.

## 8.7  Field-Strength Measurements

Radio communications service providers (telephony, broadcasting, military, traffic, and security services) use field-strength meters for propagation measurements in the planning stages and for coverage measurements during operation.

The field strength is determined by placing an antenna of known effective length in the electric field under measurement. Then, using a calibrated radio receiver, the induced voltage in the antenna is measured. Loop and halfwave doublet antennas are usually used in the high frequency (HF) and in the very high frequency (VHF)/ultra high frequency (UHF) bands, respectively. For wideband

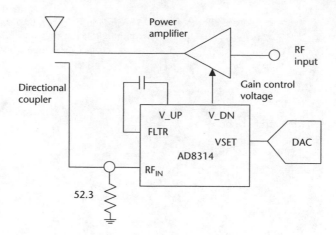

**Figure 8.5** AD 8314 logarithmic amplifier used as a transmitter power controller. (*Source:* [12]. *T&M World Magazine,* reproduced with permission of Reed Business Information.)

measurements, antennas such as log-periodic, biconical, conical and log-spiral are used. The calibrated radio receiver must have the following capabilities:

- To selectively measure a radio signal without being affected by other radio signals;
- To measure field strength over a wide range from microvolts per meter to millivolts per meter;
- To accurately measure the induced electromagnetic field (EMF) of an antenna;
- To avoid errors in measurements due to shunting of the RF sections by strong fields.

Hence, a calibrated radio receiver with a tunable amplifier in the front end with excellent linearity, selectively, stability, and sensitivity is configured as a field-strength meter. The field strength is obtained by measuring the induced EMF of the antenna and dividing the value by the effective length of the antenna.

A block diagram of a field-strength meter is shown in Figure 8.6. At the RF input stage, the signal is amplified and mixed with a local oscillator for downconversion.

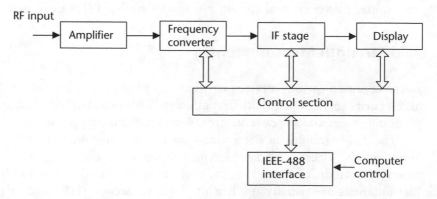

**Figure 8.6** Block diagram of a field-strength meter.

A calibration oscillator is used to calibrate the gain. The control section interacts with the frequency dividers and mixers to suitably down convert the incoming frequency to the IF. At the IF stage, a bandwidth switch is used to select the bandwidth of the signal under measurement. Step amplifiers with gain control (in inverse proportion to the input signal level) are used to keep the detector output within a fixed range. The detector stage in the IF section and the log amplifiers output the detected signal to a digital display. In modern instruments, IEEE-488 and other interfaces are used for automatic control. Field-strength measurements, in particular propagation and coverage measurements, are usually made in mobile mode. Therefore, portability and battery operation are important criteria in the choice of a test receiver.

A calibrated antenna with known effective length is stored in the instrument memory for use in direct readout of field strength. Instrument manufacturers supply calibrated antennas in general; however, other antenna elements could also be used by storing the effective length in the memory. An antenna factor is used for this purpose, which takes into account the effective length of the antenna as well other anomalies due to mismatch and loss. For more details, [1] is suggested.

## 8.8   Spectrum Observations and Frequency Measurements

With the growth of wireless LAN, cellular, WLL, satellite, and terrestrial microwave systems, spectrum observations and accurate frequency measurements have become very important. Deregulation of telecommunications services has made spectrum management a very important topic.

Frequency measurements up to about 3 GHz are based on simple frequency counters, and measurements are possible with down-conversion techniques up to about 140 GHz. Modulation domain analyzers are available for observing the signals in the modulation domain. While this chapter does not allow a detailed treatment for these techniques, [2] provides a good account.

Traditionally, spectrum measurements were based on swept-spectrum analyzers. However with advanced digital storage oscilloscopes (DSOs) and FFT techniques becoming available, three different options are available for the test engineer, depending on the frequency range of interest. Table 8.2 provides a summary of the capabilities of modern instruments. DSOs and FFT analyzers are based on sampling the signal under observation, and swept-spectrum analyzers are based on heterodyne techniques [2].

For most RF and microwave communications systems, swept-spectrum analyzers (or heterodyne type) are the instruments of choice due to the need for real-time observations and wide dynamic range. These are based on the principle of mixing the signal with a local oscillator, followed by a set of filters and detectors to extract the

**Table 8.2**   Comparison of Spectrum Instruments

| Instrument Type | Typical Bandwidth Range | Typical Sample Rates | Typical Dynamic Range of the Signal |
|---|---|---|---|
| Swept-spectrum analyzer | 100 kHz to over 26 GHz | Not applicable | 80 dB |
| FFT analyzers | 100 kHz | 256 ksamples/s | 80 db |
| DSOs with FFT capability | Over 3 GHz for nonrepetitive signals | 500 ksamples/s to 2.5 gigasamples/s | 55 db |

frequency components within a wideband signal, with the limitation of only amplitude observations (without the phase response). Figure 8.7 depicts the simplified signal-processing elements of a spectrum analyzer. For details of the swept or heterodyne spectrum analyzers and their application, the reader is referred to [2, 15]. For frequency measurements and modulation domain analyzers, [2] is suggested.

## 8.9  DSL Testing and Measurements

As discussed in Chapter 6, DSL systems exploit the high-frequency capabilities of the subscriber loop. ADSL modems establish the data rates by transmitting at the highest possible data rate at the beginning and lowering it until an error-free connection is made. This process is called *training*. During crosstalk testing, for example, two ADSL modems must first train each other at a reference noise level before increasing that noise to complete the test. In the following section, discussion is based on ADSL measurements using the ANSI standard T1. For details of other techniques, [2, 14] are suggested.

### 8.9.1  ADSL Testing and Measurements as per ANSI T1.413

The ANSI T1.413 document defines conformance tests for DMT ADSL modems. In testing, a line of twisted-pair wires (called a test loop) must be simulated, electrical impairments must be added to the lines, and the BER must be measured. ANSI T1.413 defines several tests for ADSL modems, including tests for crosstalk, impulse-noise, and POTS interference.

  ADSL modem chip sets have built-in test features similar to the analog modems. Chip sets have loop-back features that allow the testing of the device's transmitter and receiver circuits. The loop-back tests can take place within the chips or over a simulated local loop to a remote transmitter. Figure 8.8 shows a typical setup for remote loop-back testing.

#### 8.9.1.1  Lab Test Setup and Measurements

Figure 8.9 shows the lab test setup for testing an ADSL modem's conformance to ANSI T1.413. The ATU-C is the ADSL phone company's CO ADSL modem, and

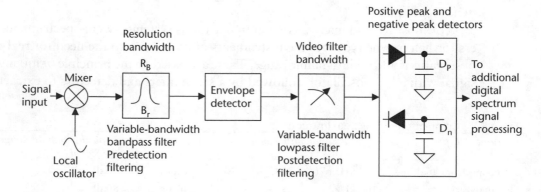

**Figure 8.7**  Simplified signal-processing elements of a swept-spectrum analyzer. (*Source:* [15]. Courtesy of *Electronic Design Magazine*.)

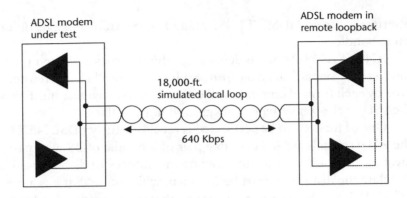

**Figure 8.8** Remote loop-back testing of ADSL loops.

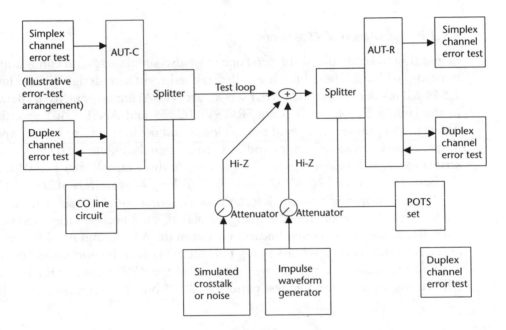

**Figure 8.9** Conformance testing on ADSL modems by injection of signals into a simulated test loop. *(Source:* [14].)

the ATU-R is the modem at the customer premises. The ATU-C is sometimes called the transmitter because it transmits at a rate of megabits per second. The ATU-R is called the receiver.

Simulation of the test loop is carried out by using a line simulator or a spool of wire. The splitters separate POTS signals (normal voice signals) from ADSL signals. During the tests, a POTS phone or modem is connected to the ATU-R splitter and a CO simulator is connected to the ATU-C splitter. It is also necessary to use a high-impedance network to introduce impairments into the test loop without altering its characteristics. Conformance testing must be done with any error correction enabled.

The first conformance test is the crosstalk test. Crosstalk testing is necessary due to bundling of 25 to 50 pairs of the subscriber loops. The tests are based on interference from HDSL, T1, ISDN, and ADSL transmissions on other lines in the same bundle.

ANSI T1.413-Annex B defines the characteristics of each type of signal. Each interference signal (called a *disturber*) has its own PSD. Therefore, the noise generated by each form of crosstalk will differ. The test engineer must also make sure that the PSD of the test system's background noise is flat.

One of the tests to be performed is the 24-disturber DSL NEXT test. In this test, the ADSL signals are fed over one pair of a bundle of 25 wire pairs. Then the PSD level of the noise is varied to simulate the number of disturbers. The noise used to simulate the disturbers must be Gaussian white noise that has a crest factor of five. (The T1E1.4 committee is investigating these specifications and may change them in the future.) Connection of the noise source to the test loop must have an impedance greater than 4 kΩ.

### 8.9.1.2   Simulation of Wire Loops

In addition to simulating disturbers, one must also simulate several configurations of twisted-pair lines. These lines are called revised-resistance design (RRD) lines (for 1.544-Mbit/s data rates) and carrier service area (CSA) lines (for 6-Mbit/s data rates). In the United States, the Bellcore SR-TSV-002275 and ANSI T.601 specifications define the lengths and gauges of the test loops and the locations of bridge taps in the loops, as these cause reflections and can distort signals. Figure 8.10 shows two subscriber loops with bridge taps used for testing. In this example, the T1.601 loop #13 has both 24-AWG and 26-AWG wires; the CSA loop #4 is entirely of 26-AWG wire.

While performing a crosstalk test, the noise source should be set to the appropriate level for the type of crosstalk being simulated. That becomes the reference level, 0 dB. Establishment of communication between the ATU-C and ATU-R is necessary at the reference noise level (assuming that the ATU-R is the unit under test).

Once training is complete, measurement of the BER with a BERT is necessary while using a pseudorandom test pattern of $2^{23}-1$ bits. By increasing the noise level

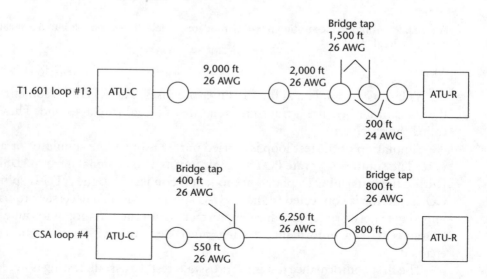

**Figure 8.10**   Simulations of subscriber loops with bridge taps. (*Source:* [14].)

in 1-dB increments, the test engineer can find the highest noise level that produces a BER that is less than $10^{-7}$. The difference between the noise level and the reference level is called the *margin*. To pass crosstalk conformance tests, recommendations specify a margin of 6 dB. Because the noise signals are random, the tests must be performed long enough to accurately measure the BER. Table 8.3 shows the minimum test times for crosstalk conformance testing.

### 8.9.1.3 Impulse—Noise Test

Impulse-noise testing subjects the ADSL products to wideband noise delivered in short bursts. The impulses specified in ANSI T1.413-Annex C are reconstructions of pulses recorded at field sites. There are two impulse test waveforms in ANSI T1.413, as shown in Figure 8.11. To generate those waveforms, an arbitrary waveform generator or a line simulator that generates impulse impairments can be used.

The impulse tests could begin after the training of the ADSL modems using the 0-dB noise level that was used for the crosstalk tests. After communication is established between the two modems, each impulse is applied 15 times at intervals of at least 1 sec. The tests are repeated by adjusting the amplitudes of the impulses until errored seconds are measured on the BER tester. The test engineer can then follow a formula in ANSI T1.413 to calculate the probability that a given amplitude will produce an errored second. To pass the test, that probability must be less than 0.14%.

### 8.9.1.4 Errors Created by POTS Equipment

These tests do not account for errors created by POTS equipment. ADSL and POTS equipment must share the same wires. Hence, testing must be carried out on the

**Table 8.3** Minimum Test Times for Crosstalk Tests

| Bit Rate | Minimum Test Time |
|---|---|
| Above 6 Mbit/s | 100 sec |
| 1.544 Mbit/s to 6 Mbit/s | 500 sec |
| Below 1.544 Mbit/s | 20 min |

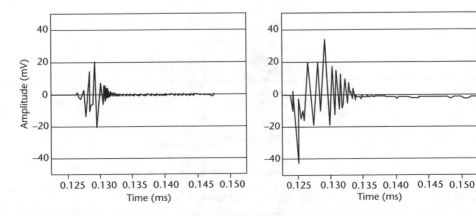

**Figure 8.11** Use of two waveforms for impulse testing on ADSL equipment. (*Source:* [14].)

ADSL equipment with interference from POTS equipment on the same line. Connection of a CO simulator to the ATU-C splitter (see Figure 8.7) and connection of a POTS phone to the ATU-R splitter allow this. The POTS equipment should not interfere with ADSL operation. During the POTS interference test, simulations are necessary on the common conditions that POTS equipment produces, such as signaling, alternating, ringing, answering, and going on- and off-hook.

## 8.10   Testing of Digital Transmission Systems

Digital transmission systems carry multiplexed bit streams. The bit durations can vary from milliseconds to picoseconds depending on the multiplexing scheme. The basics of pulse parameters, definitions, and techniques are detailed in [2]. Most measurements for SDH/PDH/SONET systems are based on observing and measuring the short- and long-term variations of timing parameters of these signals.

### 8.10.1   Jitter and Wander—Definitions and Related Standards

A practical digital signal's position in time continually moves backward and forward with respect to an ideal clock source. Jitter and wander are defined, respectively, as the short-term and the long-term variations of the significant instants of a digital signal from its ideal positions in time. Most engineers' first introduction to jitter is as viewed on an oscilloscope (see Figure 8.12). When triggered from a stable reference clock, jittered data is clearly seen to be moving.

Jitter and wander on a digital signal are equivalent to the phase modulation of the clock signal used to generate the data (see Figure 8.13). Naturally, in a practical situation, jitter will be composed of a broad range of frequencies at different amplitudes.

Jitter and wander have both an amplitude and a frequency. Amplitude and frequency indicate how much and how quickly the signal is shifting in phase, respectively. Jitter is defined in ITU Recommendation G.810 as phase variation with frequency components greater than or equal to 10 Hz, while wander is defined as phase variations at a rate less than 10 Hz (see Figure 8.14).

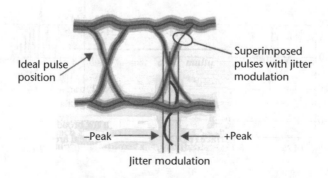

**Figure 8.12**   Jitter as viewed on oscilloscope.

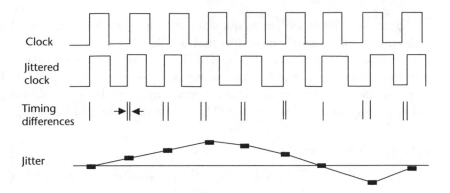

**Figure 8.13**   Phase variation between two signals.

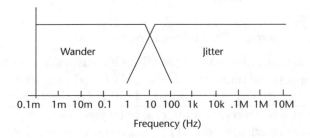

**Figure 8.14**   Frequency ranges of jitter and wander as per ITU G.810 recommendation.

### 8.10.2   Metrics of Jitter

Jitter is normally specified and measured as the maximum phase amplitude within one or more measurement bandwidths. A single interface may be specified using several different bandwidths, as the effect of jitter varies depending on its frequency as well as its amplitude.

Jitter amplitudes are specified in *unit intervals* (UIs), such that one UI of jitter is one data bit width, irrespective of the data rate. For example, at a data rate of 2,048 kbit/s, one UI is equivalent to 488 ns, whereas at a data rate of 155.52 Mbit/s, one UI is equivalent to 6.4 ns.

Jitter amplitude is normally quantified as a peak-to-peak value, rather than an RMS value, as it is the peak jitter that would cause a bit error to be made in network equipment. However, RMS values are also useful for characterizing or modeling jitter accumulation in long-distance systems using SDH regenerators, for example. The appropriate specifications use this metric instead of peak-to-peak value.

### 8.10.3   Metrics of Wander

A wander measurement requires a "wander-free" reference, relative to which the wander of the signal under observation is measured. Any PRC can serve as a reference because of its long-term accuracy (1 in $10^{-11}$ or better) and good short-term stability. A PRC is usually realized with a cesium-based clock, although it may also be realized using the global positioning system (GPS).

Because it involves low frequencies with long periods, measured wander data can consist of hours of phase information. As phase transients are of importance, high temporal resolution is also needed. To provide a concise measure of synchronization quality, three wander parameters described next have been defined and are used to specify performance limits. Formal mathematical definitions of these and other parameters can be found in ITU Recommendation G.810 [17].

### 8.10.3.1  Time Interval Error

Time interval error (TIE) is defined as the phase difference between the signal being measured and the reference clock, typically measured in nanoseconds. TIE is conventionally set to zero at the start of the total measurement period $T$. Therefore, TIE gives the phase change since the measurement began. An example is given in Figure 8.15. The increasing trend shown is due to a frequency offset—about 1 ns per 10s, or $10^{-10}$ in this case.

### 8.10.3.2  Maximum Time Interval Error

Maximum time interval error (MTIE) is a measure of wander that characterizes frequency offsets and phase transients. It is a function of a parameter $\tau$ called the *observation interval*. The definition of MTIE ($\tau$) is the largest peak-to-peak TIE (i.e., wander) in any observation interval of length $\tau$ (see Figure 8.16).

In order to calculate the MTIE at a certain observation interval $\tau$ from the measurement of TIE, a time window of length $\tau$ is moved across the entire duration of TIE data, storing the peak value. The peak value is the MTIE ($\tau$) at that particular $\tau$. This process is repeated for each value of $\tau$ desired. For example, Figure 8.18 shows a window of length $\tau = 20$ sec at a particular position. The peak-to-peak TIE (PPTIE) for that window is 4 ns. However, as the 20-sec window is slid though the entire measurement period, the largest value of PPTIE is actually 11 ns (at about 30 sec into the measurement). Therefore MTIE (20 sec) = 11ns. Figure 8.16(b) shows the complete plot of MTIE ($\tau$) corresponding to the plot of TIE in Figure 8.15. The rapid 8 ns transient at $T = 30$ sec is reflected in the value MTIE ($\tau$) = 8 ns for very small $\tau$. It should be noted that the MTIE plot is monotonically increasing with observation interval and that the largest transient masks events of lesser amplitude.

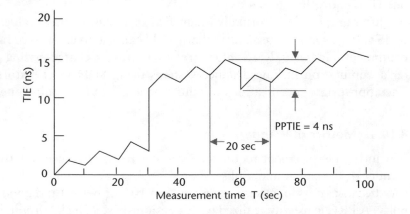

**Figure 8.15**  An example of TIE wander measurement. (*Source:* [18].)

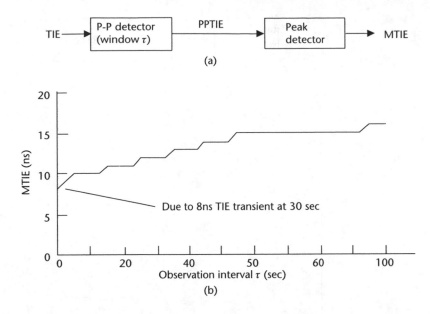

**Figure 8.16** MTIE definition and measurement: (a) functional definition of MTIE, and (b) plot of MTIE. (*Source*: [18].)

### 8.10.3.3 Time Deviation (TDEV)

Time deviation (TDEV) is a measure of wander that characterizes its spectral content. It is also a function of $\tau$, the observation interval. The definition of TDEV ($\tau$) is the rms value of the filtered TIE, where the bandpass filter (BPF) is centered on a frequency of $0.42/\tau$.

Figure 8.17(b) shows two plots of TDEV ($\tau$). The first plot (for $T = 100$ sec), corresponding to the TIE data of Figure 8.15, shows TDEV rising with . This is because, for the short measurement period $T = 100$ sec, the two transients in Figure 8.15 dominate.

If longer TIE measurements up to $T = 250$ sec are made, the effect of the two transients on TDEV would become less, assuming there are no more transients. The TDEV characteristic labeled $T = 250$ sec would be the result. It should also be noted that TDEV is insensitive to constant phase slope (frequency offset). To calculate TDEV for a particular $\tau$, the overall measurement period $T$ must be at least $3\tau$. For an accurate measure of TDEV, a measurement period of at least $12\tau$ is required. This is because the rms part of the TDEV calculation requires sufficient time to get a good statistical average.

## 8.11 BERTs

A critical element in a digital transmission system is the probability of error. This measurement is made by a BERT, which replaces one or more of the system's components during a test transmission.

The digital transmission system shown in Figure 8.18 includes a data source, such as computer memory, a voice digitizer, or a multiplexer, that originates a digital signal, $D$.

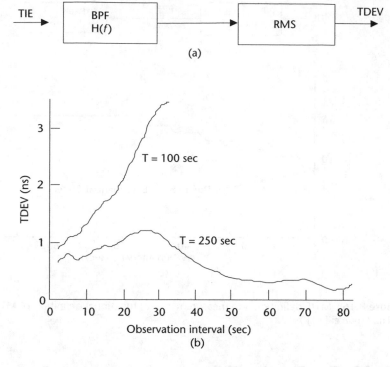

**Figure 8.17** TDEV measurement: (a) functional definition, and (b) example of TDEV wander measurement for case of Figure 8.15. (*Source*: [18].)

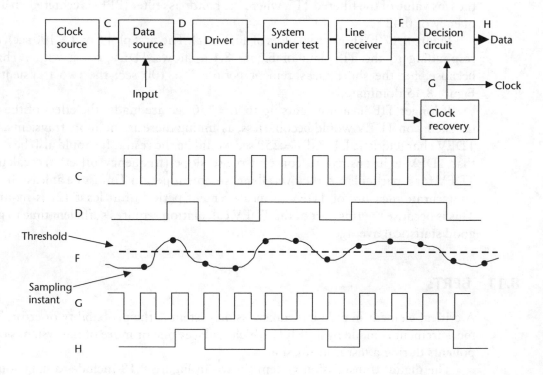

**Figure 8.18** A typical digital transmission system and its signals: (a) block representation of system, and (b) interrelations or signals. (*Source*: [18].)

A clock source produces a clock signal, $C$, which times the occurrence of each bit in the digital signal. A driver, which may be a power amplifier, a laser diode, an RF modulator, or a tape head, prepares the signal for the system under test. The system under test can be a transmission line with repeaters, an optical-fiber link, microwave radio link, or digital tape recorder. The received signal, $F$, exhibits the noise and pulse dispersion that the transmission system adds to the digital signal.

If the noise and distortion are within limits, the decision circuit can correctly decide whether the original bit was a 1 or a 0. The circuit does this by comparing $F$ (at sampling instants determined by clock signal $G$) with a threshold halfway between the two levels (see Figure 8.18). If no errors are made in the decision process, $H$ is a delayed replica of the original data signal $D$. A clock-recovery circuit generates $G$ from information in the data signal $F$.

A malfunction in any system component can also cause the recovered data to differ from the original data. The primary job of a BERT is to determine the system's error rate rather than isolate the faulty component or the cause. However, for the sake of convenience, the BERT may replace the clock source in the transmitter or receiver. In this case, some fault isolation may be possible by comparing the performance of the system clock sources with that of the BERT. But for the comparison to be meaningful, users must understand the timing jitter specifications of both units.

### 8.11.1   Detection of Data Errors

The simplest measuring technique, as shown in Figure 8.19, is to replace the system's data source with the BERT's data-pattern generator $D'$ and have the BERT receiver monitor the recovered signal for errors.

The data pattern generator can mimic typical traffic by creating pseudorandom patterns, or it can stress the system by outputting fixed patterns stored in memory. To monitor the transmission, the BERT receiver generates its own data pattern $H'$, which is the same as the desired data $D'$. The BERT receiver compares the received signal $H$ with $H'$ and looks for errors. The tester records the total number of errors, the ratio of errors to bits (the BER), the number of *errored seconds* (ES), and the ratio of ES to total seconds. To make a valid comparison, the BERT receiver must synchronize $H'$ with $H$. Accomplishing synchronization depends on whether the data is a fixed or pseudorandom pattern.

### 8.11.2   Supplying Clock

Sometimes it is convenient for the BERT to supply its own clock signals for its transmitter or receiver. For instance, the system clock may be unavailable in a field situation, or the test engineer may want to avoid providing and phasing the clock at the BERT receiver. In this case, the BERT's transmitter clock is $C'$ and its receiver is $G'$, as shown in Figure 8.19(b). In laboratory applications, it is common for the BERT to provide a wide range of clock frequencies.

The BERT's clock source and clock-recovery circuit must be as good as their counterparts in the system under test. The source must introduce negligible timing

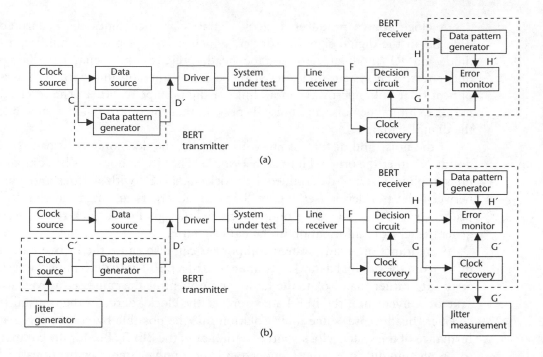

**Figure 8.19**  A simple application of a BERT: (a) BERT generating a known data pattern, and (b) BERT supplying both clock and data. (*Source*: [18].)

jitter, because phase jitter in *C'* causes phase jitter in the recovered clock signal, *G*, relative to the received data signal, *F*. Likewise, the BERT's clock-recovery circuit must tolerate at least as much jitter as the system's recovery circuit without causing errors.

### 8.11.3   Stressing with Jitter

Although the BERT clock source should be essentially jitter free to test the digital transmission system under normal conditions, to stress the system at times, the BERT must generate controlled jitter. For this, some BERTs have a jitter generator that can sinusoidally modulate the phase of the clock source, as shown in Figure 8.19(b).

On the receive end, the BERT monitors the effect of the controlled jitter in two ways. First, it looks for an increased error rate. Second, it measures the jitter remaining in the recovered data, which yields the system's jitter transfer function. The jitter measurement circuit can also be used without the jitter generator to measure the system's own jitter. For more details of BERT applications, [19] is suggested.

## 8.12   Broadband Network Measurements

Based on broadband networks in Chapter 7, some essentials of broadband network measurements are presented next. More details can be found in [2, 18].

### 8.12.1  Jitter and Wander Testing on SDH Systems

#### 8.12.1.1  Jitter Tests

Jitter testing has been an established part of the telecom industry for a long time, and several tests are popular and well known from their use in PDH systems. These are output jitter, input jitter tolerance, and jitter transfer function.

Several new jitter tests required for SDH systems are providing new challenges for test equipment. These new sources of jitter are so significant that 50% of the allowed jitter at a demultiplexed PDH output is allocated to them. The sources are *pointer jitter* and *mapping jitter*. Pointer jitter is considered to be a more significant source of timing impairment than mapping jitter.

#### 8.12.1.2  Wander Tests

The important wander tests are output wander, input wander tolerance, wander noise transfer, phase transient response, and holdover performance.

#### 8.12.1.3  ITU-T Test Equipment Recommendations

ITU-T Recommendations 0.171 and 0.172 are two specifications for test equipment for PDH and SDH, as per the following headings:

- 0.171: Timing jitter and wander measuring equipment for digital systems based on the PDH;
- 0.172: Jitter and wander measuring equipment for digital systems based on the SDH.

Recommendation 0.172 has been developed in order to specify measurement requirements for the whole SDH network (i.e., both SDH line interfaces operating at PDH bit rates, simply referred to as *tributaries*). For details, [17] is suggested.

## 8.13  Use of Modern Digital Storage Oscilloscopes for Physical Layer Testing or Tributary Signals

### 8.13.1  Telecom Testing and Pulse Marks

Telecom standards dictating the shapes of pulses that represent bits include marks that define the tolerances of timing and the amplitude of the pulses. Telecom receiving equipment must properly interpret pulses that fall within the mask limits, and transmitters must drive the signals along a medium so they arrive within tolerance. For example, pulse masks in Figure 8.20(a) set the limits for physical parameters, such as rise time, fall time, pulse width, amplitude, overshoot, and undershoot. The entire pulse must fall within the mask for it to comply with a standard. Engineers use pulse masks as "sanity checks" for transmitter and receiver designs. Later, the masks are used in tests to verify that a product complies with industry standards. Masks may also be used in pass-fail tests at production.

Some digital scopes include mask options that let you measure the pulses. Figure 8.20(b) shows the mask for a 155.52-Mbps STS-3E binary-1 pulse. The

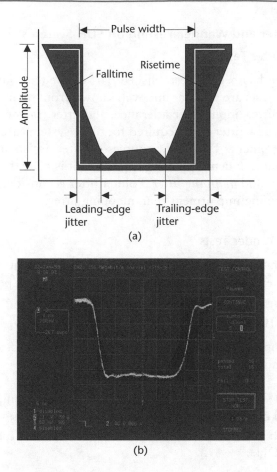

(a)

(b)

**Figure 8.20**   Pulse marks and oscilloscope measurement: (a) masks defining amplitude, rise/fall times, and jitter, and (b) a typical oscilloscope measurement.

mask is typical of that for other electrical signals, such as T1, T3, and DS3, that travel over coaxial and twisted-pair wires. Binary 1s may have two marks with inverse polarities. The polarity depends on the type of bit encoding and on the bit pattern used in the transmission system. For details, [20] is suggested.

### 8.13.2   BERTs Versus Modern Fast Sampling DSOs

BERTs can be very effective in finding errors that occur very infrequently, but only if the communication device under test is working well enough to accept inputs and transmit data. If the design is not working, a BERT cannot help a designer determine the source of the problem. When a transmitter is not working or is partially working, an oscilloscope is invaluable for debugging the design. The graphical display of the digital data makes finding the problem easier. By contrast, the BERT typically has a numeric display that shows BER only.

In many cases, the advanced triggering available in DSOs [2] allows quick location of signal errors. The digital phosphor oscilloscopes from Tektronix is a good example of suitable DSO for this purpose. Digital phosphor oscilloscopes offer five different categories of advanced triggers, including communications, logic, and pulse width triggering. Many of these advanced triggers can be used to quickly debug a communications problem. For more details on use of digital oscilloscopes for telecom tributary signals, [21] is suggested.

# References

[1]    Kularatna, N., *Modern Electronic Test and Measuring Instruments*, London: IEE Press, 1996.

[2]    Kularatna, N., *Digital and Analogue Instrumentation Testing and Measurement*, London: IEE Press, Chapters 7 and 9, 2003.

[3]    Tant, M. J., *The White Noise Book*, Hertz: UK: Marconi Instruments, 1974.

[4]    Bellamy, J., Digital Telephony, 3rd Edition, New York: John Wiley, 2000.

[5]    Reeve, W., *Subscriber Loop Signaling Handbook—Analog*, New Jersey: IEEE Press, 1992.

[6]    Reeve, W., *Subscriber Loop Signaling Handbook—Digital*, New Jersey: IEEE Press, 1995.

[7]    Grizmala, F., " New Methods Cut Envelope Delay Test Time," *Test and Measurement World*, November 1994, pp. 59–63.

[8]    Mazda, F. F., *Electronic Instruments and Measurement Techniques*, Cambridge, England: Cambridge University Press, 1987.

[9]    Bryant, G. H., *Principles of Microwave Measurements*, 2nd Ed., London: Peter Perigrinus, 1993.

[10]   Loser, A., "New Thermocouple Power Sensors NTV-Z51 and NRV-Z52 Based on Semiconductor Technology," *News from Rohde and Schwartz*, Vol. 32, No. 139, 1992/IV, pp. 34.

[11]   Rohde and Schwartz, "Power Meter NRVS," Technical Brochure/Application Note.

[12]   Mayer, J. H., "Re-Evaluating RF Power Measurements," *Test and Measurement World*, November 2000, pp. S3–S6.

[13]   Israelsohn, J., " Make Short Work of RF Power Measurements," *EDN*, August 3, 2000, pp. 54–64.

[14]   Rowe, M., "ADSL Products Must Confirm and Perform," *Test and Measurement World*, February 1997, pp. 39–48.

[15]   Engelson, M., " Knowing the Signal Classes Greatly Simplifies Spectrum Analysis," *Electronic Design*, November 6, 2000, pp. 127–134, 2000.

[16]   Tektronix, Inc., "New ITU-T 0.172 Defines Jitter and Wander Test Equipment for SDH Systems," Technical Brief TD/XBS 2GW-13178-0, June 1999.

[17]   G. 810 recommendation (1996)- ITU.

[18]   Tektronix, Inc., "Performance Assessment of Timing and Synchronization in Broadband Networks," Application Note FL 5336/XBS 2GW-11145-0, April 1997.

[19]   Tektronix, Inc., "Ensure Accuracy of Bit-Error Rate Testers," Technical brief.

[20]   Rowe, M., "Pulse Marks Define Signal Limits," *Test and Measurement World*, September 1999, pp. 15–18.

[21]   Tektronix, Inc., "Testing Telecommunications Tributary Signals," Application Note TD/XBS55W-12045-0, May, 1998.

## CHAPTER 9

# Embedded Systems Design for Telecommunications: A Case Study

By Ethan Bordeaux, DSP Software Engineer, Analog Devices, Inc., Wilmington, MA

## 9.1 Introduction

Over the last decade (1990–2000), the semiconductor industry advanced rapidly, providing designers with powerful DSP and microprocessor chips that can handle very complex real-time software applications. The telecommunications industry was one of the important sectors that profited greatly from these developments. Standardization bodies and special-interest forums were active in grouping and developing the expertise required for newer applications.

In the previous chapters, we discussed the fundamentals and applications of modern telecommunication systems, gradually transforming into the digital domain with intensive DSP applications. The contents here provide the reader with a practical overview of the development approach to a typical software-intensive product, such as a voice codec for a cellular phone. In particular, the chapter provides an overview on the software development for an adaptive multirate (AMR) codec for a GSM cellular phone, including the development process, hardware usage, and the use of reference code from a standardization body. This chapter covers the analysis and steps followed in porting the AMR speech codec to Analog Devices' AD6526 multicore processor. While the discussion is specific to this algorithm and architecture, the basic steps followed in porting and testing this code apply to virtually all embedded algorithms and hardware platforms.

The reader is expected to have a background on the processor-based product developments. Reference [1] provides a useful overview. For an overview of DSP processor architectures, [2] is recommended.

## 9.2 AMR Speech Codecs and the GSM Environment

By the beginning of the new millennium, GSM subscriptions grew to over 600 million, with speech as a prime service. The challenge for the operators is to enhance the speech quality while optimizing spectrum efficiency, particularly in urban areas with dense subscriber populations. By the latter part of the 1990s, GSM operators and the developers created a speech-processing standard to deal with nonoptimum partitioning of the source and channel coding rates. With fixed channel coding rates, optimum use of error correction was not possible. When the channel is bad, the redundancy inserted by the channel coder may be not sufficient to correct transmission errors. Similarly, when the channel is good, speech quality can be improved if more bits are spent in source coding. The AMR standard helps solve this problem.

Figure 9.1 depicts the basic concept of the AMR speech transmission system. Both the base station and the mobile station consist of the following entities.

- A speech codec with variable bit rate (speech encoder and speech decoder);
- A channel codec with variable error-protection rate, matching to the bit rate of the speech codec (channel encoder and channel decoder);
- A channel estimation entity;
- A control unit for rate adaptation.

The base station is the master and determines the appropriate rates (modes) in both uplink (UL) and downlink (DL) directions. A channel-quality parameter is derived from the soft output generated by the equalizer on the MS and is used to control the codec rates. In the UL direction, the mobile transmits the mode bits (channel rate information) on a traffic channel, and the DL channel metric (DL channel quality information) is transmitted to the channel decoder using a control channel slow associated control channel (SACCH). At the receiver (base station), channel decoding is done first, followed by speech decoding. In parallel, a measurement of the UL channel is carried out by the base station channel estimator. Measured UL channel quality and the DL channel metric (received from the mobile station) are fed to the base station control unit, which determines the current DL rate as well as the required UL rate.

In the DL direction, the current DL mode and the required UL rate are transmitted inband within the traffic channel to the mobile station. Similar to the UL, a DL channel quality measurement is carried out by the estimator of the mobile station, and the required UL rate is decoded from the received bit stream. From the measured DL channel quality a DL channel metric is calculated by the MS control unit and then transmitted via a SACCH to the base station. The speech encoder now operates at the newly requested UL rate. The AMR concept allows operation with near-landline speech quality for poor channel conditions and better quality for good channel conditions. For more details of AMR and variable rate codecs, [3, 4] are suggested.

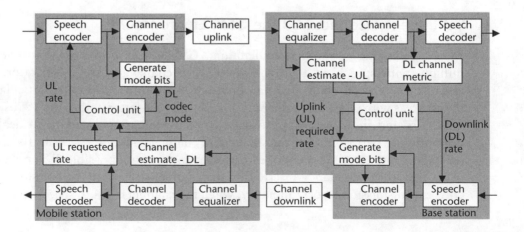

**Figure 9.1**   Block diagram of the GSM AMR concept.

## 9.3   An Overview of Digital Signal Processor Architectures

Microprocessors and microcontrollers are generally used in applications where a common sequence of activities are handled, with the possibility of subroutine calls to deviate into not-so-common sequences. For example, in a cellular phone, when a number is dialed, the number-processing algorithm inside the microprocessor sub-system takes care of the activity and handles the process. When a phone call is received, the phone's microprocessor activates various processes with the base station and starts the voice conversation.

A DSP is a special type of microprocessor with an architecture specifically tuned for real-time signal processing algorithms and applications within embedded systems. This is different from a microprocessor/microcontroller architecture, which is typically designed for handling a wide variety of applications and whose systems can typically handle a greater level of indeterminacy in total execution time. For instance, the difference between 19.9 ms and 20.1 ms in completing a user interface–related task on a microprocessor is indistinguishable and, generally speaking, unimportant. However, this difference in computational delay on a DSP can mean the difference between successfully processing a frame of data and buffer overflow, leading to corrupted output or a dropped communications channel.

While DSPs must be designed to handle a wide variety of mathematical operations in a fast and flexible manner, the operation that defines a DSP is a multiply/accumulate with simultaneous dual data accesses inside of a loop. The multiply/accumulate instruction is the kernel of such algorithms as autocorrelation, finite impulse response (FIR) filtering, and convolution, all of which are critical for signal modification and analysis. A simple example of a multiply/accumulate instruction inside a loop (written in C code) is given in Figure 9.2.

This function multiplies the *input* and *coeff* arrays together and places the accumulating sum into the variable named *result*. On a DSP, the multiply/accumulate instruction plus the code maintaining the loop counter would execute in a single cycle (ignoring extra cycles that might be necessary to initialize the hardware loop counters). DSPs normally have an accumulator that is much larger than the values being multiplied in order to support multiple multiply/accumulates placed in a loop without overflowing the output register.

Of course a general-purpose microprocessor can also compute the result of the code in Figure 9.2. However, microprocessors were not originally designed to handle signal processing–intensive applications. Earlier generations of microprocessors performed quite poorly on signal-processing algorithms. Recently, there have

```
long correlate(short *input, short *coeff, int corrlen) {

    long result = 0;
    int i;

    for (i=0; i<corrlen; i++) {
        result=result+(((*input)++)*((*coeff)++));
    }
    return result;

}
```

**Figure 9.2**   A simple example of a looped multiply/accumulate instruction in C code.

been improvements in DSP processing capabilities for mainstream desktop processors, such as the introduction of single instruction multiple data (SIMD) and streaming SIMD extensions (SSE) in the Pentium® line of processors and AltiVec in the PowerPC® architecture. However, even with these enhancements, DSPs still often execute signal-processing algorithms more efficiently than their microprocessor counterparts, particularly in low-power applications.

## 9.4   Overview of Target Hardware for the AMR Speech Codec—The AD6526 Multicore Processor

Within the latter part of the 1990s, microprocessor designers and manufacturers were able to integrate both the microprocessor cores and DSP cores on the same die or in hybrid form, leading to reduced cost, power consumption, and system integration effort. The AD6526 is such a multicore IC, which contains both an Advanced RISC Machines ARM7® microcontroller and an Analog Devices ADSP-218x-based DSP core, along with a wide variety of peripherals designed specifically for wireless handset applications. The DSP core is also enhanced to improve its performance in wireless algorithms and applications. In particular, the DSP includes additional functionality for performing memory accesses to a wide address bus by using the read-write state machine (RWSM) page registers and the DSP cache. Figure 9.3 depicts a simplified block diagram of the AD6526.

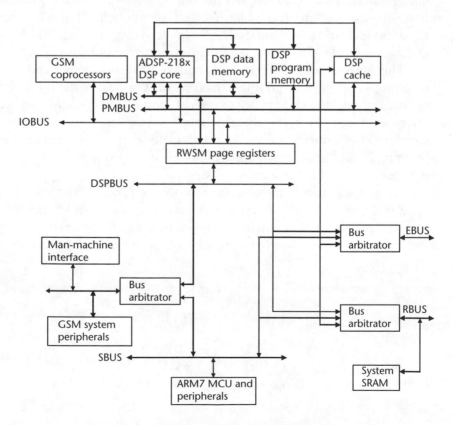

**Figure 9.3**   Block diagram of the AD6526. (*Courtesy of:* Analog Devices, Inc.)

The AD6526 DSP has 8K words of internal program memory and 8K words of internal data memory. These memory segments do not require any wait states and typically hold the most important portions of a number of different DSP algorithms. However, the DSP must have access to a much greater amount of memory in order to perform all of the tasks required of a wireless terminal. To allow the DSP to access a 32-bit address space (the base ADSP-218x core only supports a 14-bit address space), two functional blocks were added to the AD6526: the RWSM and the DSP cache. These blocks are briefly described in the following sections.

### 9.4.1   RWSM

The RWSM allows the DSP to have a series of 1K-word windows into the 32-bit system memory. This is handled by configuring a set of RWSM page registers to the appropriate base address and then executing DSP instructions that perform accesses into the 1K window. Many DSPs do not have the wide address bus required to access a large external memory, as needed in many wireless applications. The DSP has a total of 14 RWSM pages for accesses to data through both the program memory bus and the data memory bus (see Figure 9.4).

Note how the processor can access any region of system memory with a 14-bit address, along with the values in the page registers. However, there are only 14 1K-word windows into memory available at any one time. When porting an

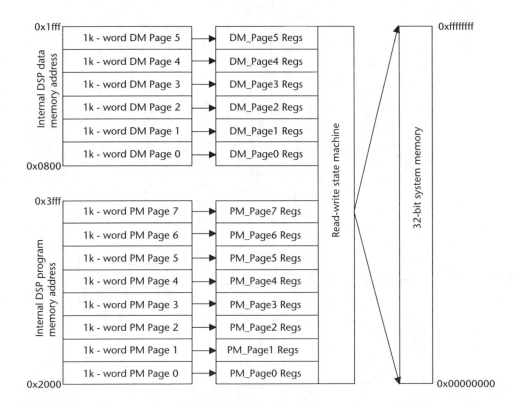

**Figure 9.4**   Diagram of RWSM/system memory accesses on the AD6526. (Reproduced with permission of Analog Devices, Inc.)

algorithm to the AD6526, it must either set the RWSM inside the algorithm or use less than 14K words of data memory.

While this memory mapping scheme allows for access to the full 32-bit address space without modifying the existing DSP architecture, it does add some complexity when handling DSP algorithms that require large amounts of data. If more than 14K words of system memory are required by an algorithm, the RWSM registers must be managed inside the algorithm. This is undesirable, as it makes a portion of the algorithm's contents architecture dependent, complicating porting and maintenance. Therefore, one design goal of the ported AMR speech codec was to avoid modifying the RWSM page registers during execution of the algorithm.

### 9.4.2 DSP Cache

The AD6526 also includes a caching mechanism to speed accesses to external program memory. The DSP has a configurable 8K-word window for program memory accesses into the full 32-bit address space. There is a 4K-word cache, which loads eight instructions into the cache when a cache boundary is crossed. Because the cache is 4K words long and the total window into memory is 8K words long, the cache is mirrored at the 4K boundary (see Figure 9.5). For example, if an instruction is executed at address 0x2000 (the base address of the cache from the perspective of the DSP), code in the cache from address 0x3000 would be automatically invalidated.

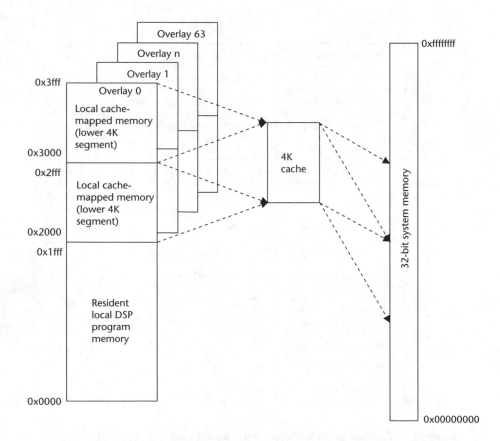

**Figure 9.5**  Caching mechanism as implemented on the AD6526.

Additionally, because there is only a single 4K-cache memory, accessing code on a different overlay will also overwrite the previously stored instructions and lead to cache invalidation. The implications of using the cache in a DSP algorithm and methods of reducing the impact of using the cache are covered later in this chapter.

Note how all overlays map into the same 4K-word cache. If two overlay pages are accessed at the same address (or at a 4K offset of that address), the cache is invalidated.

## 9.5   Overview of the AMR Speech Codec

The AMR speech codec was introduced at the end of 1998 as a new codec for both the GSM wireless network and as the mandatory speech codec for third generation wideband code division multiple access (WCDMA) systems. At the beginning of 2003, it is being widely introduced into existing GSM systems. Some of the defining features of this speech codec include:

- Eight adaptive data rates between 4.75 kbit/s and 12.2 kbit/s, based on channel characteristics and system loading [5];
- VAD, silence description (SID), CNG, and discontinuous transmission (DTX) capabilities for lowering data rates during silences in speech;
- The ability to substitute and mute lost traffic frames due to errors in the traffic channel;
- Efficient implementation on fixed point DSP architectures.

The specification for the AMR speech codec was generated by the Special Mobile Group, a subgroup of ETSI®, a nonprofit organization that produces and maintains the telecommunications standards that are used specifically in Europe and potentially throughout the world. All of the technical specifications for the AMR speech codec can be found at ETSI's Web site [6].

### 9.5.1   Technical Overview of the AMR Speech Encoder

The AMR speech encoder is designed to take 160 16-bit words as an input and produce between 95 and 244 bits of encoded parameters as an output, depending on the encoder operating mode. The required ADC/DAC sample rate of an AMR-based application is 8 kHz. Therefore, the AMR speech codec continuously processes 20 ms ($160 \times 125\,\mu s$) frames of data. The processing path can be loosely divided into the following functional blocks: preprocessing, linear predictive analysis and quantization, open-loop pitch and gain analysis (adaptive codebook search), and algebraic codebook search. The basic principles of these functions are described in Chapter 2. Some of the functional blocks of the AMR speech codec are calculated on 20-ms frames of data, while other algorithms are performed on 5-ms or 10-ms subframes of the sampled data (often referred to as the subframe loop).

The AMR speech encoder also performs a VAD computation to determine whether the current frame contains speech or background noise, and a DTX mode in decoding to lower aggregate transmission requirements when speech is not present in the encoder. The data flow through the functional blocks of the AMR speech encoder is shown in Figure 9.6.

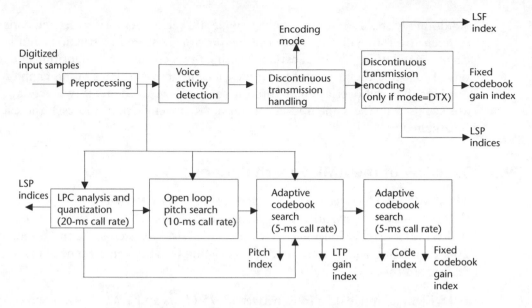

**Figure 9.6**   Simplified block diagram of the AMR speech encoder.

This summary is here to give a sense of the types of computations and algorithms needed for AMR encoding and is not an exhaustive explanation of the algorithm. For the sake of simplicity, only the 4.75-Kbps encoding mode is explained. While there is a great deal of similarity of computations between all eight operating modes, some encoding rates significantly add or remove from the computational steps covered here. For full technical information on AMR speech encoder operation, refer to the specifications listed in Table 9.1 (which are available for download on ETSI's Web site).

### 9.5.1.1   Preprocessing

Once the input buffer of 160 samples is filled, the AMR speech codec performs preprocessing on the data. The first step is stripping the three least significant bits (LSBs) from the input samples. Normally this loss of data is not audible due to the low SNR of telecom ADCs and the inherent distortions introduced in the

**Table 9.1**   ETSI® Specifications for the AMR Speech Codec

| Standard Name | Description |
| --- | --- |
| GSM 06.73 | ANSI-C code for the AMR speech codec |
| GSM 06.74 | Test sequences for the AMR speech codec |
| GSM 06.90 | AMR speech transcoding |
| GSM 06.91 | Substitution and muting of lost frames for AMR speech traffic channels |
| GSM 06.92 | Comfort noise aspects for AMR speech traffic channels |
| GSM 06.93 | DTX for AMR speech traffic channels |
| GSM 06.94 | VAD for the AMR speech codec |

compression algorithm. The input data is then passed through a second-order infinite impulse response (IIR) HPF with a cutoff frequency of 80 Hz. This filtering process is acceptable, as the base bandwidth in telecom systems is 300–3,400 Hz. Additionally, the input is scaled down by a factor of two to reduce the chance of errors or distortion due to saturation or overflow from performing 16-bit fixed-point computations.

### 9.5.1.2    VAD

Before the speech encoder processes the input speech to derive the appropriate parameters, a VAD algorithm is run on the incoming speech to determine if the current sample buffer contains active speech or background noise. If the input data is determined to be speech, the encoding procedure is followed. Otherwise, the DTX algorithm is run to determine if the buffer should still be fully encoded as speech or if a simplified frame of data (known as a *silence descriptor* or SID frame) will be transmitted.

The AMR speech codec includes two distinct VAD algorithms, known as VAD1 and VAD2. Only one VAD algorithm is compiled into a complete AMR codec. In either case, the output of the VAD algorithm is the same—a flag indicating whether the current input data contains speech. It is up to the system designers to determine which algorithms are appropriate for their applications. In GSM wireless communications, it is expected that the VAD1 algorithm will be the more common choice; therefore, only this algorithm is described next.

The VAD1 algorithm performs a power analysis of the input signal across nine frequency bands. Additionally, it uses pitch delay (gathered from the open- and closed-loop pitch analysis) from previous encoder calculations to detect additional periodic signals and tone detection (again using data from the previous pitch analysis) to determine if there is an informational tone, such as DTMF. Complex signal detection is also performed using data from the open-loop pitch analysis to search for complex correlated signals, such as music. The VAD1 algorithm takes this set of information, along with an estimate of the background noise, to find the ratio of the power of the background noise in nine frequency bins to the power of the signal in those same bins. This ratio is compared against an adaptive threshold, and if it crosses this value, then an intermediate decision that the input is noise is made. The final VAD decision depends on an analysis of the complexity of the noise signal and whether there should be hangover in the decision such that the speech encoder continues interpreting the signal as speech at the end of a speech burst.

### 9.5.1.3    DTX

The DTX processing and control block determines whether the current frame should be processed as either speech or noise. Note that this is different from the VAD processing block, which determines if the current frame *is* speech or noise. A series of sequential noise frames must be acquired before the AMR speech encoder switches from processing data as speech to processing it as noise.This is done to avoid fast switching between two different modes of the encoder, leading to annoying audible artifacts. Once enough frames of data are identified as noise, the speech encoder enters a simplified mode, where fewer parameters are encoded and transmitted. The encoder attempts to match the spectral qualities of the sampled noise and sends this through a SID_FIRST frame. If the signal continues to be noise, a SID_UPDATE frame is sent every eighth frame to update the noise model, starting with the third frame after the SID_FIRST frame.

DTX processing is essentially a state machine that controls whether the full encoded parameters are calculated (SPEECH), a subset of these parameters are calculated (SID_FIRST or SID_UPDATE), or no parameter need be calculated in the current frame (NO_DATA). DTX processing must balance the desire for lower bandwidth requirements with the need for intelligible speech, especially at onsets (the start of a new burst of speech). DTX processing makes particular sense in wireless applications where it is highly desirable to avoid transmitting signals continuously.

### 9.5.1.4   Linear Predictive Analysis and Quantization

As described in Chapter 2, LPC is a common method of representing and compressing speech into a series of parameters. Briefly, LPC algorithms model the behavior of the vocal tract as a series of resonating tubes and airflow across the vocal cords as either quasiperiodic impulses or white noise (for creating voiced and unvoiced signals, respectively). The LPC model in AMR attempts to calculate the resonant frequencies of the vocal tract with a tenth-order all-pole filter (often referred to as a formant filter). The coefficients of this filter are calculated via autocorrelation and a Levinson-Durbin recursion algorithm [7–11].

Once the coefficients are computed, they are converted to line spectral pairs. This transformation is done for three reasons. First, line spectral pairs are easily evaluated to determine if the corresponding LP filter will have any resonances. Second, errors are minimized when quantizing the line spectral pairs versus quantizing the LP coefficients, as they have a lower spectral sensitivity. Additionally, because the line spectral pairs have a direct relationship to the resonant frequencies of the formant filter, they are ideal for interpolating coefficients from frame to frame. A first-order moving average predictor is applied to the line spectral pairs, and the residual vectors are quantized via split matrix quantization [12]. The quantization indices are transmitted as a part of the encoded parameters.

### 9.5.1.5   Open- and Closed-Loop Pitch Analysis (Adaptive Codebook Search)

Open-loop pitch analysis is used to find an estimate for the pitch of the input vector. This estimate is used in the closed-loop pitch analysis to allow for searching fewer pitch lags. The closed-loop pitch analysis calculates an integer and fractional lag. This lag is calculated by processing the synthesized speech for all lags within a certain range and then picking the pitch delay with the smallest mean squared error between the original and synthesized speech. A pitch gain is also calculated here, by computing the ratio of the gains of the power of the filtered adaptive codebook to the ratio of the correlation of the filtered adaptive codebook and the target signal. The lag and pitch are both encoded and are a part of the transmitted parameters.

### 9.5.1.6   Algebraic Codebook Search

The algebraic codebook is searched by minimizing the mean squared error between the weighted input signal and the weighted synthesized speech signal. The basic idea of the codebook search is to find a series of pulses, which, when passed through the synthesis filter and added to the previous estimate of the speech signal, will minimize the error between this new signal and the target input signal. In the 4.75-kbit/s mode of AMR, two pulses per subframe are used for excitation. The location and sign of the pulses within the subframe are encoded in such a way that 36 bits are required to represent the position and sign of the pulses for all four subframes.

### 9.5.2   Technical Overview of the AMR Speech Decoder

The AMR speech decoder follows a similar computational model as the AMR speech encoder, except that it has a lower computational requirement. The basic processing blocks in the AMR decoder are LP filter decoding, decoding the adaptive and algebraic codebook vector and gains, and computing the synthesized speech (see Figure 9.7). The AMR speech decoder also includes a DTX processing algorithm.

### 9.5.2.1   DTX Processing

The first step in decoding the received parameters from the AMR encoder is to determine whether the incoming frame contains speech or is a part of a DTX frame. This is done by analyzing the current frame type, along with parameters generated

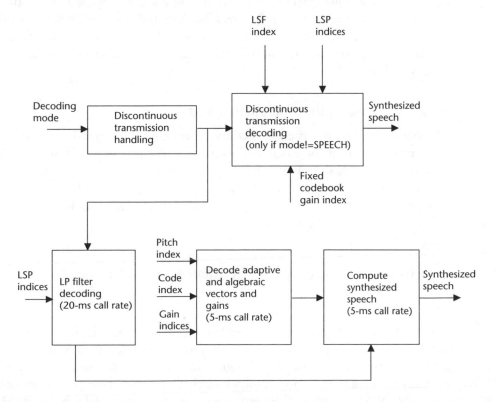

**Figure 9.7**   Simplified block diagram of AMR speech decoder.

from previous frames. If the input vector is determined to not be speech, the parameters are passed through a DTX decoder. The DTX decoder acts as a simplified speech synthesizer, generating an estimate of the comfort noise from the received signal in the absence of speech. DTX processing can occur for a number of reasons. If the DTX handler determines that the input data is a SID frame, it is processed by the DTX decoder, and comfort noise is generated. Additionally, if the incoming data is corrupted, it could also trigger DTX processing to avoid annoying garbled output from the speech decoder.

### 9.5.2.2    LP Filter Decoding

The line spectral pair coefficients are interpolated across four subframes and are then converted into LP filter coefficients. These coefficients are used to create the synthesis filter for reconstructing the output speech vector.

### 9.5.2.3    Decoding the Algebraic and Adaptive Codebook Vectors and Gains

After calculating the LP filter coefficients, the adaptive codebook vector is found by decoding the received pitch index. Decoding this index gives the integer and fractional pitch lags. Using this pitch delay, the adaptive codebook vector is found by interpolating the past excitation. The algebraic codebook vector is found by extracting the pulse locations and signs using the algebraic codebook index input parameter.

The adaptive and algebraic codebook gains are calculated from the received codebook index, which directly provides the adaptive codebook gain and the algebraic codebook gain correction factor. The algebraic codebook gain is estimated and scaled by the correction factor. The codebook gain and vector then go through post-processing routines. The codebook gain passes through a smoothing filter to avoid unnatural fluctuations, and an antisparseness filter is also run in the low-bit-rate modes of AMR.

### 9.5.2.4    Computing the Synthesized Speech

The synthesized speech is computed by subtracting the filtered previous outputs of the synthesizer from the calculated excitation. If the synthesis filter overflows, the excitation is scaled down and the output is once again synthesized. This output is then fed into an adaptive post filter and finally through an HPF to remove any inadvertent outputs below 60 Hz. At this point, speech decoding is complete, and the output can be passed to a DAC.

## 9.6    Algorithmic Complexity of the AMR Speech Codec

### 9.6.1    MIPS

The MIPS of the AMR speech codec are estimated through the usage of weighted millions of operations per second (WMOPS), a methodology developed by ETSI for counting basic mathematical and logical operations along with data accesses, and weighting the total operation counts such that they reflect the performance expected on a DSP. WMOPS are a nice convention for estimating MIPS because they can be adapted through time as processor architectures change, or they can even be tweaked by the system architect if it is known how many cycles it takes to perform a specific operation. Table 9.2 lists the WMOPS operations (often referred to as *basic-ops*) and their default weighting within the AMR speech codec reference code.

**Table 9.2**    ETSI Defined Basicops Along with Default Cycle Weightings

| WMOPS Operation | Description | Weighting |
|---|---|---|
| Add | Sixteen-bit by 16-bit add with saturation | 1 |
| Sub | Sixteen-bit by 16-bit subtract with saturation | 1 |
| abs_s | Sixteen-bit absolute value | 1 |
| shl | Sixteen-bit arithmetic shift left with saturation | 1 |
| shr | Sixteen-bit arithmetic shift right with saturation | 1 |
| extract_h | Sixteen-bit extract low from 32-bit value | 1 |
| extract_l | Sixteen-bit extract high from 32-bit value | 1 |
| mult | Sixteen-bit by 16-bit multiply, 16-bit result with saturation | 1 |
| L_mult | Sixteen-bit by 16-bit multiply, 32-bit result with saturation | 1 |
| negate | Sixteen-bit negation with saturation | 1 |
| round | Sixteen-bit rounding of a 32-bit input | 1 |
| L_mac | Sixteen-bit by 16-bit multiply with 32-bit accumulate, 32-bit result with saturation | 1 |
| L_msu | Sixteen-bit by 16-bit multiply with 32-bit subtract, 32-bit result with saturation | 1 |
| L_macNs | Sixteen-bit by 16-bit multiply with 32-bit accumulate, 32-bit result with no saturation | 1 |
| L_msuNs | Sixteen-bit by 16-bit multiply with 32-bit subtract, 32-bit result with no saturation | 1 |
| L_add | Thirty-two-bit by 32-bit addition with saturation, 32-bit result | 2 |
| L_sub | Thirty-two-bit by 32-bit subtraction with saturation, 32-bit result | 2 |
| L_add_c | Thirty-two-bit by 32-bit addition with no saturation, 32-bit result and carry bit | 2 |
| L_sub_c | Thirty-two-bit by 32-bit subtraction with no saturation, 32-bit result and carry bit | 2 |
| L_negate | Thirty-two-bit negation with saturation | 2 |
| L_shl | Thirty-two-bit arithmetic shift left with 32-bit result with saturation | 2 |
| L_shr | Thirty-two-bit arithmetic shift right with 32-bit result with saturation | 2 |
| mult_r | Sixteen-bit by 16-bit multiply with rounding and a 16-bit result | 2 |
| shr_r | Sixteen-bit arithmetic shift right with saturation and rounding and a 16-bit result | 2 |
| mac_r | Sixteen-bit by 16-bit multiply with 32-bit accumulation and 16-bit result | 2 |
| msu_r | Sixteen-bit by 16-bit multiply with 32-bit subtraction and 16-bit result | 2 |
| L_deposit_h | Deposit 16-bit value into upper half of 32-bit result, zero lower 16 bits | 2 |
| L_deposit_l | Deposit 16-bit value into lower half of 32-bit result, sign-extend upper 16 bits | 2 |
| L_shr_r | Thirty-two-bit arithmetic shift right with 32-bit result with saturation and rounding | 3 |
| L_abs | Thirty-two-bit absolute value with saturation | 3 |
| L_sat | Thirty-two-bit saturation | 4 |
| norm_s | Calculate number of left shifts needed to normalize a 16-bit value | 15 |
| div_s | Sixteen-bit by 16-bit division, 16-bit result | 18 |
| norm_l | Calculate number of left shifts needed to normalize a 32-bit value | 30 |
| move16 | Sixteen-bit immediate value write to register or 16-bit register to register move | 1 |
| move32 | Thirty-two-bit immediate value write to register or 16-bit register to register move | 2 |
| logic16 | Sixteen-bit logical operation (AND, OR, XOR) | 1 |
| logic32 | Thirty-two-bit logical operation (AND, OR, XOR) | 2 |
| test | Comparison with branch | 2 |

Table 9.2 shows the expected number of cycles to perform a particular operation. For instance, a 32-bit absolute value (*L_abs*) is expected to take three cycles. By running a test vector through the AMR speech codec, a baseline WMOPS (and consequently MIPS) value is determined. The total WMOPS for the AMR speech codec running in the 4.75-kbit/s mode is approximately 10.7 for the encoder and 2.0 for the decoder. Of course, if the target processor executes any of these operations at a slower or faster rate, the weighting table must be adjusted accordingly to provide a better estimate of the total WMOPS. It should be noted that this WMOPS figure does not take into consideration any architectural limitations of a target processor, such as a limited register set or deep pipelines, or whether the processors' computational units do not handle these operations in a bit-exact fashion. Additionally, these tests do not consider additional cycles added due to fetching data for algorithmic calculations. The reference code assumes that when data fetches are required for mathematical operations, they can be loaded without any additional cycles. This is potentially not true for two reasons: either the processor does not support simultaneous loading of all of the required data with the mathematical operations or the variable being loaded is located in nonzero wait state memory. However, the WMOPS value can be a valuable guide in determining the absolute best-possible performance of a particular algorithm on a processor.

### 9.6.2 Memory

When coding for an embedded system of this nature, it is important to break down the total memory usage into a number of types for both program memory (PM) and data memory (DM). These categories and a brief description are listed in Table 9.3.

The final implementation of the AMR speech codec on the AD6526 was approximately 14,000 assembly language instructions. The AD6526 uses a 24-bit opcode size; therefore, the program memory size of the algorithm is approximately 42,000 bytes.

The DM state is found in the structure definitions *Speech_Encode_FrameState* and *Speech_Decode_amrState* in the reference C code. These structures must retain their values from frame to frame and be placed in read/write memory. The encoder required 2,744 bytes of state (assuming it is using VAD1), and the decoder requires 1,628 bytes of state.

In the AMR reference code, the DM scratch is primarily held on the stack. While this may be a common and normal occurrence in personal computer–based applications, in embedded systems it is quite common to reduce stack-based scratch and allocate it via an alternate methodology. This will be covered in detail in the Section 9.8.

**Table 9.3**  Basic Memory Types Required by a DSP Algorithm

| Category | Description |
| --- | --- |
| PM code | Executable opcodes |
| DM state | Computed data that must be saved from frame to frame |
| DM scratch | Data that is only needed within a single frame |
| DM tables | Fixed data provided by the algorithm |

The DM tables are spread among many of the C files found in the AMR reference code. The total amount of tables used in the AMR speech codec is 14,729 16-bit words. This data is read only and can be spread among internal DSP memory and external system memory.

## 9.7 Implementing the AMR Speech Codec—Trade-offs and Considerations

When porting or coding a speech codec (or any algorithm), it is important to consider the whole system that the algorithm is working within and the requirements of the processor and application. These constraints can come from the DSP/system architecture and the end-application requirements and can greatly complicate the porting process. In this example, the designers used the reference code provided by ETSI, who originally developed the case for an AMR. ETSI provides a full C implementation of the AMR speech encoder and decoder, along with documentation explaining how the code works.

### 9.7.1 MIPS Versus Memory Usage

One of the most common trade-offs in embedded programming is determining the correct balance between the allowable algorithmic execution time and total memory usage. As discussed earlier, speech codecs must run in real time, processing its current data before the next set of data is scheduled to arrive. In Section 9.4.1, we discussed the need to process 160-word blocks of 8-kHz sampled voice in 20-ms time with full-duplex facility.

In general, it is usually possible to decrease the total MIPS by modifying where an algorithm is placed in memory. Many modern DSPs contain multilevel caches or multiple memory types with varying wait states for each access. DSPs also typically support means for placing data in a secondary memory space to enable dual data accesses. On some DSP architectures, this could be a portion of program memory, while on others it could be a small scratchpad space. Regardless, careful choices need to be made here to ensure that the secondary memory is used efficiently. Methods used to improve MIPS through memory optimizations are covered in detail later.

### 9.7.2 Application Constraints on the Algorithm

The AMR speech codec was designed specifically for wireless handsets, which implies that certain system constraints helped determine baseline algorithmic requirements. Because a cellular phone handles one conversation (and, therefore, speech codec) at a time, this rules out the need for multiple instances of the speech codec running concurrently. This means that the AMR speech codec was ported such that it was not *reentrant*. A reentrant algorithm requires that all references to state and scratch memory are handled through pointers rather than absolute memory addresses. An algorithm that uses directly addressed DM cannot be reentrant because the scratch and state from one instance is corrupted by the processing from another instance. The ADSP-218x architecture is less efficient for pointer-based memory accesses as compared to absolute memory accesses. Therefore, a single instance version of AMR, which saved both MIPS and memory, was implemented.

In addition to this, because AMR is running inside a wireless device, it must not use much internal program or data memory due to the variety of algorithms that must be running simultaneously to support a phone call (e.g., channel codec, equalizer, operating system, user applications). This means that memory analyses should be run on all variables and arrays to determine the appropriate split of data between "fast memory" (such as internal SRAM located on the AD6526) and "slow memory" (such as external flash memory). The methods used for performing this analysis are covered in the following sections.

## 9.8   Porting the Reference AMR Speech Codec C Code to Target Hardware

Obviously, the first step in porting the AMR speech codec is to acquire the reference C code. However, the design team has to check if any change requests were made and ratified after the completion date of the reference code. Such change requests for the AMR speech codec are found on the Third Generation Partnership Project (3GPP®) Web site [13]. This group manages and updates many of the protocols used in second and third generation wireless networks, including the GSM network. In addition, 3GPP typically provides explanations and source code for all the required changes.

Once these updates are made, the reference C code is ready to be modified to better support the unique characteristics of the specific processor within the embedded environment. For the example here using the AD6526, the modifications include the following tasks.

### 9.8.1   Evaluating and Developing a Bit-Exact Version of Basicops

One of the most important issues to deal with when converting the reference code to an optimized embedded algorithm is to properly handle the *basicop* functions. The reference C code inefficiently uses function calls for most mathematical calculations, while many *basicop* functions are translatable into a single DSP instruction. The most common way of optimizing *basicops* is to use the *inline* C preprocessor directive, which tells the compiler not to make a function call but rather to include the called-function code directly in the file. On the AD6526, most simple *basicops* became simple assembly language instructions, and the more complex *basicops* (such as *div_s*) remained function calls.

### 9.8.2   Counting and Analyzing Memory Accesses to Variables and Tables

One very important part of analyzing and optimizing the AMR speech codec is to determine the number of accesses to all tables and variables during the processing of a frame of data. Different tables and variables are accessed with widely varying frequency, and some accesses are unique to particular modes of AMR. Because of this variable nature, it is important to characterize the number of accesses to each table and consider how memory architecture and wait states will affect the total required MIPS.

If there is no support in hardware emulation or software simulation for counting accesses to a region of memory, the reference C code can be probed to determine the number of accesses to each variable and array. One method of counting memory accesses is to place a counter macro next to each place the variable or table is accessed and run one of the test vectors through the codec. This will provide the number of accesses made during each frame. If the total amount of memory available and associated wait states are known, it is possible to get a good estimate of the impact of placing variables and tables into slow memory on the system.

When performing this test, it is very important to run a test vector for all eight modes of the speech codec. Knowing the frequency of each memory access for every operating mode is useful information, as each AMR mode executes in a widely varying amount of MIPS, and not all tables are accessed in each processing mode. MIPS-intensive and frequent data accesses can be moved to faster memory segments in order to equalize performance across all modes of operation.

### 9.8.3   Minimizing Stack Usage and Determining a Suitable Memory Architecture

The reference code for the AMR speech codec places all of its temporary variables (also known as scratch variables) on the stack. A stack is a common data structure where data and addresses are stored and retrieved sequentially in memory. Variables that are declared and used inside a C function are placed on a stack, and when the function ends, the memory is freed for future usage. While this is a simple method of managing memory, it is very inefficient in embedded designs. Placing all temporary variables on the stack limits efficient use of slow and fast memory. Stack usage also forces the placing of variables in sequence and does not allow for full use of multiple memories for dual data accesses (which is one important aspect of DSP architectures and algorithms). Finally, in some DSPs, stack-memory use is very inefficient compared to directly addressed arrays, as DSPs often do not handle pointer-based accesses as well as direct-memory accesses.

Given these reasons, the following methods are used for removing stack manipulations from C functions. One common method is *dynamic memory allocator,* such as the ANSI C standard function *malloc.* Malloc allocates memory on a heap, which is quite similar to a stack. One advantage is that most compilers allow the location of the heap to be controlled, enabling the system designer to place it in an appropriate memory segment. Additionally, some implementations of malloc allow for multiple heaps, providing a method of placing variables in different memory segments based on how frequently they are accessed. Another method is to create a *scratch memory structure* by analyzing how memory is allocated in each function and then placing those allocations in a C structure. In this case, the high-level API interface then calls malloc to allocate enough space. A simple example of how a program can be converted from a standard stack-based memory allocation to one that uses memory allocation and a scratch structure is given in Figures 9.8 and 9.9.

The data memory requirements between the two listings are the same. Because the two subfunctions have their scratch arrays unioned together, they take up only as much memory as the larger of the two arrays. This is similar to how the stack would overlay the local memory allocation of two functions at the same call level. If the algorithm had another level of calls within it (*subfunc1* or *subfunc2* made function calls), another level of structures (and potentially unions based on the exact details of the function calls) would be placed within the *scr_sf1* and *scr_sf2* structures.

The implementation of the AMR speech codec on the AD6526 did not use a passed pointer and indirect memory accesses to fetch data from the state and scratch

```
#include <stdio.h>

int subfunc1();
int subfunc2();

int main() {

        int sf1_out, sf2_out, total;

        sf1_out = subfunc1();
        sf2_out = subfunc2();

        total = sf1_out+sf2_out;
        printf("total:  0x%x", total);

        return(1);

}

int subfunc1() {

        int sf1_scratch1[5];
        int i;
        for (i=0; i<5; i++) {
                sf1_scratch1[i] = i*i;
        }

        return(sf1_scratch1[2]);

}

int subfunc2() {

        int sf2_scratch1[50];
        int i;
        for (i=0; i<50; i++) {
                sf2_scratch1[i] = i*i;
        }

        return(sf2_scratch1[4]);

}
```

**Figure 9.8** The main function and the two subfunctions all use stack to store their variables.

```
#include <malloc.h>
#include <stdio.h>

void subfunc1(scrmemptr);
void subfunc2(scrmemptr);

struct scr_scrmem {
        int sf1_out, sf2_out, total;
        union {
                struct scr_sf1 {
                        int sf1_scratch1[5];
                } sf1;
                struct scr_sf2 {
                        int sf2_scratch1[50];
                } sf2;
        } u;
};

typedef struct scr_scrmem* scrmemptr;

int main() {

        scrmemptr p_scrmem;

        p_scrmem = malloc(sizeof(struct scr_scrmem));

        subfunc1(p_scrmem);
        subfunc2(p_scrmem);

        p_scrmem->total=p_scrmem->sf1_out+p_scrmem->sf2_out;
        printf("total:  0x%x", p_scrmem->total);

        free(p_scrmem);

}

void subfunc1(scrmemptr p_scrmem) {

        int i;
        for(i=0; i<5; i++) {
                p_scrmem->u.sf1.sf1_scratch1[i]=i*i;
        }

        p_scrmem->sf1_out=p_scrmem->u.sf1.sf1_scratch1[2];
        return;

}

void subfunc2(scrmemptr p_scrmem) {

        int i;
        for(i=0; i<50; i++) {
                p_scrmem->u.sf2.sf2_scratch1[i]=i*i;
        }

        p_scrmem->sf2_out=p_scrmem->u.sf2.sf2_scratch1[4];
        return;

}
```

**Figure 9.9**   The main function and two subfunctions all place a portion of their scratch memory into a structure called scr_scrmem, which is allocated at the start of main and deallocated at the end of main. While this code uses the standard memory allocator, in a DSP application additional parameters would be added to the allocation function to allow for placing the structure in a specific memory type.

structures. However, the structures created here were used to create the assembly code for allocating space for the structures. A small personal computer–based C program that created an instance of the AMR state and scratch structures and output a line of ADSP-218x assembly code, which corresponded to the name and size of that structure element, was written. By doing this, the C structure was converted to an equivalent DSP assembly language file, which would statically allocate the appropriate amount of data memory.

### 9.8.4   Grouping Tables into Subsets of Files

Another important task to take care of when porting the AMR speech codec is consolidating all of the tables from the various C files into a small number of files. A table is any variable that is initialized and never changed in the code. It is important to bring all of these constants into a small number of files because it may be necessary to place tables into different memory segments based on the frequency they are accessed. In the original ETSI source code, tables are simply included in the source files that they are accessed in. Depending on the software-development tools designed for a particular DSP architecture, it may not be simple (or even possible) to arbitrarily place different tables from a single file into different memory segments.

   The simplest method of handling this situation is to determine the frequency that all tables are accessed and then, depending on the available memory, place as many of the most-often accessed tables into a single file. This will then be linked into that particular memory segment. Determining the proper grouping is handled through analysis of the data gathered from the memory analyses and placement of the most-commonly accessed elements into a single table. This is then targeted for fast internal memory, while less commonly accessed table data is placed in slower system memory.

### 9.8.5   Optimizing Cache Usage and Creating Unused Zones in Cache for Deterministic Operation

Due to the size and complexity of common DSP algorithms, it is often necessary to use a caching mechanism to handle the increased memory requirement. The main problem with using a cache in DSP algorithms is that it adds a level of indeterminacy in execution time. There is no longer a simple one-to-one correspondence between the code being executed and the time it will take for it to execute. Additionally, interruptions or preemptive executions of code can flush a previously full cache, leading to further uncertainty in execution time. Determining and minimizing the worst-case scenario in execution time is very important in DSP algorithms, especially those that have hard real-time constraints. Based on the information in Section 9.3.2, the following decisions were made in how to partition the DSP code.

   First, the most-commonly-called functions were placed into internal program memory. These functions never used the cache and would execute most quickly. Next, a small zone of program memory that is mapped into the cache was configured such that only a single function would be mapped into this memory segment. There would be a cache penalty the first time this function was called, but in subsequent accesses the code would be stored in the cache and would execute efficiently. Third, the code from the encoder subframe loop was configured such that it would be mapped inside a single 4K-word segment of memory. This was done to avoid multiple cache misses inside a single encode of a frame of data. Because the cache on the AD6526 is only 4K words while a page of program memory is 8K words, if some portions of the subframe loop were placed into the upper 4K while other portions were placed into the lower 4K, the cache would be continually reloaded inside this critical portion of the AMR algorithm (see Figure 9.10).

The region of DSP memory from 0x0000-0x1fff is internal to the processor and has zero wait states. The region from 0x2000-0x3fff goes through the instruction cache. A small portion of the cache has only one piece of DSP code mapped into it, ensuring that that code will always be present in the cache after it is loaded the first time. The subframe loop is designed to reside in a single block of memory, which can be loaded into the cache without forcing any misses (i.e., the total size must be less than 4K words minus the size of the fast reserved cache zone). If the subframe loop could not fit in this segment, there would be at least four cache misses that must be reloaded into the cache per frame for each segment.

## 9.9   Testing and Packaging of the AMR Speech Codec

ETSI provides a suite of test data for proving basic functionality of the AMR speech codec, including DTX processing. Because of the many modes and general complexity of the algorithm, literally hundreds of test vectors are provided to prove

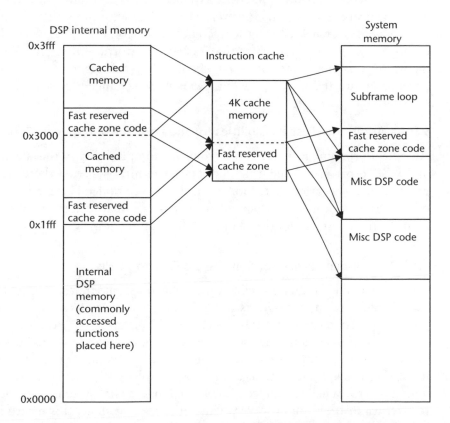

**Figure 9.10**   Memory mapping for optimal performance in AMR speech codec.

conformance with the original C code. Once a speech codec passes these vectors, it is considered compliant with the reference code. However, despite the quantity of vectors provided by ETSI, full coverage of the speech codec is not provided by simply running these tests. It is often wise to generate additional vectors to further prove compatibility between the original code and the ported implementation. Any test vector, whether it is test tones generated from an audio editor or sampled speech, is a valid source of data for testing the algorithm. Once additional tests are run, it is possible that differences will be found in the output generated by the reference code versus what is generated by the ported code. One common cause of the difference is that the *basicops* are handled slightly differently in the ported code versus the reference code. ETSI's *basicops* have certain unique features (particularly in how they handle rounding and saturation) that make slight incompatibilities in boundary cases a possibility. Often, these minute differences do not cause an error in the base test vectors, but they could create a problem in a new test vector. The AMR speech codec uses a number of thresholds throughout the algorithm that can cause a dramatic change in output if the computed value is just above or below this value. Additionally, the speech codec also holds a great deal of state, which can cause a small difference in one frame to create wildly divergent operation for long periods of time after an error originally occurs.

Once the exact location and cause of the error is determined, whether this error needs to be fixed is a matter for debate. Generally speaking, it is advisable to remain completely compatible with the reference C code; otherwise, the reference code becomes increasingly difficult to use as a method for generating new test vectors and verifying proper operation. However, if the difference is not objectionable, it is not a requirement to conform with the reference C code on tests not created by ETSI. If a coding trade-off was made such that a particular portion of the AMR speech codec algorithm was computed more quickly at the cost of losing compatibility over all possible inputs, this may be a valid reason to not change the code.

### 9.9.1   Additional Test Possibilities for the Ported AMR Speech Codec

Along with running functional tests to prove compatibility of the speech codec, it is also wise to perform some general tests to determine whether there are any coding errors that might affect behavior when the speech codec is forced to operate with other DSP algorithms or when enhancements are made after initial code delivery. Some of these tests are outlined in the following sections.

#### 9.9.1.1   Assuming Processor State Across Function Calls

A good assembly language coding practice is to assume that the state of a large and well-defined set of registers (so-called *scratch registers*) can be safely modified between function calls. If this is done, the DSP programmer need not be concerned with the path of calls that led to a particular function and can make any modifications to any function without worrying about effects on its parent functions. One method of proving that state is not held is to replace all function return instructions with a macro. When the codec is integrated into a final system, the macro compiles to a return instruction; however, while in the debug stage, this macro is defined to corrupt all scratch registers. If this test is run and the algorithm continues to function properly, the programmer gains great assurance that any function can be modified without worrying about the effects on its parents.

### 9.9.1.2   Finding and Removing Rogue Memory Accesses

A rogue memory access is defined as a memory read or write that occurs outside the bounds of the defined arrays of an algorithm. These are the sorts of bugs that do not affect an algorithm when it is running on its own, but they can cause subtle and difficult-to-solve problems when they are placed in the final application.

Rogue memory writes are obviously dangerous because they could potentially modify the state of another algorithm, causing it to misbehave in any number of ways. Rogue memory reads may at first appear innocuous, but, depending on the processor they occur on, they can cause even stranger and more subtle errors in a final application. In the AD6526, there are system-control registers mapped into the data memory space of the DSP, which are affected by memory read and can lead to unexpected behavior in a DSP interface.

On the AD6526, there are hardware breakpoints, which greatly simplify the task of finding these types of improper memory accesses. Hardware breakpoints are used when controlling the processor with an in-circuit emulator (ICE). They are similar to the standard breakpoints, where code execution stops once the program counter reaches a certain address. However, hardware breakpoints are much more powerful in that they allow for halting code execution when there is a memory access inside or outside a certain range of either program or data memory.

If the target processor does not have hardware breakpoints, it is still possible to find many rogue memory accesses. One method is to load all memory with a known value before downloading algorithms to the target hardware. After executing the code, the target processor memory is read back, and, by comparing memory contents before and after execution, it is possible to see whether there were any memory writes outside of the expected range. Unfortunately, finding rogue memory reads cannot be as easily checked for as memory writes. Setting all memory might uncover some rogue reads, as the algorithm might have functioned properly if the read accessed one particular value but not if it accessed another. Another method that might help uncover rogue reads (and writes) is to rearrange how arrays are laid out in memory. By moving the location of various arrays, bugs can potentially be uncovered by forcing the rogue memory access to a different address that cannot be safely read, such as a memory mapped peripheral.

While rogue memory accesses (especially writes) should be avoided, there may be times where performing one additional access on an array can help improve performance of an algorithm. Often, software pipelines make it advantageous to include an extra memory read before a loop starts to improve overall performance. However, it is important to be sure that the system designer knows that an additional memory access was made in a part of the algorithm. Much embedded programming involves a trade-off between increasing performance and improving reliability and portability. When working on algorithms with hard real-time requirements, it is sometimes necessary to increase integration complexity to gain significant improvements in performance. If a coding decision that leads to additional reads of an array is made, it should be carefully documented in the source code and release notes.

### 9.9.1.3   Checks for Untested Code

It is very useful to determine a methodology for finding code that is not verified with the provided test vectors. One such methodology provided by some DSP vendors is statistical profiling. Statistical profiling captures and processes data on which functions are called and arrays are accessed. Because the ETSI test vectors do not guarantee full coverage of the entire AMR speech codec, it is advisable to determine exactly which paths through functions are not followed with the base vector set and then generate new vectors that force execution through these paths of the algorithm. If statistical profiling is not available, macros can be used to determine the call tree, and the location of untested code can be deduced from this data.

## 9.10   Final Packaging and Documentation of the AMR Speech Codec

After completing the steps of code porting and development, debugging, and testing, the algorithm is ready for integration. Along with the actual source code, clean and complete documentation is an absolute requirement in the delivery. It is particularly necessary for an algorithm in an embedded application. This is because performance within a larger application is often dependent on knowledge of how the algorithm was created, and compatibility can only be guaranteed if the system state assumptions are known by the system architect. A base set of information that should be provided includes:

- An overview of the directory structure;
- An explanation of the API, including all necessary information on input and output data types and sample drivers;
- Any assumed processor state, including processor modes of operation, registers that may never be used or modified, and registers that are assumed to be modifiable across function calls;
- Observed MIPS, including any assumptions on how the value was calculated (e.g., wait states and cache usage);
- Total memory usage, including program memory, scratch memory, state memory, and tables;
- Testing methodology, including any tests that were not specified by ETSI and additional methodology tests, such as the ones suggested earlier.

## 9.11   Conclusions

This chapter is an attempt to indicate the rigors in porting reference C codes for reliable and efficient implementation. The reference code provided by a standards body is not always written for easy use in embedded hardware. While the AMR speech codec was written from the perspective that it would be run on wireless handsets, and the actual algorithm can run efficiently on modern DSPs, the initial C implementation must be modified in many ways to make it suitable for an embedded application and DSP architecture. The total time required to finish this project was approximately 40 person-months. The steps indicated in this chapter could be used as a general guide for developing real-time voice-coding algorithms on any DSP hardware.

# References

[1]   Kularatna, N., *Modern Component Families and Circuit Block Design*, Boston, MA: Butterworth-Heinemann, 2000.

[2]   Laplsey, P., et al., *DSP Processor Fundamentals*, Berkely, CA: Berkeley Design Technology, 1994–1996.

[3]   Heinen, S. et. al. "A 6.1 to 13.3 kb/s Variable Rate CELP Codec (VR-CELP) for AMR Speech Coding," *49th IEEE-ICASSP '99 Conference Proceedings*, 1999, pp. 9–12.

[4]   Corbun, O., M. Almgren, and K. Svanbro, "Capacity and Speech Quality Aspects Using Adaptive Multirate (AMR)," *9th IEEE Int'l. Symp. on Personal, Indoor and Mobile Radio Communications,* September 1998, pp. 1535–1539.

[5]   Hindelang, T., et. al., "Improved Channel Coding and Estimation for Adaptive Multirate (AMR) Transmission," *Vehecular Technology Conf. Proc.,* Tokyo, 2000, pp. 1210–1214.

[6]   http://www.etsi.org.

[7]   Rabiner, L. R., and R. W. Schafer, *Digital Processing of Speech Signals*, Englewood Cliffs, NJ: Prentice Hall, 1978, p. 82.

[8]   Rabiner, L. R., and R. W. Schafer, *Digital Processing of Speech Signals*, Englewood Cliffs, NJ: Prentice Hall, 1978, pp. 398–404.

[9]   Rabiner, L. R., and R. W. Schafer, *Digital Processing of Speech Signals*, Englewood Cliffs, NJ: Prentice Hall, 1978, pp. 411–413.

[10]   Jayant, N. S. and P. Noll, *Digital Coding of Waveforms: Principles and Applications to Speech and Video*, Englewood Cliffs, NJ: Prentice Hall, 1984.

[11]   Kondoz, A. M., *Digital Speech*, John Wiley and Sons, 1994.

[12]   Xydeas, C. S., and C. Papanastasiou, "Split Matrix Quantization of LPC Parameters," *IEEE Transactions on Speech and Audio Processing,* Vol. 7 No. 2, March 1999, pp. 113–125.

[13]   http://www.3gpp.org.

# List of Acronyms

**AAL**     ATM adaptation layer

**ACEL**     Algebraic Codebook Excited LP

**ACI**     Adjacent channel interference

**ADC**     Analog to digital conversion

**ADM**     Add-drop multiplexer

**ADPCM**     Adaptive differential pulse code modulation

**ADSL**     Asymmetric DSL

**AFE**     Analogue front end

**AGC**     Automatic gain control

**ALiDD**     ADSL-Lite Data DSP

**AM**     Amplitude modulation

**AMI**     Alternative mark inversion

**AMPS**     Advanced mobile phone system

**AMR**     Adaptive multirate

**ANSI**     American National Standards Institute

**AP**     Application parts

**APC**     Adaptive predictive coding

**API**     Application programming interface

**ARIB**     Association of Radio Industries and Businesses

**ARQ**     Automatic repeat request

**ASCII**     American Standard Code for Information Interchange

**ASIC**     Application-specific ICs

**ASK**     Amplitude shift keying

**ATM**     Asynchronous transfer mode

**ATU**     ADSL Transmission Unit

**ATU-C**     ADSL Transmission Unit—Central Office site

**ATU-R**     ADSL Transmission Unit—Remote site

**AWG**     Additive white Gaussian noise

**AWG**     Arrayed waveguide grating

**AWGN**     Additive white Gaussian noise

**BER**     Bit error rate

**BHCA**     Busy hour call attempts

**B-ISDN**     Broadband ISDN

**BOD**     Bandwidth on demand

**BPSK**    Binary phase shift keying

**BRA**    Basic rate access

**BRAN**    Broadband radio access

**BS**    Base station

**CP**    Central processors

**BWA**    Fixed wireless access systems

**CAP**    Carrierless amplitude and phase

**CAS**    Channel-associated signaling

**CBR**    Constant bit rate

**CCI**    Cochannel interference

**CCS**    Common channel signaling

**CDF**    Cumulative distribution function

**CDMA**    Code division multiple access

**CELP**    Code excited LPC

**CH-Est**    Channel estimation entity

**CIR**    Committed information rate

**CL**    Convergence layers

**CLI**    Calling line identification

**CLP**    Cell loss priority information

**CNG**    Comfort noise generation

**CO**    Central office

**CORBA**    Common object request broker architecture

**CPR**    Call processors

**CRC**    Cyclic redundancy check

**CS**    Control store

**CSA**    Carrier service area

**CSMA/CA**    Carrier sense multiple access/ collision avoidance

**CT**    Cordless telephony

**CTI**    Computer telephony integration

**DAA**    Data access arrangement

**DAC**    Digital to analog conversion

**DASP**    Digital ADSL signal processor

**DCA**    Dynamic channel allocation

**DCC**    Digital colour code

**DCE**    Data circuit terminating equipment

**DCME**    Digital circuit multiplexing equipment

**DECT**    Digital European cordless telephony

**DFE**    Decision feedback equalization

**DHCP**    Dynamic host configuration protocol

**DID**    Direct inward dialing

**DiffServ**    Differentiated services

**DL** Downlink

**DLC** Data link control

**DM** Data memory

**DMBUS** Data memory bus

**DMR** Digital microwave radio

**DMT** Discrete multitone

**DP** Distribution points

**DPO** Digital phosphor oscilloscopes

**DQDB** Distributed queue dual bus

**DQDC** Distributed queue dual bus

**DQPSK** Differential quaternary phase shift keying

**DS0** Digital signal-level zero

**DSI** Digital speech interpolation

**DSL** Digital subscriber loop

**DSLAM** DSL access multiplexer

**DSO** Digital storage scopes

**DSP** Digital signal processing

**DS-SS** Direct-sequence spread spectrum

**DTE** Data terminal equipment

**DTMF** Dual tone Multifrequency

**DTX** Discontinuous transmission

**DuSLIC** Dual channel subscriber line interface concept

**DWDM** Dense wavelength-division multiplexing

**DWMT** Discrete wavelet multitone

**EBCID** Extended binary coded decimal interchange code

**EC** Echo cancellation

**EC-ADSL** Echo-canceled ADSL

**EDFA** Erbium doped fiber amplifier

**EDGE** Enhanced data rates for GSM evolution

**EIR** Excess information rate

**EM** Extension modules

**EMB** Extension module bus

**EOC** Embedded operations channel

**ES** Errored seconds

**ESN** Electronic serial number

**ETC** Exchange terminal circuits

**ETSI®** European Telecommunications Standards Institution

**FAL** Frame alignment loss

**FAU** Fixed access unit

**FAW** Frame alignment word

**FCC** Federal Communications Commission

**FDD**     Frequency division duplexing

**FDD-ADSL**     Frequency division duplexing-ADSL

**FDDI**     Fiber distributed data interface

**FDM**     Frequency division multiplexing

**FDMA**     Frequency division multiple access

**FEC**     Forward error correction

**FEXT**     Far-end crosstalk

**FFT**     Fast Fourier transform

**FH-SS**     Frequency-hopped spread spectrum

**FIC**     Fabric interface chip

**FIR**     Finite impulse response

**FM**     Frequency modulation

**FP**     Fixed part

**FR**     Frame relay

**FRAD**     FR access device

**FRAU**     Fixed radio access unit

**FSK**     Frequency shift keying

**FWA**     Fixed wireless access

**GFC**     Generic flow control

**GFR**     Guaranteed frame rate

**GFSK**     Gaussian FSK

**GI**     Guard interval

**GMPLS**     Generalized multiprotocol label switching

**GMSK**     Gaussian minimum shift keying

**GOS**     Grade of service

**GPRS**     General packet radio service

**GPS**     Global positioning system

**GSA**     Geographic service area

**GSM**     Global System for Mobile

**HDL**     Hardware description language

**HDLC**     High level data link

**HDSL**     High-bit-rate-DSL

**HEC**     Header error control

**HEMT**     High electron mobility

**HLR**     Home location register

**HPF**     Highpass filter

**HSCSD**     High-speed circuit-switched data

**ICE**     In-circuit emulator

**IDN**     Integrated digital networks

**IDT**     Integrated Device Technology Inc.

**IEEE**     Institution of Electrical and Electronic Engineers

**IETF**    Internet Engineering Task Force

**IFFT**    Inverse fast Fourier transform

**IIR**    Infinite impulse response

**IMT**    International mobile telephony

**IMTS**    Improved mobile telephone system

**INAP**    Intelligent network application part

**IntServ**    Integrated services

**IR**    Infrared

**IROS**    Internet radio operating system

**ISDN**    Integrated services digital network

**ISI**    Intersymbol interference

**ISM**    Industrial, scientific and medical

**ISP**    Internet service providers

**ISUP**    ISDN user part

**ITDF**    Integrated test and diagnostic functions

**ITeX**    Integrated telecom express

**ITU**    International Telecommunications Union

**ITU-T**    International Telecommunications Union—Telephony

**IWU**    Interworking unit

**JTACS**    Japanese Total Access Communication System

**LAN**    Local area network

**LE**    Local exchange

**LIC**    Line interface circuits

**LMCS**    Local multipoint communication service

**LMDS**    Local multipoint distribution service

**LOS**    Line of sight

**LPAS**    Linear prediction based analysis by synthesis

**LPC**    Linear predictive coding

**LPF**    Lowpass filter

**LPR**    Line processors

**LSBs**    Least significant bits

**MAC**    Mobile attenuation code

**MAN**    Metropolitan area network

**MAP**    Mobile application part

**MAU**    Maintenance unit

**MCM**    Multicarrier modulation

**MCU**    Multipoint control unit

**MDF**    Main distribution frame

**MDS**    Multichannel distribution service

**MEMs**    Microelectromechanical systems

**MFC**    Multifrequency-compelled

**MIN**     Mobile identification number

**MIPS**     Millions of instructions per second

**MMA**     Multimedia mobile access

**MMAC**     Multimedia mobile access communication system

**MMDS**     Multichannel multipoint distribution service

**MMSE**     Minimize the mean-square error

**MOEMS**     Micro-optical electromechanical systems

**MOS**     Mean opinion scores

**MPE**     Multipulse excited LPC

**MPLS**     Multiprotocol label switching

**MPR**     Main processor

**MS**     Mobile station

**MSB**     Most significant bit

**MSC**     Mobile switching center

**MSRN**     Mobile station roaming number

**MSU**     Message signal units

**MTIE**     Maximum time interval error

**MTP**     Message transfer part

**MTSO**     Mobile telecommunications switching office

**MUP**     Multiple positions

**MuSLIC**     Multichannel subscriber line interface circuit

**NEXT**     Near-end cross talk

**N-ISDN**     Narrowband ISDN

**NLOS**     Nonline-of-sight

**NMT**     Nordic Mobile Telephone

**NSP**     Network service part

**NTT**     Nippon Telephone and Telegraph

**O/E**     Optical to electrical

**OADM**     Optical add/drop multiplexing

**OAM&P**     Operations, administrations, maintenance, and provisioning

**OBS**     Optical burst switching

**OC**     Optical carrier

**OCX**     Optical cross connect

**OFDM**     Orthogonal frequency division multiplexing

**OFDMA**     Orthogonal frequency division multiple access

**OMAP**     Operation and maintenance application part

**ONU**     Optical networking unit

**OOK**     On-off keying

**OPS**     Optical packet switching

**OSI**     Open systems interconnect

**OSM**     Optical signal (spectral) monitoring

**OXCs**    Optical cross-connects

**PABX**    Private automatic branch exchange

**PACS**    Personal access communications system

**PAD**    Packet assembly and disassembly

**PAM**    Pulse amplitude modulation

**PAN**    Personal area networks

**PANS**    Pretty amazing new services

**PAR**    Peak-to-average ratio

**PCB**    Printed circuit board

**PCM**    Pulse code modulation

**PCP**    Primary cross-connection points

**PCS**    Personal communications systems

**PDH**    Plesiochronous digital hierarchy

**PGA**    Programmable gain amplifier

**PHS**    Personal handyphone system

**PLL**    Phase locked loop

**PLMN**    Public land mobile network

**PM**    Program memory

**PMBU**    Sprogram memory bus

**PMD**    Physical medium dependent

**POTS**    Plain old telephone system

**PP**    Portable part

**PRC**    Primary reference clock

**PRA**    Primary rate access

**PSD**    Power spectral density

**PSDN**    Packet switched data network

**PSK**    Phase shift keying

**PSPDN**    Packet switched public data network

**PSTN**    Public switched telephone network

**PTI**    Payload type identifier

**PVC**    Permanent virtual circuit

**PWT**    Personal wireless telecommunications

**PWT-E**    Personal wireless telecommunications—enhanced

**QAM**    Quadrature amplitude modulation

**QoS**    Quality of service

**QPSK**    Quadrature phase shift keying

**QSAFA**    Quasi-static automatic autonomous frequency assignment

**RA**    Raman amplifiers

**RADSL**    Rate-adaptive DSL

**RELP**    Residual excited linear predictive coding

**RF**    Radio frequency

**RFI**   Radio frequency interference

**RFP**   The radio fixed part

**RLL**   Radio local loop

**RMS**   Root mean square

**RNC**   Radio node controller

**ROBOs**   Regional Bell Operating Companies

**RP**   Regional processors

**RPB**   Regional processor bus

**RPE**   Regular pulse excited LPC

**RRD**   Revised-resistance design

**RSS**   Remote subscriber station

**RSS**   Remote switching system

**RSVP**   Resource reservation protocol

**RWSM**   Read-write state machine

**SA**   Smart antennas

**SAM**   Scalable ADSL modem

**SAT**   Supervisory audio tone

**SCCP**   Signaling connection control part

**SCN**   Switched communication networks

**SCP**   Service control points

**SDH**   Synchronous digital hierarchy

**SDL**   Signaling data link

**SDSL**   Single-line DSL

**SID**   System identification number

**SIP**   Session initiation protocol

**SL**   Signaling link

**SLIC**   Subscriber line interface circuit

**SME**   Small and medium enterprise

**SMS**   Short message service

**SNR**   Signal-to-noise ratio

**SoC**   Systems-on-a-chip

**SOHO**   Small-office/home-office

**SONET**   Synchronous optical network

**SPC**   Stored program controlled

**SP-Dec**   Speech decoder

**SP-Enc**   Speech encoder

**SPM**   Space switch modules

**SQNR**   Signal-to-quantization noise ratio

**SS**   Speech store

**SS7**   Signaling systems No. 7

**SSUs**   Synchronization supply units

ST      Space-time

ST      Signaling terminals

STM      Synchronous transport module

STP      Shielded twisted pair

STP      Signal transfer point

STS      Space-time-space

SVCs      Switched virtual circuits

TACS      Total access communication system

TC      Transmission convergence

TCAP      Transaction capabilities application part

TCH      Traffic channel

TCP      Transmission control protocol

TCP/IP      Transmission control protocol / Internet protocol

TC-PAM      Trellis-coded pulse-amplitude modulation

TDD      Time division duplexing

TDEV      Time deviation

TDM      Time division multiplexing

TDMA      Time division multiple access

TEQ      Time domain equalizer

THD      Tight total harmonic distortion

THD      Total harmonic distortion

TIE      Time interval error

TLP      Transmission level point

TRX      Transceiver

TS      Time-space

TSI      Time switch interchange

TSM      Time switch modules

TST      Time-space-time

TUP      Telephony user part

UDP      User datagram protocol

UI      Unit intervals

UL      Uplink

UNI      User network interface

UNII      Unlicensed National Information Infrastructure

UP      User parts

UTP      Unshielded twisted pair

UTRA      UMTS terrestrial radio access

VAD      Voice activity detection

VBR      Variable bit rate

VC      Virtual circuits

VCC      Virtual channel connection

**VCI**     Virtual channel identifier

**VCXO**     Voltage controlled crystal oscillator

**VDSL**     Very high bit rate DSL

**VHSIC**     Very high speed Integrated circuit

**VLR**     Visitor location register

**VLSI**     Very large scale integrated circuits

**VOD**     Video on demand

**VoDSL**     Voice over DSL

**VoIP**     Voice over IP

**VPI**     Virtual path identifier

**VSWR**     Voltage standing wave ratio

**WACS**     Wireless access communications system

**WAN**     Wide area network

**WCDMA**     Wideband code division multiple access

**WDM**     Wavelength division multiplexing

**WLAN**     Wireless local area network

**WLL**     Wireless local loop

**WMOPS**     Weighted millions of operations per

**WRSs**     Wavelength routing switches

**WS**     Workstation

**WWW**     World Wide Web

# About the Authors

Nihal Kularatna, former CEO of the Arthur C Clarke Institute for Modern Technologies in Sri Lanka (2000–2001), has over 28 years of experience in professional and research environments as an electronics engineer. He is the author of several books, *Modern Electronic Test & Measuring Instruments* (IEE, 1996), *Digital and Analogue Instrumentation: Testing and Measurement* (IEE, 2003), *Power Electronics Design Handbook: Low-Power Components and Applications* (Butterworth, 1998), and *Modern Component Families and Circuit Block Design* (Butterworth, 2000). He was the principal author of the Datapro/NBI report *Sri Lanka Telecoms—An Industry and Market Analysis* (1997). He was an active consultant and a trainer for many Sri Lankan organizations and a few U.S. companies, including the Gartners Group.

In his early career as an electronics engineer involved with civil aviation and telecommunications, he experienced many specialized trainings with equipment manufacturers and other training organizations in the United States, the United Kingdom, and France, including the FAA and CIT Alacatel. He has contributed over 50 papers to many international journals and international conferences. A Fellow of the IEE (London), a Senior Member of IEEE (United States) and an honors graduate from the University of Peradeniya, Sri Lanka, he is currently active in research and development work in power electronics, processor-based hardware, and sensor systems. During his research career in Sri Lanka, he was a winner of the Presidential Awards for Inventions (1995), The Most Outstanding Citizens Awards–1999 (Lions Club), and a TOYP Award for academic accomplishment (Jaycees) in 1993.

He is presently working as a senior lecturer in the Department of Electrical and Computer Engineering at the University of Auckland. His hobby is gardening cacti and succulents.

Dileeka Dias is a professor in the Department of Electronic and Telecommunication Engineering at the University of Moratuwa, Sri Lanka. She holds an M.S. and Ph.D. from the University of California, Davis, specializing in signal processing and modulation related to digital mobile communications. She has over 12 years of experience in teaching and research at undergraduate and postgraduate levels and in continuing professional development of practicing engineers.

She was the head of the Department of Electronic & Telecommunication Engineering, at the University of Moratuwa, Sri Lanka, from 1999 to 2003, during which time she was extensively involved in curriculum and course development in electronic and telecommunication engineering at undergraduate and postgraduate levels.

Her main research interests are in digital signal processing and its applications to communications. Presently, she is actively engaged in developing novel telecommunications applications, particularly in the mobile communications arena.

She has over 20 research publications to her credit, and is also a chapter coauthor of *Power Electronics Design Handbook: Low-Power Components and Applications* (Butterworth,1998), *Modern Component Families and Circuit Block Design* (Butterworth, 2000), and *Digital and Analogue Instrumentation: Testing and Measurement* (IEE, 2003), authored by Nihal Kularatna.

# Index

For further information on these and other Artech House titles,
including previously considered out-of-print books now available through our In-Print-Forever® (IPF®) program, contact:

Artech House
685 Canton Street
Norwood, MA 02062
Phone: 781-769-9750
Fax: 781-769-6334
e-mail: artech@artechhouse.com

Artech House
46 Gillingham Street
London SW1V 1AH UK
Phone: +44 (0)20 7596-8750
Fax: +44 (0)20 7630-0166
e-mail: artech-uk@artechhouse.com

Find us on the World Wide Web at:
www.artechhouse.com